LIFE-CYCLE ASSESSMENT OF BIOREFINERIES

LIFE-CYCLE ASSESSMENT OF BIOREFINERIES

EDGARD GNANSOUNOU
Bioenergy and Energy Planning Research Group (BPE), Ecole Polytechnique Fédérale de Lausanne (EPFL), Lausanne, Switzerland

ASHOK PANDEY
Center of Innovative and Applied Bioprocessing, Mohali, Punjab, India

ELSEVIER

elsevier.com

Elsevier
Radarweg 29, PO Box 211, 1000 AE Amsterdam, Netherlands
The Boulevard, Langford Lane, Kidlington, Oxford OX5 1GB, United Kingdom
50 Hampshire Street, 5th Floor, Cambridge, MA 02139, United States

Library of Congress Cataloging-in-Publication Data
A catalog record for this book is available from the Library of Congress

British Library Cataloguing-in-Publication Data
A catalogue record for this book is available from the British Library

ISBN: 978-0-444-63585-3

For information on all Elsevier publications
visit our website at https://www.elsevier.com/

Publisher: John Fedor
Acquisition Editor: Kostas Marinakis
Editorial Project Manager: Christine McElvenny
Production Project Manager: Maria Bernard
Cover Designer: Greg Harris

Typeset by SPi Global, India

Contents

6. Life Cycle Assessment of Thermochemical Conversion of Empty Fruit Bunch of Oil Palm to Bio-Methane

P. MONTAFIA, E. GNANSOUNOU

7. Life Cycle Assessment of Algal Biorefinery

E. GNANSOUNOU, J.K. RAMAN

8. Life Cycle Assessment and Land-Use Changes: Effectiveness and Limitations

W. BEN AOUN, B. GABRIELLE

9. Modeling Land-Use Change Effects of Biofuel Policies: Coupling Economic Models and LCA

L. PANICHELLI, E. GNANSOUNOU

10. Towards an Integrated Sustainability Assessment of Biorefineries

D. TUAZON, E. GNANSOUNOU

Contributors

W. Ben Aoun Université Paris-Saclay; Terres Innovia, Thiverval-Grignon, France

K. Dufossé Université Paris-Saclay, Thiverval-Grignon, France

B. Gabrielle Université Paris-Saclay, Thiverval-Grignon, France

E. Gnansounou Bioenergy and Energy Planning Research Group (BPE), Ecole Polytechnique Fédérale de Lausanne (EPFL), Lausanne, Switzerland

P. Montafia Ecole Polytechnique Fédérale de Lausanne (EPFL), Lausanne, Switzerland

E. Ruiz Pachón Ecole Polytechnique Fédérale de Lausanne (EPFL), Lausanne, Switzerland

A. Pandey Center of Innovative and Applied Bioprocessing, Mohali, Punjab, India

L. Panichelli International Consultant (Freelance), Milano, Italy

J.K. Raman Ecole Polytechnique Fédérale de Lausanne (EPFL), Lausanne, Switzerland

D. Tuazon Ecole Polytechnique Fédérale de Lausanne (EPFL), Lausanne, Switzerland

P. Vaskan Ecole Polytechnique Fédérale de Lausanne (EPFL), Lausanne, Switzerland

Preface

Biomass is available in a variety of raw materials composed of interwoven compounds that can be fractionated for valorizations into energy, as well as intermediate and final products such as food/feed, chemicals, and biomaterials. It is a feedstock suitable for multiple conversions, including chemical, physical, thermal, biological, and biotechnological, and several combined processes. Due to the potential competition between alternative uses, biomass is at the crossroads of several debates related to food security, substitution for depleting fossil resources, reducing dependency on the limited number of fossil fuel exporting countries, improving energy security by diversifying energy imports, reducing greenhouse gas (GHG) emissions, improving quality of air, soil and water, and avoiding deforestation.

The concept of a biorefinery, due to its potential flexibility, has opportunely emerged as a relevant strategy to efficiently valorize the services provided by biomass while minimizing burdens on the environment, creating new opportunities for social and economic development and adapting their expansion and operation to economic conjunctures. Sustainability is one of the core issues of biorefineries. The biorefinery is a metaphor for the petroleum refinery, in which biomass is considered to be a possible substitute for petroleum and which is capable of generating as many products as a petroleum refinery. Production of a wide spectrum of biobased products, which are fully competitive with their conventional equivalents, is one of the key objectives of biorefineries.

This book comprises 10 chapters. Chapter 1 is on the classification of biorefineries. Biorefineries have been classified by considering sustainability potential and flexibility as additional criteria to feedstock, conversion, platforms, and final products. Chapter 2 describes the fundamentals of the life cycle assessment (LCA). The use of methods to assess environmental performance of industrial products became popular in the 1960s, with growing environmental concerns. This chapter introduces LCA methodology based on the four-stage approach popularized by ISO. The theoretical background is presented as well. The strengths and weaknesses of the methodology are discussed and a new allocation approach is proposed. Extension of environmental LCA to social, economic and organizational LCA is outlined. Chapter 3 deals with the LCA of agricultural feedstocks for biorefineries. This chapter reviews the main challenges associated with the application of LCA to biomass supply chains from agricultural residues, arable crops or perennial lignocellulosic grasses, and addresses them through state-of-the-art approaches involving spatially explicit feedstock modeling. Chapter 4 describes LCA of sugar crops or starch-based integrated biorefineries and presents a comprehensive analysis of the existing approaches and key issues in the LCA methodology applied to evaluate environmental performance of sugar- or starch-based integrated biorefineries. Moreover, this chapter includes an economic analysis and details of process design and technological achievements in integrated biorefineries.

Chapter 5 is on LCA of vetiver-based biorefineries with production of bioethanol and furfural. A prospective LCA was performed to compare the environmental impacts of vetiver-based biorefinery systems with conventional systems. Chapter 6 describes LCA of thermochemical conversion of empty fruit bunch of oil palm to biomethane. The chapter evaluates and compares effects of different allocation methods of raw materials and it points out a rigorous approach for economic allocation. Furthermore, a section is devoted to the end of the plantation life and its consequences on carbon stock and GHG emissions. Chapter 7 is on LCA of an algal biorefinery in India for biodiesel and protein production from algae, and biodiesel, protein, and succinic acid production, which were compared with their respective fossil-based reference system. Chapter 8 deals with the LCA and land-use changes (LUC) and focuses on how to address the issue of LUC in the environmental assessment of biofuels, by introducing the different types of underlying mechanisms while emphasizing their complexity, and also by reviewing the methodologies put forward to include LUC effects in LCAs of biofuel chains. Chapter 9 is on modeling LUC effects of biofuel policies: coupling economic models and LCA. This chapter concludes that the LUC effects of biofuel policies need to be addressed in the broader perspective of sustainable agriculture and land-use planning. The last chapter of the book (Chapter 10) provides a broad discussion of salient sustainability aspects of biofuel production. Within the framework of biorefineries, a scenario is proposed that promotes accountability and embraces uncertainty and adherence to sustainability guiding principles and frameworks.

The chapters in this book provide state-of-art general information, basic data and knowledge on the fundamental aspects and principles of LCA of biomass-based biorefineries, which can greatly help in understanding and developing technoeconomic feasibilities of biorefineries. The book provides key features for the education and understanding of integrated biorefineries, describing models to cope with LUC effects of biorefineries; it also presents case studies. We hope that readers will find it useful.

We thank the authors of all the chapters for their cooperation and their preparedness in revising the manuscripts in a time-framed manner. We also acknowledge the help received from the reviewers who, despite their busy professional activities, helped us by evaluating the manuscripts and gave their critical inputs to refine and improve the articles. We warmly thank Dr. Marinakis Kostas, the acquisition editor, Christine McElvenny and the Elsevier team for their cooperation and efforts in producing this book.

Edgard Gnansounou, Switzerland
Ashok Pandey, India

CHAPTER

1

Classification of Biorefineries Taking into Account Sustainability Potentials and Flexibility

E. Gnansounou, A. Pandey†*

*Bioenergy and Energy Planning Research Group (BPE), Ecole Polytechnique Fédérale de Lausanne (EPFL), Lausanne, Switzerland †Center of Innovative and Applied Bioprocessing, Mohali, Punjab, India

1.1 INTRODUCTION

Biomass is available in a variety of raw materials composed of interwoven compounds that can be fractionated for valorization into energy, intermediate and final products, such as food/feed, chemicals, and biomaterials. It is a feedstock suitable for multiple conversions including chemical, physical, thermal, biological, and biotechnological, and several combined processes. Due to the potential competition between alternative uses, biomass is at the crossroads of several debates related to food security, substitution for depleting fossil resources, reducing dependency on the limited number of fossil fuel–exporting countries, improving energy security by diversifying energy imports, reducing greenhouse gas (GHG) emissions, improving quality of air, soil and water, and avoiding deforestation. With regard to the complex relationships between these debates, the issues of biomass conversion into energy, materials, and chemicals have been in the spotlight during the last three decades, with changing perspectives depending on the levels of international prices of petroleum oil and fossil fuels.

During periods of soaring petroleum oil prices, environmental policies enhance incentives in favor of renewable energy, including biomass. However, at the same time issues related to food security and burdens on the environment retain attention, raising questions about which

http://dx.doi.org/10.1016/B978-0-444-63585-3.00001-2

relevant trade-offs should be the focus in order to develop a sustainable biobased economy. When oil prices decrease, the efforts in favor of bioprocesses tend to slow down.

The concept of a biorefinery, due to its potential flexibility, has opportunely emerged as a relevant strategy for efficiently valorizing the services provided by biomass while minimizing burdens on the environment, creating new opportunities for social and economic development and adapting the biomass services expansion and operation to economic conjunctures. Sustainability is one of the core issues of biorefineries. That is why the International Energy Agency (IEA) considers biorefining to benefit *sustainable utilization* of biomass [1]. Therefore, Task Group 42 of the IEA defined the biorefinery concept as "the *sustainable processing* of biomass into a spectrum of marketable products and energy." The definition provided by the IEA is close to the one from the US Department of Energy (DOE): "A biorefinery is an *overall concept* of a processing plant where biomass feedstocks are converted and extracted into a spectrum of valuable products." However, before being a defined concept, the term biorefinery was a metaphor for the petroleum refinery, where biomass is considered as a possible substitute for petroleum and capable of generating as many products as a petroleum refinery does. Most of the definitions of biorefineries emphasize the technological or industrial challenges. For the US National Renewable Energy Laboratory (NREL), "a biorefinery is a *facility* that integrates biomass *conversion processes and equipment* to produce fuels, power, and value-added chemicals from biomass. The biorefinery concept is analogous to today's petroleum refineries which produce multiple fuels and products from petroleum." [2]. De Jong and Jungmeier [3] compared a petroleum refinery with a biorefinery and pointed out more heterogeneous feedstock and processes, a higher number of process types (e.g., chemical and biotechnological) in the case of biorefineries. The main drivers of the development and market deployment of biorefineries are energy and environmental policy. In the European Union, for instance, the Directive of Renewable Energy required at least 20% renewable energy in the final energy of each member state by 2020 and at least 10% for the transportation sector. Second-generation biofuels in particular are promoted by a double counting system [4] and by limiting the share of first-generation biofuels to 7%. In 2011 a large consortium of research organizations proposed a European Biorefinery Vision for 2030 [5]. That vision includes the following key points:

- Production of a wide spectrum of biobased products which are fully competitive with their conventional equivalents
- Versatile biomass supply chains
- A revitalized, competitive, and knowledge-intensive rural economy based on biorefineries
- Growing integration of biobased industrial sectors
- A focus on sustainable bioproducts which gives Europe a competitive edge
- Versatile biorefinery development routes

The European Commission as well as the EU member states funded several projects on biorefineries. In the United States, an ambitious biomass program is being developed with a voluntary goal of petroleum oil partial substitution. The program is encouraged by several federal policies and incentives including the Energy Independence and Security Act of 2007 (EISA) that mandated at least 136 billion liters of biofuels by 2022. The Food Conservation and Energy Act of 2008 also promoted biobased market programs. Other countries or regions such as Brazil, China, and India are also developing biofuels with the perspective of biorefineries. A few plants started operating at a commercial level in recent years worldwide and a large number of plants are at pilot or demonstration stages. Other facilities are rather potential biorefineries that would emerge from integrating novel additional processes into existing paper mill, sugar mill, starch, and oilseed industrial plants. Benchmarking different facilities would give relevant insights for monitoring the research-development-deployment procedures. However, biorefineries can differ a great deal from each other. There is a need for a rigorous classification system that will allow the grouping of similar plants and projects and comparing their achievements. After analyzing the classification system developed by the IEA Bioenergy Task Group 42 [3], complementary criteria are proposed and discussed considering the perspectives of energy and environmental policy.

1.2 CLASSIFICATION SYSTEMS

1.2.1 Main Features

The main purpose of a classification system for biorefineries is to homogenize their nomenclature in order to allow comparisons within each class and from one class to the other. The Task Group 42 of IEA [3] proposed a classification system based on the following criteria:

- Platforms
- Product groups
- Feedstock groups
- Conversion processes

Fig. 1.1 shows the background network of the IEA classification. The main requirements with which this classification attempts to comply are the following:

- Be unambiguous for all stakeholders in the biorefinery field
- Be explicit with regard to the feedstocks, the platforms and final products
- Reflect the complexity of the biorefinery facility
- Be specific enough in naming each biorefinery

This classification is useful for naming existing biorefineries in an explicit way. The structure of the name of a biorefinery as proposed by the IEA's classification is:

Platforms—Portfolio of final products—Feedstocks—Conversion processes (Fig. 1.2).

The term *platform* means an intermediate product that can generate a series of building blocks, precursors of the final biobased products. The term *final* refers to the products in the outlet of the biorefinery, and as such depends on the boundary of the biorefinery system. The platforms often need to be cleaned up and conditioned to comply with the requirements of the final products.

A few of the biorefineries studied in Chapters 5–7 are described in the next sections using the classification system of the IEA.

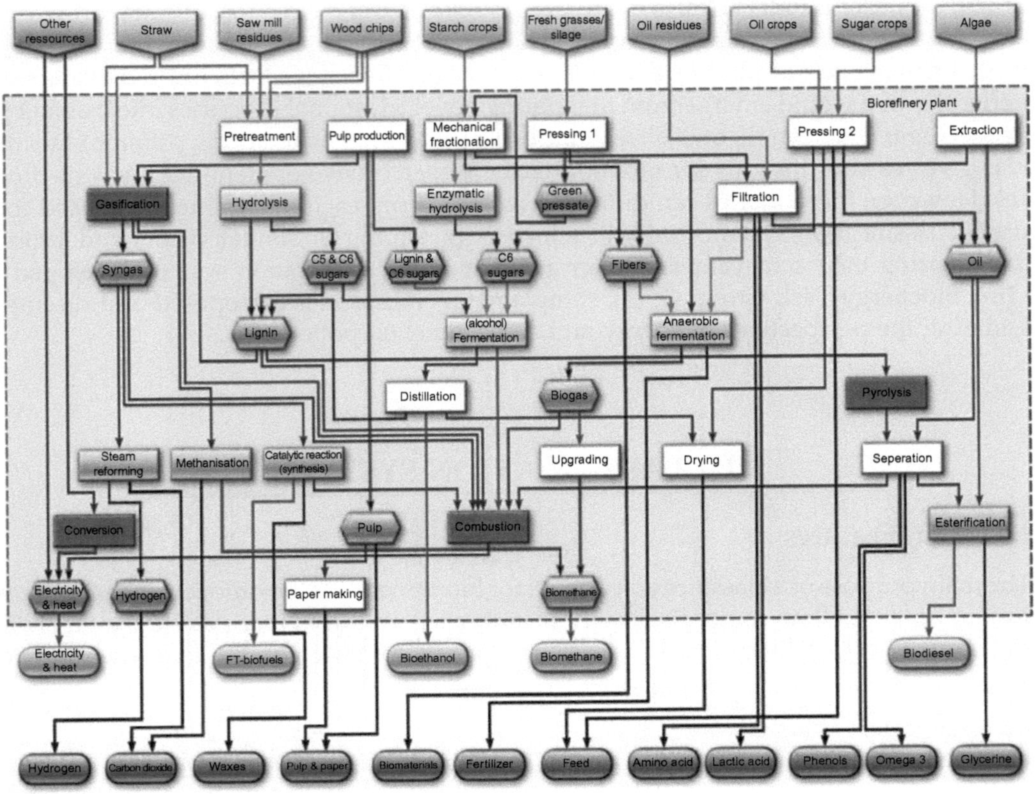

FIG. 1.1 Background network of the IEA classification system [3].

1.2.1.1 *Case 1: A C5/C6 Plant Producing Bioethanol From Vetiver Leaves Using a Biochemical Route*

This plant aims at valorizing wastelands in India for planting vetiver, a perennial grass. The vetiver plant would have two functions: storage of carbon in the soil through roots and use of the leaves as feedstock for producing bioethanol. The production of the bioethanol proceeds with dilute acid pretreatment followed by saccharification, cofermentation of the monomers sugars, distillation, and dehydration. Optimizing the efficiency of the cofermentation of C5/C6 sugars into ethanol would necessitate the use of performant yeasts such as engineered *Saccharomyces cerevisiae*. The lignin fraction of biomass enters a cogeneration plant to fuel the process. Does this second-generation ethanol plant deserve the name of biorefinery? Its unique purpose is to produce bioethanol (Fig. 1.3).

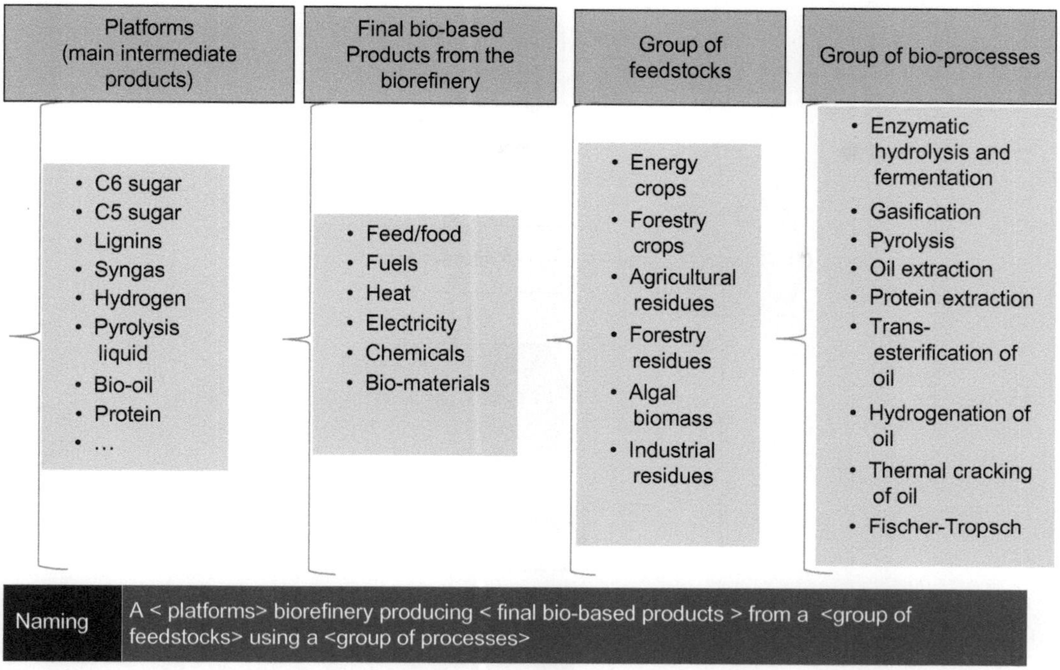

FIG. 1.2 Illustration of the IEA naming system.

Depending on the amount of lignin, a surplus of electricity could be sold to the grid. Adding a biomass-based electrical generator that would work in a synergetic way with the ethanol plant could enhance the generation of electricity. In that case, a biorefinery would emerge based on bioethanol and electricity generation through a cluster of two synergetic bio-industries. Another possibility is to separate the C5 and C6 platforms and valorize each, specifically as in Case 2. Consequently, the second-generation ethanol plant is a *potential* biorefinery.

1.2.1.2 Case 2: A C5/C6 Biorefinery Producing Bioethanol and Furfural From Vetiver Leaves Using a Biochemical Route

Case 2 is a diversification of Case 1 by routing the xylose sugar (C5 platform) to furfural production (Fig. 1.4).

The aim of this biorefinery as described in Chapter 5 is to diversify the biomass used to produce furfural. This is achieved by utilizing part of the vetiver leaves to produce furfural jointly with bioethanol production. The expected benefit is to decrease competition between feed and chemical production since mainly fodder crops presently provide feedstock for furfural production. In Figs. 1.3 and 1.4, the C6 platform appears only implicitly since C6 sugars, the main components of the cellulose, are fermented into ethanol after the saccharification stage, within a stream that also contains lignin and other insoluble compounds.

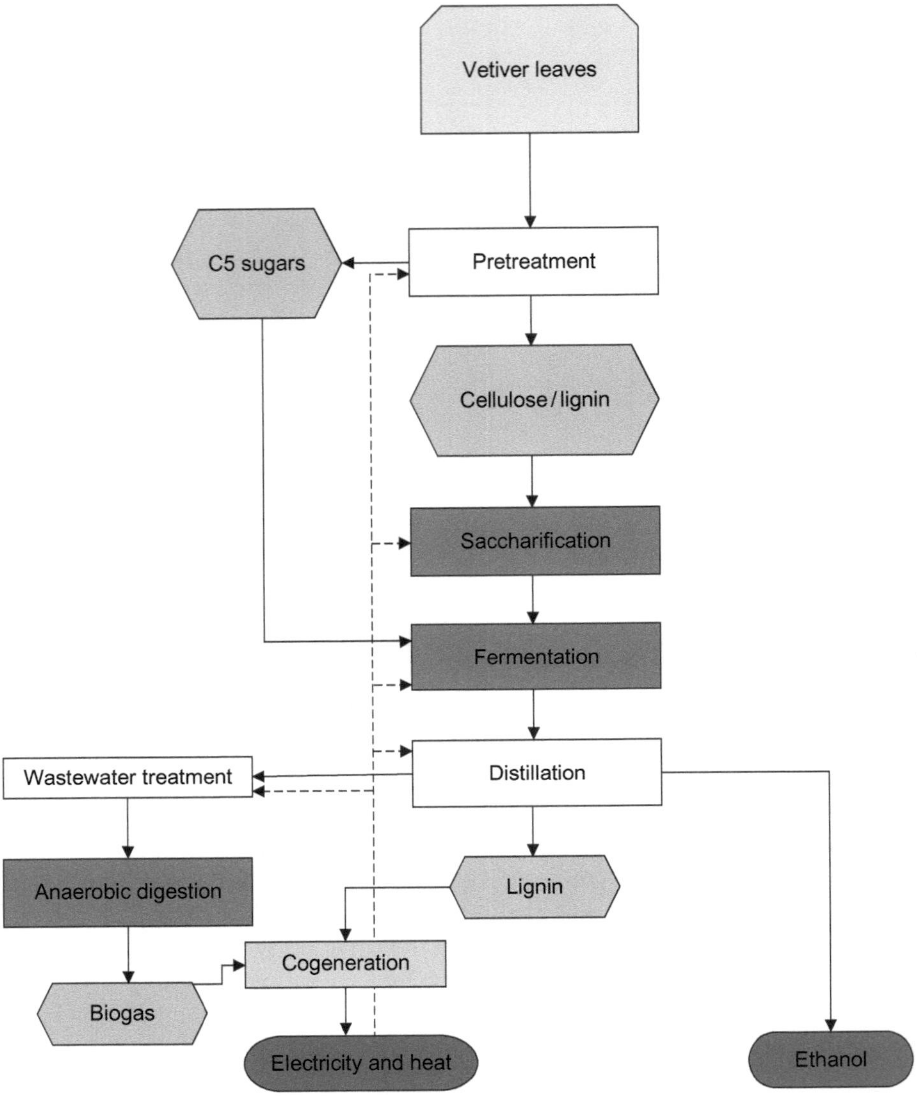

FIG. 1.3 Illustrative Case 1.

1.2.1.3 *Case 3: An Oil Biorefinery Producing Biodiesel, Glycerin, and Protein From Microalgae by a Biochemical Route*

This *prospective* biorefinery is one of the two options presented in Chapter 7. Its main aims are the following (Fig. 1.5):

- Diversify and increase the availability of biomass resources in order to comply with the biofuel mandate in India by developing and processing microalgae in the southern region (Tamil Nadu), a coastal area
- The plant would produce 100,000 tons per year of biodiesel along with protein and glycerin

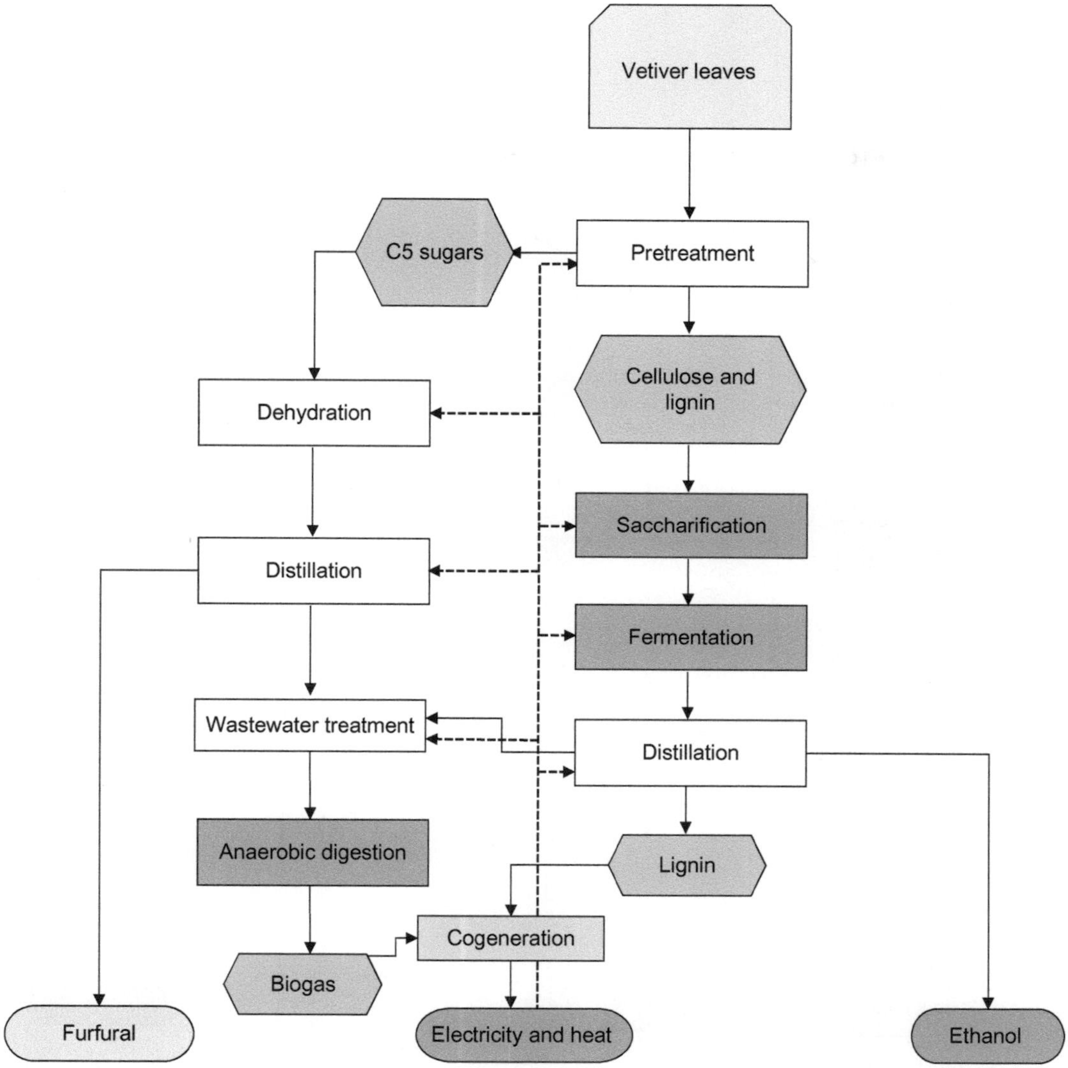

FIG. 1.4 Illustrative Case 2.

- The products of the biorefinery would partially replace fossil diesel, soybean protein, and conventional glycerin

The remaining residues after an anaerobic digestion produce biogas which enters a cogeneration to fuel the process.

1.2.1.4 Case 4: An Oil/C5/C6 Biorefinery Producing Biodiesel, Glycerin, Protein and Succinic Acid From Microalgae Using a Biochemical Route

This biorefinery is the second option presented in Chapter 7. Following the extraction of oil and protein, the residual carbohydrate-rich biomass is fermented into succinic acid (Fig. 1.6).

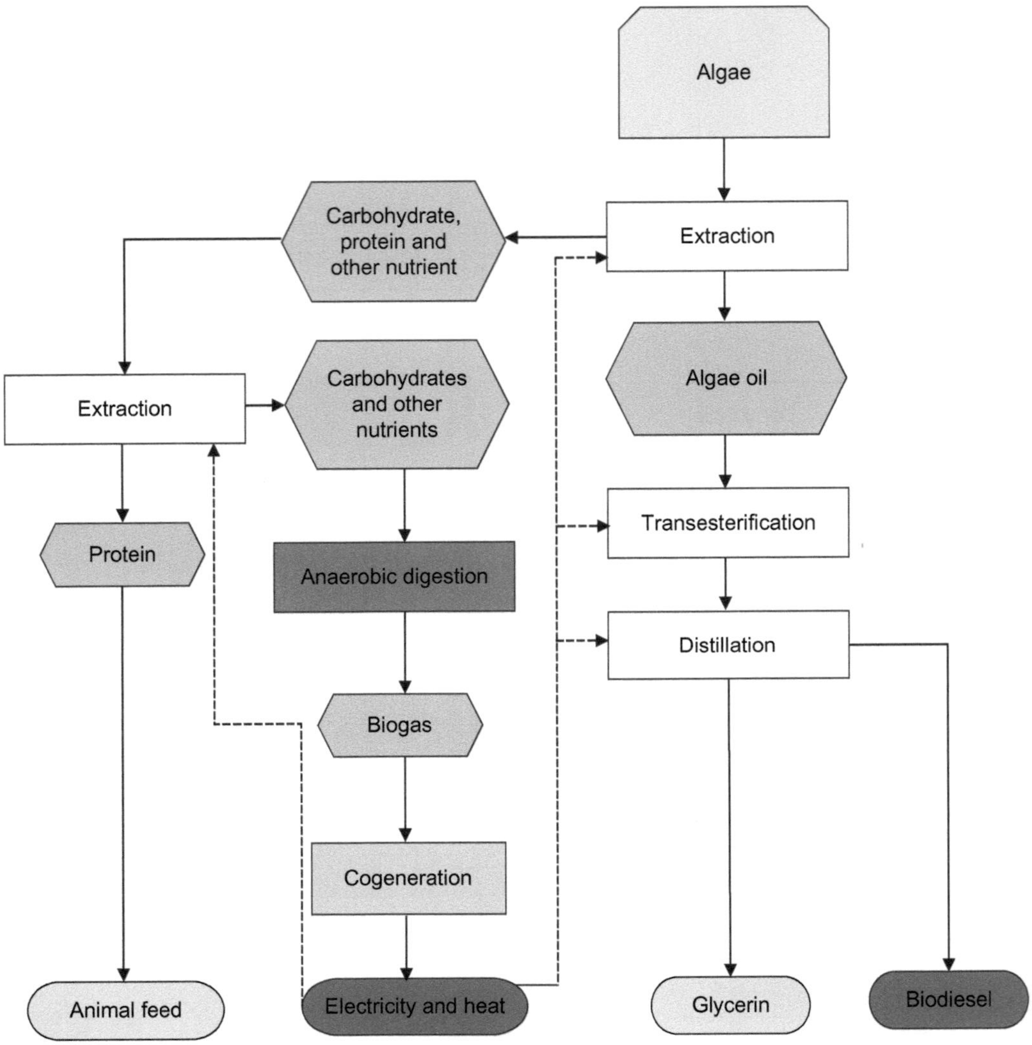

FIG. 1.5 Illustrative Case 3.

The choice between the two options (Cases 3 or 4) would depend on the composition of the microalgae.

In addition to the products that are substituted in Case 3, the conventional succinic acid produced from fossil fuel would be replaced by bio-succinic acid, which would make this biorefinery more environmentally performant.

1.2.1.5 Case 5: A Synthesis Gas Biorefinery Producing Biomethane and Fertilizer From Oil Palm Empty Fruit Bunches Using a Hydrothermal Gasification Route

In this illustrative case, oil palm empty fruit bunches (EFB) in Pará State, Northern Brazil, is supposed to be converted into biomethane and salt brine, using a hydrothermal process.

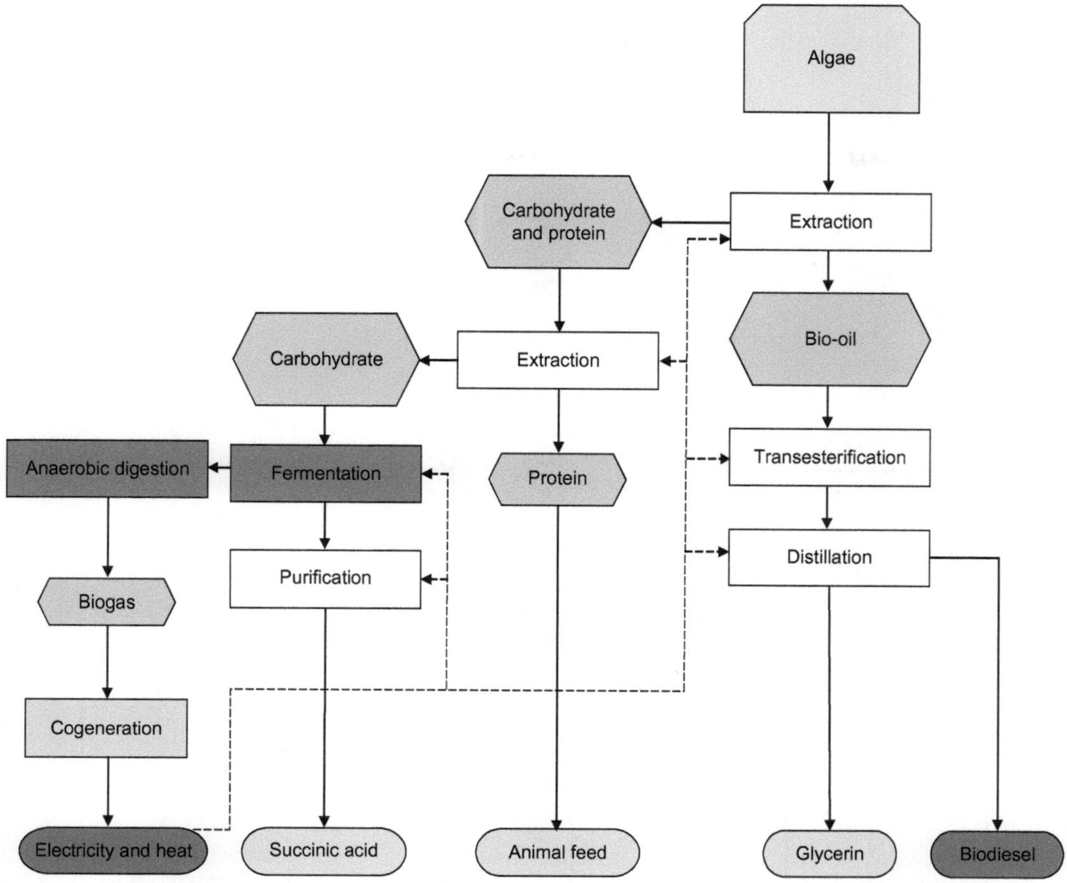

FIG. 1.6 Illustrative Case 4.

The biomethane would replace natural gas imported from Bolivia and the salt brine would substitute for mineral fertilizer consisting of a mix of nitrates, sulfates, phosphates, and magnesium (Fig. 1.7).

1.2.2 Main Platforms of Biorefineries

1.2.2.1 C5/C6 Sugars

C6 sugars are present in a number of feedstocks and can be more or less easy to extract. Extraction is easy from sugar crops such as sugarcane, sugar beet, and sweet sorghum, where glucose and fructose are the main constituents of the juice. They are stored in the form of disaccharides ($C_{12}H_{22}O_{11}$), also referred to as sucrose. As an example, in Brazil the saccharose of the sugarcane is converted into refined sugar and ethanol. The share of sugar crops and molasses for world ethanol production was 31% during the period 2012–14 [6]. Sucrose conversion to bioethanol shows the highest efficiency compared to starch and lignocellulose.

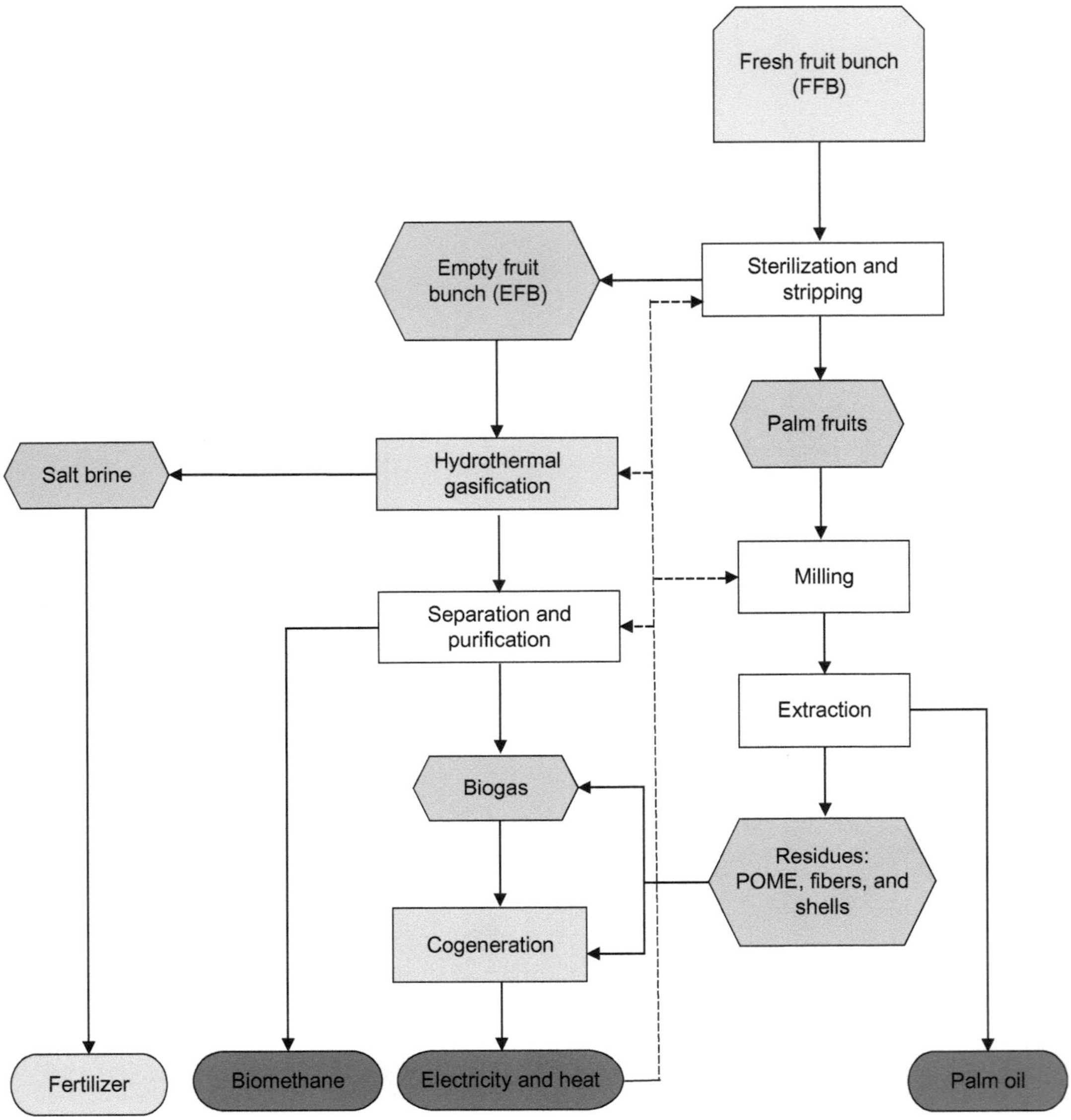

FIG. 1.7 Illustrative Case 5.

However, the price of sucrose is unstable due to the competition with refined sugar. On the other hand, the contribution of starch crops to the world ethanol production during the period 2012–14 was 57%, corn being the starch crop that most contributed, especially in the United States, the first ethanol producing country worldwide. Starch is a polymer of glucose. It is decomposed into monomers using enzymatic hydrolysis with the α-amylase enzyme family (Fig. 1.8A).

In addition to ethanol, a corn wet mill can coproduce corn gluten meal and oil. Indeed, contrary to the dry milling process where the main output is bioethanol and dried corn

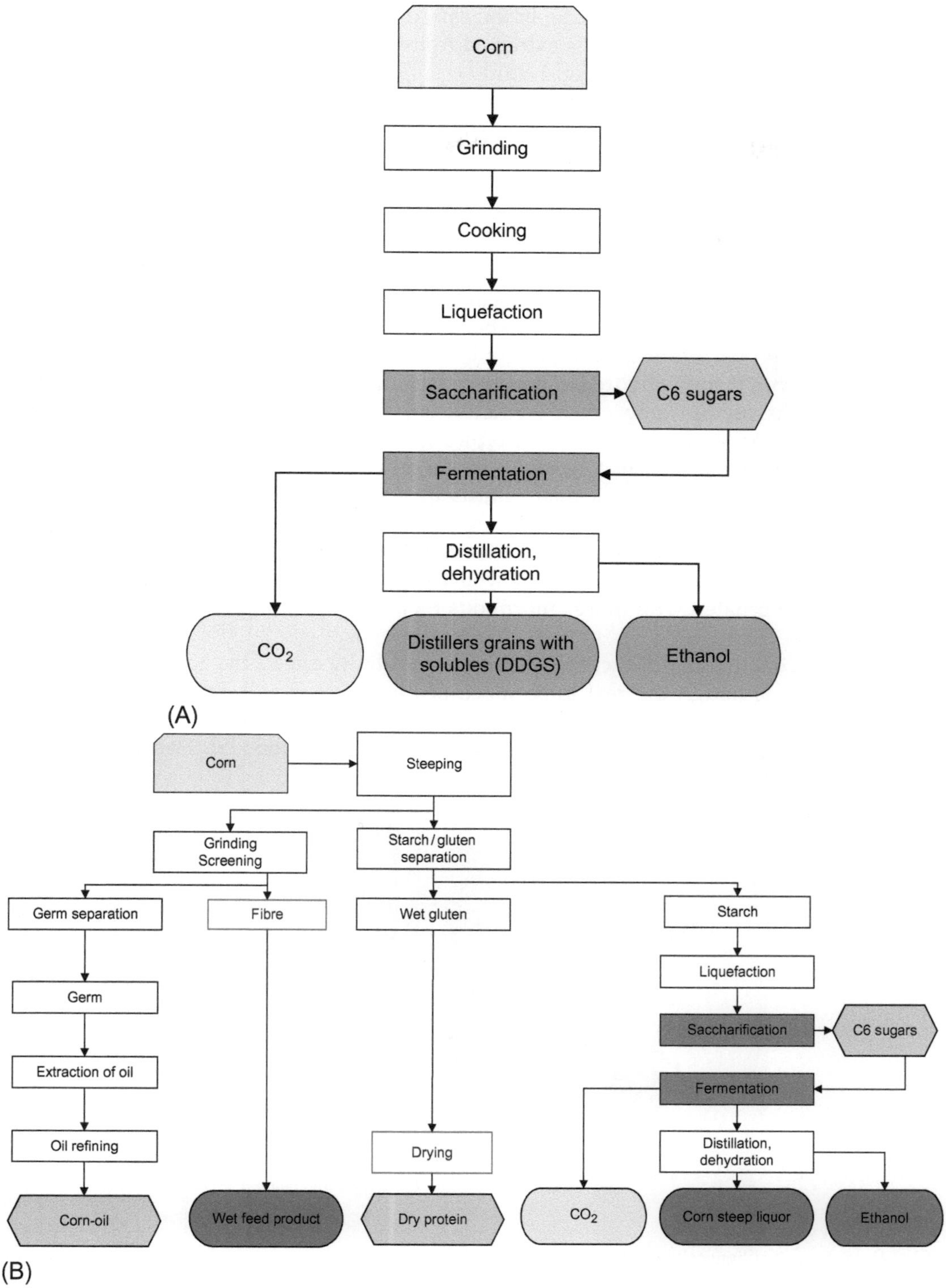

FIG. 1.8 (A) Dry milling process of corn and (B) wet milling process of corn.

distillers' grains with solubles (DDGS), in the case of wet mills the corn is fractionated into starch, fiber, gluten, and germ. Oil is extracted from the germ and gluten is converted into corn gluten meal for animal feed (Fig. 1.8B).

The use of starch for production of bioethanol is severely criticized due to its potential competition with food. Lignocellulosic biomass and particularly agricultural and forestry residues are seen as potential biomass resources that can allow the avoidance of fuel-food competition.

Lignocellulose is one of the most complex sources of C5/C6 sugar platforms. It is mainly composed of cellulose, hemicellulose, and lignin. Cellulose is an unbranched homopolymer constituted of glycosyl units in the pyranose configuration. The linkage connecting the molecules is a β-1,4 glycosidic. Cellulose molecules are long linear chains that are stable and insoluble in water. They are the ideal backbone for a composite fiber design, the one that is actually adopted by cell walls. Single cellulose molecules have the specificity of binding together in bunches through hydrogen bonds. In fact, numerous H-bonds are present among the glucose units of the same chain (intramolecular) and between two adjacent chains (intermolecular). As a result, groups of typically 30–50 cellulose chains are assembled together to give rise to a fiber-like structure of roughly 3–5 nm diameter called micro-fibril. While in the primary wall, cellulose micro-fibrils are randomly oriented, while in the secondary wall they create parallel sheets that are then superimposed on other sheets. This configuration gives particular strength to the wall. However, native cellulose is not totally ordered: the regions where microfibrils are well arranged in a geometrical crystalline pattern are sometimes alternated with more amorphous regions where long-range order is not preserved (Fig. 1.9).

With respect to ethanol production, cellulose crystallinity affects the accessibility to cellulolytic enzymes, although the real importance of this effect is still debated. Furthermore, the spacing between cellulose chains creates a porous network that similarly affects the activity of enzymes: this aspect is also subject to investigations [9]. Hemicelluloses are

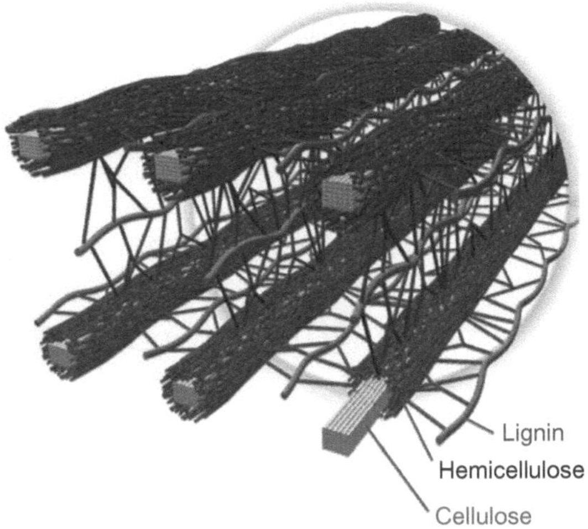

FIG. 1.9 Structure of lignocellulose [7,8].

a family of polysaccharides that have a common structure, being characterized by an equatorial β-1,4 linked backbone and a branched structure. Despite this common point, hemicelluloses are a vast group; sometimes the name is used in the singular for simplicity. However, there is no single hemicellulose molecule with a unique structure. Sometimes, operational definitions based on extractability with a particular solvent are used to define hemicelluloses. Depending on the source, this family might include different molecules. Hemicelluloses can be divided into xyloglucans, xylans, mannans, glucomannans, and β-1,3 1,4-glucans. These names reflect the presence of one or two particular saccharides in the main chain. Different families of plants show a characteristic composition, which is however strongly variable according to the exact species, the particular tissue examined, and the different geographical conditions. Xyloglucans are constituted of a backbone analogous to that of cellulose, with glycosyl residues disposed in a linear fashion: the backbone, however, is branched and bears different residues including xylosyl, galactosyl, and fucosyl. Lignins that are extremely complex and variable phenolic heteropolymers are present in the secondary walls. Their main biological role is to impart toughness and resistance to the plant. Lignin polymers are hydrophobic and thus impermeable to water; they also serve as a barrier to pathogens and are difficult to break. A remarkable fact is that lignin's primary structure is not regular, being rather dictated by the random gathering of different monomeric units. Lignin is hence a very complex polymer to study. The current knowledge about lignin is limited to substructures extracted with different techniques. However, these fragments vary according to the procedure employed to extract them, and caution should be observed in extending a structure model to native lignin, namely the one found in the plant cell wall. High variability is present between different plants as well as in the same plant, according to the tissue. In addition, due to their chaotic structure and complex stereochemistry, it is unlikely that two lignin molecules would have exactly the same constitution even in the same plant.

In addition to the main components (cellulose, hemicellulose, and lignin), a typical compositional analysis would consider materials such as extractives and ash. Extractives are part of the biomass that can be easily dissolved into solvents such as water and ethanol. Examples of extractives include fatty acid, phenols, phytosterols, and soluble sugars. Ash consists mainly of mineral compounds in very variable percentages involved in the biomass itself or in undesirable compounds from soil that are attached to the biomass during the harvest. If the fraction of ash is significant, this is unfavorable to thermochemical conversions due to slagging that affects the operations and shortens the lifetime of the facilities.

There are several possible classification systems for lignocellulosic biomass, such as a botanic one: grass, softwood, hardwood. Another classification is related to the type of land use with a distinction between agricultural lands and forestry lands, and between crops and residues, which gives the three following categories:

- Lignocellulosic agricultural crops
- Forestry crops
- Agricultural and forestry residues

An outline is given for these different categories of lignocellulosic feedstocks. In Chapter 3 of the book, more information is available, particularly on the first category.

LIGNOCELLULOSIC AGRICULTURAL CROPS

Dedicated nonfood crops offer an opportunity for diversifying agricultural production in the case where there is no pressure on the agricultural lands. Perennial lignocellulosic energy crops are of particular interest due to their adaptability to a variety of soils including marginal and degraded ones. A large number of investigations on switch grass undertaken in North America and to a lesser extent in Europe for several decades have shown its capability to provide significant amount of solid biocombustible or feedstock for biofuels [10]. Another promising perennial crop that can contribute significantly to the C5/C6 platforms is miscanthus. Furthermore, Chapters 4 and 5 deal respectively with the cases of sugarcane and vetiver. The contribution of lignocellulosic agricultural crops to the production of C5/C6 platforms in the future will depend on availability of land for food.

FORESTRY CROPS

Short rotation coppice (SRC) and short rotation forestry (SRF) aim at planting fast-growing trees such as eucalyptus, willow, and poplar in dense cultures and harvesting them more regularly and rapidly compared to a conventional forestry management. Interests in SRC and SRF are multiple:

- Reduction of wood uptakes on naturally regenerated forests
- Reforestation and rehabilitation of degraded forest lands
- Integration with agriculture, especially in areas where surplus agricultural lands exist
- Provision of feedstocks for several services and activities such as construction, bioenergy, and biorefineries

Forestry crops are used in the framework of "planted forests," defined by the United Nations' Food and Agricultural Organization (FAO) as "those forests predominantly composed of trees established through planting and/or deliberate seeding of native or introduced species" [11]. The importance of planted forests for wood production is growing rapidly. Carle and Holmgren [11] forecasted that the planted area in the world would increase from 261 Mha in 2005 to 303–345 Mha in 2030. However, biorefineries would compete for feedstock procurement with the well-established timber industries. Due to the expected high delivery cost, the contribution of forestry crops to the biorefineries feedstock would be marginal.

AGRICULTURAL AND FORESTRY RESIDUES

The contribution of agricultural and forestry residues to biorefineries feedstocks is more promising compared to forestry crops. While lignocellulosic agricultural crops would contribute in areas where there is no competition with food for land procurement, residues can be a useful complement if there is a large enough amount left on the field for soil regeneration. Smil [12] estimated the annual production of agricultural residues during the mid-1990s to about 1.4 times the amount of annual harvest of crops. However, the low density of agroforestry residues strongly limited the possible harvestable amount. The agricultural residues production RP_{ij} (kg year^{-1}) for a country i and a crop j is:

$$RP_i = \sum_{j=1}^{J} \left(A_{ij} \times RPR_{ij} \times Y_{ij}\right) \tag{1.1}$$

where A is the harvested area (ha), RPR is the ratio to product, and Y is the annual yield of the product per year (kg ha^{-1} $year^{-1}$).

Eq. (1.2) gives the estimation of RPR as a function of the *harvest index* (HI) that is the ratio of crop yield to the total aboveground phytomass of the crop [12].

$$\mathrm{RPR} = \frac{1-\mathrm{HI}}{\mathrm{HI}} \tag{1.2}$$

For most of the agricultural crops, HI is around 0.40–0.60. However, there is a high imprecision and uncertainty on the data used to estimate HI. Forestry residues show a significant potential as well. These residues include direct wastes from forest management as well as residues from wood-based industries such as sawmill plywood, and particleboard production. The strengths of using agroforestry residues for production of C5/C6 platforms are their large availability, the possibility of avoiding competition with food, and the possibility of achieving a low environmental footprint. The main weakness is their low energy density, leading to an expensive cost of logistics. Finally, the availability of agricultural residues should be carefully estimated, taking into account all competing uses, the existing barriers, and the need to secure the whole supply chain. The assumption that lignocellulosic residues will remain cheap in the long term is not relevant for a sustained vision of biorefinery development. Besides the cost of logistics, which would be significant, economic assessment must anticipate an increase of the farm gate price of feedstocks.

The potential of given feedstocks to be converted into C5/C6 platforms depends on the share of holocellulose in their composition. In this respect, Table 1.1 shows typical compositions of selected lignocellulosic feedstocks in cellulose, hemicellulose, and lignin fractions. It also gives other useful properties such as elemental composition and lower heating value [13–29]. The lignocellulosic feedstocks in Table 1.1 represent categories such as herbaceous crops (miscanthus, switchgrass, and energy cane—a hybrid of sugarcane composed of more fiber and less sucrose), and one woody crop (eucalyptus).

FRACTIONATION AND SACCHARIFICATION OF LIGNOCELLULOSIC FEEDSTOCKS

Fractionation of the feedstock into its main components is the major step of the conversion into C5/C6 platforms. Fractionation consists of pretreatment and saccharification. The goal of pretreatments is to recover a high percentage of holocellulose. As far as biorefineries are concerned, mild conditions are required in order to limit degradation of part of the holocellulose and the lignin and formation of inhibitors for the subsequent phases. Pretreatment of lignocellulosic materials can be classified into the following categories (Table 1.2) [30–33].

- Physical pretreatments

  This category comprises combinations of methods such as chipping, grinding, and milling aimed at reducing the size of the lignocellulose. After grinding, for instance, the size of the particles is typically in the range of 0.2–2 mm. Physical pretreatments allow increasing the specific surface area of the particles, and also contribute to reducing the crystallinity of the cellulose, enhancing its digestibility.

TABLE 1.1 Typical Composition of Selected Lignocellulosic Feedstocks

		Herbaceous Crops			Agricultural Residues		Woody Plant
Composition and Properties	**Units**	**Miscanthus**	**Switchgrass**	**Energy Cane**	**Corn Stover**	**Wheat Straw**	**Eucalyptus**
Productivity	Mg ha^{-1} $year^{-1}$	1–44	7–10	26.7	1.2–3.6	0.1–1.75	33
Moisture	%WM	4–5	13–65	73	15–35	5–8	59–63
Components							
Cellulose	%DM	40–60	31–35	43	31–38	32.6	47–50
Hemicellulose	%DM	20–40	24–28	24	19–25	22.6	13–14
Lignin	%DM	10–30	17–22	22	17–21	16.9	27–28
Extractives	%DM	0.3–2.2	5–17	nr	3–12	13.0	0.6–4.2
Ash	%DM	2.2–3.5	5–6	0.8	10–14	10.2	0.6–1.2
Juice	%WM	n.a.	n.a.	54	n.a.	n.a.	n.a.
Juice sugars	% Juice	n.a.	n.a.	10	n.a.	n.a.	n.a.
Proximate analysis							
Volatile matters	%DM	72.6–78.2	76.6–77.9	78.6–81.1	70.2–74.8	69.4	79.5–82.5
Fixed carbon	%DM	15.1–20.4	17.0–22.1	17.0–18.8	20.0–22.6	21.5	17.2–19.0
Ultimate analysis	%DM						
C		47.1–49.7	46.6–48.5	48.2–49.0	44.8–48.0	43.9	49.5–49.9
H		5.4–5.9	5.3–5.7	5.5–6.0	5.4–6.1	5.3	5.7–6.3
O		41.4–44.6	38.2–42.1	39.2–41.6	37.0–41.4	38.8	42.0–43.5
N		0.1–1.0	0.4–0.7	0.2	0.6–0.7	0.6	0.1–0.5
S		0.07–0.19	0.07–0.11	0.02–0.03	0.06–6.1	0.2	0.01–0.04
Energy content	MJ kg^{-1}						
LHV		17–20	18–19	19.0	18–19	17.4	20
Sources		[13]	[14,15]	[14,16]	[14,17,18]	[14,19,20]	[14,21,22]

WM, wet mass; DM, dry mass; n.a., not applicable; nr, not reported.

- Physico-chemical pretreatments
  The methods classified under this category are based on the combined action of high variation of pressure and chemical reactions on the structure of the lignocellulose. A few examples of physico-chemical pretreatments are steam explosion [34], ammonia fiber explosion (AFEX), and carbon dioxide explosion.

TABLE 1.2 Selected Pretreatment Methods of Lignocellulosic Biomass

Pretreatment	Description	Advantages	Drawbacks
Physical			
Chipping, grinding Milling	Mechanical reduction of the materials size: 10–30 mm after chipping, 0.2–2 mm after milling or grinding	-. Reduction of the crystallinity of the biomass -. Increase of the contact surface for enzymatic hydrolysis	High energy requirement; the energy consumption may be higher than the energy content of the biomass
Physico-chemical			
Steam explosion	The biomass is exposed to a high pressure steam and explodes after a sudden decompression	Low energy requirement compared to physical pretreatment	Less effective for softwood, need for additional catalyst in that case
Ammonia fiber explosion (AFEX)	The same compared to steam explosion but in the case of AFEX, the biomass is exposed to liquid ammonia	Good results in the cases of herbaceous and grass feedstocks	Less effective results when the feedstock contains high percentage of lignin
Carbon dioxide explosion	The same process compared to steam explosion and AFEX but the biomass is exposed to carbon dioxide at a supercritical temperature	Good results reported in cases such as alfalfa and bagasse of sugarcane	Less effective compared to steam explosion and AFEX
Chemical			
Acid hydrolysis	Hydrolysis in the case of biorefineries is performed in mild conditions with, e.g., sulfuric acid concentration often below 4%. The optimal temperature depends on the solid loadings and the type of reactor (batch vs. continuous-flow)	Good susceptibility towards enzymatic hydrolysis was reported for a number of feedstocks including hardwoods, grasses, and agricultural residues	Depending on the severity of the acid hydrolysis, inhibitors may be produced leading to low performances of saccharification and fermentation
Alkaline hydrolysis	Sodium, ammonium hydroxides have been used for dilute alkaline hydrolysis and resulted in delignification of the biomass	Increase of the internal contact surface, depolymerization, dislocation of the structural linkages between the lignin and the holocelluloses	Less effective for softwoods with high lignin content
Organosolv	Use of mixtures or organic and inorganic solvents	Good delignification of the biomass	Necessity to remove the solvents to avoid inhibition of the saccharification and the fermentation stages
Biologic			
Use of microorganisms	Microorganisms such as fungi are used to decompose the biomass	Environmentally friendly pretreatment	The decomposition process is low

- Chemical pretreatments

  Several chemical pretreatments can be used to decompose the structure of the lignocellulose, including the following: liquid hot water, dilute acid hydrolysis, dilute alkaline hydrolysis, Organosolv, and oxidative delignification.
- Biological pretreatments

  These pretreatments proceed with degradation of hemicellulose and lignin with microorganisms such as brown, white, and soft-rot fungi, which liberate the cellulose for further hydrolysis.

SACCHARIFICATION

After the pretreatment stage, most of the hemicellulose fraction is decomposed in monomers. Using enzymatic hydrolysis, the cellulose fraction is then saccharified into hexoses. To achieve this depolymerization, a cocktail of enzymes is used where endoglucanase randomly cleaves the cellulose fraction in the amorphous zones, fragmenting the material; exoglucanase cleaves at the extremities of the reducing and nonreducing parts. Finally, β-glucosidase breaks the cellobiose molecules. Degraded compounds from the pretreatment as well as cellobiose and to a lesser extent the produced glucose can inhibit cellulose activity. Thus, the design optimization of the pretreatment and saccharification stages should take into account the particular structure of the lignocellulosic structure that changes from one biomass to another.

1.2.2.2 Lignin Platform

Lignin is the second most abundant fraction of the lignocellulose behind cellulose. Despite the importance of its availability, lignin is mostly combusted for process steam and electricity generation. Only a marginal part of the lignin produced worldwide is presently valorized into other products such as dispersant in cement and gypsum, emulsifier and chelating agent in treatment of industrial effluents [35]. One of the main difficulties in converting lignins into value-added products is the high variability of its composition. Since the works of Creighton et al. in the 1940s, lignins are classified according to the importance in their composition of the three basic unit structures: 4-hydroxphenylpropene (H), guaiacylpropane (G), and syringylpropane (S) [36–39]. These investigations identify three types of lignin giving rise to the following products, respectively: vanillin (G type), vanillin and syringaldehyde (GS type), and 4-hydroxy-benaldehyde, vanillin, and syringaldehyde (HGS type). Several researchers have contributed to refining this typology, including Faix [40] who proposed a classification that subdivides the three basic ones into more homogeneous groups. Despite this effort, the issue of categorization of lignins is still under investigation. One factor that increases the complexity of lignins categorization is that the substrate in which they are and the extraction methods may affect their composition and properties. The first suppliers of lignins worldwide, the pulp and paper industries, extract them from black liquor. This substrate includes, besides lignins, a large variety of chemical compounds dependent on the technology used by the pulp and paper industry. In that respect, different types of lignins are distinguished, from a technical point of view. Kraft lignin refers to the use of alkaline degradation as the wood process, while sulfonated lignin stems from the use of a sulfite pulping process. Other technical categories of lignins are Organosolv lignins, ionic liquid lignins, soda lignins, and hydrolysis lignins. Furthermore, the lignin recovery methods affect the characteristics, particularly the molecular weights of the recovered lignins. Basic recovery methods include the following: differential precipitation, use of selective solvents, use of enzyme-based oxidation methods, biomimetic oxidation methods, and organometal catalysts. García et al. [41] used differential precipitation to isolate lignin from black liquor that resulted from pulping of *Miscanthus sinensis* with 7.5% soda. Different lignin-rich permeates appeared when decreasing the pH of the black liquor from 12.64 to 0.72 by adding sulfuric acid. Toledano et al. [36] obtained different lignin-rich permeates from black liquor by ultrafiltration using membrane technology followed by acid precipitation. Passoni et al. [42] used three different solvents (an alcohol, a cyclic ether, and a ketone) to fractionate a softwood kraft lignin. Lange et al. [43] reviewed oxidative valorization of lignins. Depending on its native structure, the substrate in which it lays down, the isolation method, the structure of lignin, and consequently its properties and its possible valorizations vary significantly, which makes a lignin platform a treasure mine.

1.2.2.3 *Bio-syngas Platform*

Synthesis gas (syngas) derives from gasification of hydrocarbons (e.g., coal) or carbohydrates (e.g., biomass) followed by gas cleaning and gas conditioning (Fig. 1.10).

Gas conditioning mainly consists of adjusting the ratio H_2/CO. The water-gas-shift process converts CO into CO_2 by reacting with hot water in the presence of a catalyst. CO_2 is removed using different techniques, such as adsorption on solid sorbents.

GASIFICATION PROCESS

The gasification process includes three main steps [44]. The first step, which occurs below 125°C, consists of dehydration; during the second stage a pyrolysis of hemicellulose, cellulose and partly of lignin occurs at temperatures that are up to 500°C; finally, during the third stage, the temperature being above 500°C, the residual lignin is pyrolyzed. The efficiency and composition of the product gas depend on a few factors including biomass characteristics, operating parameters, and types of gasifier.

INFLUENCE OF THE BIOMASS CHARACTERISTICS

First, it is worth distinguishing between wet biomass (e.g., 50% weight basis or higher moisture) and dry biomass. In the case of the former, the energy needed to dry the biomass during the gasification process becomes significant. Thus, the use of hydrothermal gasification is recommended (Chapter 6). Conventional gasification is concerned with dry biomass. The moisture content is of high importance; it is the key factor of weight loss during the first stage of the gasification process. For biomasses with a high content of holocellulose, the second phase is faster [45]. Actually, during the second phase there are interactions between cellulose and hemicellulose and between cellulose and lignin, which governs the formation of gas, tar and char and the composition of tar [46,47].

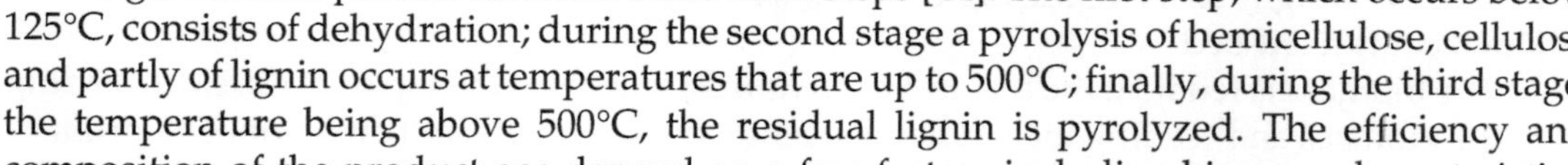

FIG. 1.10 From biomass to bio-syngas.

TEMPERATURE AND OTHER INFLUENTIAL OPERATING FACTORS

The process temperature is the operating factor that mostly determines the composition of the crude bio-syngas. The ratios between hydrogen, carbon monoxide, carbon dioxide, and methane change according to the process temperature. In general, with high temperatures (900–1000°C or above), the content of hydrogen increases. High temperatures are also favorable to the destruction and reforming of the tars. Types of oxidizing agents (e.g., air, steam) and their flow rates, biomass flow rate and the operating pressure affect the efficiency of the gasification and the composition of the crude syngas. Optimal operating factors are dependent on the gasification technology.

GASIFICATION TECHNOLOGIES

Based on criteria such as: the way the biomass is fed into and the flows are moved inside, the oxidizing agent, the operating temperature and pressure, the mode of fueling, gasifiers can be classified into various categories [48,49].

- Updraft fixed bed

  In this type of gasifier, the biomass enters at the top, whereas the oxidizing agent (air, oxygen, or steam) is fed in at the bottom. The gases leave at the top of the gasifier at a temperature around 500°C. The ash falls down and is collected at the bottom. Updraft fixed bed gasifiers are also referred to as counter-flow gasification. Their technology is simple and well proven. Their main drawback is that the remaining tar in the syngas is in the range of 10–20% weight basis, requiring cleaning up.
- Downdraft fixed bed

  The biomass is fed in at the top and the oxidizing agents at the top or at the sides leading to the biomass and feeding the oxidizing agent moving in the same direction. The primary produced gas flows downward, reacts with the produced char at a temperature in the range of 800–1200°C and generates hydrogen and carbon monoxide. The process is also referred to as concurrent-flow gasification. It converts almost all produced tar into gases. The process is simple and clean. However, it must be fed in with dry biomass <20% moisture. The temperature of the syngas exiting the gasifier is still high and about 6% of the carbon remains unconverted.
- Entrained flow

  The biomass is powdered and is fed into the gasifier at the top with pressurized oxidizing agents. At the top of the gasifier, a turbulent flame ignites part of the biomass and generates heat at high temperature in the magnitude of 1200–1500°C that converts the biomass into hydrogen and carbon monoxide. The syngas exits at the bottom as well as the molten slag resulted from melting ash.
- Bubbling fluidized bed

  The biomass is fed into the gasifier from the side in the form of fine materials. Then it lies in the bottom where it combusts with the oxidizing agents that are blown in the bottom upwards at a velocity enough to agitate the biomass material. They can be pressurized and the operating temperature is close to 900°C. The syngas leaves the gasifier at the top. The advantages include the yield of a uniform product gas, possible high conversion with low tar and unconverted carbon percentage.

- Circulating fluidized bed
  Biomass is fed into the gasifier from the side, and is suspended and entrained by blowing oxidizing gases fed into the gasifier in the bottom upwards. The syngas formed from the combustion entrains part of the biomass particles which are separated from the syngas using a cyclone. The operating temperature is below 900°C to avoid ash melting. The process can be pressurized.
- Dual fluidized bed
  The gasifier is composed of two chambers, the combustor and the gasification chamber. The oxidizing agent is fed to the gasification chamber at the bottom. The char is burnt in the combustion chamber and the hot flue gas is used in the gasification chamber to improve the gasification process. The operating temperature is below 900°C to avoid ash melting. The process can be pressurized. The produced syngas exits the gasification chamber at the top.
- Plasma
  This very high temperature gasifier can be used to convert mixed feedstocks composed of biomass and inorganic materials (e.g., auto shredder wastes, construction, and demolition wastes). The temperature is in the range of 1500–5000°C. The process uses plasma torches. The process vitrifies the inorganic materials while the organic ones produce a high-quality syngas.

GAS CLEANING

After the gasification process, the product gas is mainly composed of hydrogen, carbon monoxide, carbon dioxide, methane, nitrogen, tar, ash, other compounds and impurities. As mentioned in the gasification section, the composition of the product gas depends on that of the biomass feedstock and on the characteristics and the operating conditions of the gasification process: temperature, pressure, reaction time, oxidizing agents, flow rates, catalysts, types of gasifier. Depending on the projected use of the syngas, the requirements of the composition are different. In general, the required content of hydrogen is higher than 50%. Independent of the uses, the reduction of content of tar and inorganic impurities is one of the requisites for having a syngas that responds to basic specifications.

TAR REDUCTION

Tar is all the organic compounds present in the product gas with a molecular mass larger than benzene [50]. That complex mixture of condensable hydrocarbons condenses below a given temperature, termed as the dew point, the magnitude of which depends on the composition of the tar. There are several sampling and analytical tools for measurement of the tar content in product gases, which leads to a high uncertainty in the results [51]. The common measurement tool is the gas chromatograph (GC). According to IEA T33 [52], there are several classification systems for tars:

- Classification according to temperatures
  - Low temperature: Complex composition, fraction of polar compounds. Tars formed typically in updraft fixed gasifiers.
  - High temperature: Composed mainly of nonpolar aromatic compounds. These tars are typically formed in downdraft fixed beds and in fluidized beds.

- Classification according to the reactivity [53,54]
  - Primary tar compounds, e.g., acetic acid, acetol, phenol, guaiacol, furfural
  - Secondary tar compounds, e.g., phenol and olefins
  - Alkyl tertiary tar compounds, e.g., toluene, indene
  - Condensed tertiary tar compounds, e.g., poly-aromatic hydrocarbons (PAH)
- Classification according to water solubility/condensation
  - Class 1: gravimetric tars, not GC-detectable tars
  - Class 2: heterocyclic (high solubility), e.g., cresol, phenol, pyridine
  - Class 3: light aromatic, e.g., xylene, toluene
  - Class 4: light PAHs (2–3 rings), e.g., naphthalene, phenanthrene
  - Class 5: heavy PAHs (4–5 rings), e.g., fluoranthene, pyrene

Processes for tar reduction are classified into two categories:

- Primary methods

  They consist of reducing the tar formation during the gasification. These include the design of gasifier and choice of the operation parameters, such as temperature, pressure, oxidizing agent, residence time, and in-bed catalysts and additives.
- Secondary methods

  These methods include physical processes such as cyclones, electrostatic precipitators, scrubbers, filters and thermal-catalytic processes, e.g., cracking, partial oxidation, catalytic reforming and plasma processes.

REDUCTION OF INORGANIC COMPOUNDS

Depending on the applications for which the syngas serves, the content of inorganic compounds is limited to a certain level. The concerned inorganic materials are mainly sulfur, nitrogen compounds, halogens, and volatile metals. In the case of nitrogen compounds, the content can be reduced by choosing an appropriate oxidizing agent, e.g., oxygen in place of air. In general, various cleaning methods exist to achieve the application requirements. For instance, in the case of sulfur content, the cleaning processes are in two categories: wet processes and dry processes [55].

- Wet processes
  - Monoethanolamine process
  - Diethanolamine process
  - Methylethanolamine process
  - Alkaline hydroxide processes
  - Alkaline carbonate processes
  - Oxidative processes
- Dry processes
  - Adsorption using metal oxides
  - Adsorption using zeolites
  - Adsorption using activated carbons

1.2.2.4 *Pyrolysis Liquid Platform*

Pyrolysis liquid results from the thermochemical decomposition of solid biomass in partial or complete absence of oxygen in order to avoid combustion. In certain conditions of the thermochemical reactions, the solid biomass gives rise to dark brown organic liquids that are the result of a rapid depolymerizing of the holocellulose and the lignin in the solid biomass. Pyrolysis liquids are complex mixtures of water, guaiacols, catechols, syringols, vanillins, furancarboxaldehydes, isoeugenol, pyrenes, acetic acid, formic acid, and other carboxylic acids [56]. Pyrolysis liquids are a complex and heterogeneous product that can involve several hundreds of compounds [57–59]. These compounds can be classified into the following categories:

(1) Hydroxyaldehydes
(2) Hydroxyketones
(3) Sugars and dehydroaldehydes
(4) Carboxylic acids
(5) Phenolic compounds

Depending on the focused uses, the liquids can be refined for getting the required product.

1.2.2.5 *Bio-oil Platform*

Bio-oils mean oils derived from biomass. They include a variety of organic products such as vegetable oil, microalgae oil, tall oil, fatty, grease, used oil. Thus, their naming should depend on the biomass from which they derive and the technique of production. The main constituents of bio-oils are in the group of lipids. Oil and fat are esters formed between glycerol and fatty acids (FA). With regard to their physical properties (e.g., melting point), it is worth distinguishing saturated fatty acids (SFA) and unsaturated fatty acids (UFA). An SFA is a long-chain carboxylic acid that contains only carbon-carbon single bonds (e.g., palmitic acid), whereas a UFA fatty acid is a long-chain carboxylic acid that contains one or more carbon-carbon double bonds (e.g., linoleic acid). Polysaturated fatty acids contain more than one carbon-carbon double bond.

VEGETABLE OILS AND FATS

Vegetable oils and animal fats are the most common lipids in nature. They are mainly complexes of triesters of glycerol called triacylglycerols (TAGs) or triglycerides. They are nonpolar and hydrophobic mixtures of TAGs. Vegetable oils are mixtures of TAGs that are liquid because of their high proportion of UFA: e.g., corn oil (88% UFA), soybean oil (87% UFA), and olive oil (90% UFA). Contrary to vegetable oils, fats contain a higher proportion of saturated fatty acids: e.g., butter (47% SFA), human fat (37% SFA). Polar lipids such as nonacylglycerols may be present in fats and oil as well.

MICROALGAE OIL

The content of lipids depends on the species of the microalgae. However, several research works report that, under stress conditions, many microalgae are capable of altering their lipid metabolisms and accumulating TAGs. Nitrogen starvation is an efficient strategy for increasing lipids in general and particularly in TAG [60]. Sun et al. [61] studied the effects of different nitrogen, phosphoric concentrations and light intensities on growth and oil accumulation and TAG increase for nine *Chlorella* strains. The growth conditions that enhance saturated or unsaturated lipids are subject to investigation as well. Lin and Wu [62] studied the growth of *Chlorella* sp. Y8-1 under different cultivation conditions including carbon source and showed the influence of the presence of saturated and unsaturated lipids. Microalgae provide a promising oil container that can be subject to flexible design throughout many strategies. In the case of microalgae as well as in the oil crops case, an oil platform can be associated with other platforms, such as C5/C6 or protein platforms.

1.2.3 Products of Biorefineries

The different platforms presented in this chapter are not exhaustive. Furthermore, they can be associate or combined. Many types of products result from these biorefineries, such as feed/food products, energy, materials, chemical products, and biopolymers.

1.2.3.1 *Energy Products*

C5 and C6 sugars can ferment and produce bioethanol. They can also feed algae and convert into oil and then into biodiesel or green diesel through a hydro-treating process. Lignins can fuel cogeneration and can be converted into steam and electricity. Pyrolysis liquid can be used in a heating system or be refined into transport fuel. Syngas can be converted into biomethane and can be used as a biocombustible for heat and electricity. It can be used directly in an electricity plant. It can also feed a Fischer-Tropsch process that can yield diesel, gasoline, and a fuel similar to liquefied petroleum gas (LPG). Syngas can also be a building block for hydrogen production. Oil platforms can be used for biodiesel or hydrogenated diesel (green diesel).

1.2.3.2 *Chemical Products*

Many biobased chemicals from the biorefineries are already near or in the market. Others are at pilot or demonstration phase. These products can be, to different degrees, building blocks as well. The US Department of Energy (DOE) selected in 2004 the top 10 biobased chemicals that would appear on the market in future [63]. Bozell and Petersen revised this list in 2010 [64] and proposed nine criteria to evaluate their opportunities (Table 1.3).

Then they characterized the new top chemical opportunities according to these criteria. Ethanol as a precursor for several chemicals, glycerol and its derivatives and sorbitol perform well against each of the nine criteria. Table 1.4 shows the performance of the selected promising biobased chemicals.

1.2.3.3 *Biopolymers*

In addition to the biochemicals cited in the previous section, biopolymers constitute secondary chemicals that can be synthesized from precursors such as lactic acid and succinic acid. The strongest market dynamics are envisioned for polylactic acid, polyhydroxyalkanoate, as well as bio-ethylene that can be used to produce biobased plastics.

TABLE 1.3 Criteria for Evaluating Opportunities of Biobased Chemicals From Carbohydrates [64]

1.	*The compound or technology has received significant attention in the literature.* A high level of reported research identifies both broad technology areas and structures of importance to the biorefinery
2.	*The compound illustrates a broad technology and is applicable to multiple products.* As in the petrochemical industry, the most valuable technologies are those that can be adapted to the production of several different structures
3.	*The technology provides direct substitutes for existing petrochemicals.* Products recognized by the chemical industry provide a valuable interface with existing infrastructure and utility
4.	*The technology is applicable to high volume products.* Conversion processes leading to high volume functional equivalents or utility within key industrial segments will have particular impact
5.	*A compound exhibits strong potential as a platform.* Compounds that serve as starting materials for the production of derivatives offer important flexibility and breadth to the biorefinery
6.	*Scaleup of the product or a technology to pilot, demo, or full scale is underway.* The impact of a biobased product and the technology for its production is greatly enhanced upon scaleup
7.	*The biobased compound is an existing commercial product, prepared at intermediate or commodity levels.* Research leading to production improvements or new uses for existing biobased chemicals improves the utility
8.	*The compound may serve as a primary building block of the biorefinery.* The petrochemical refinery is built on a small number of initial building blocks: olefins, BTX, methane, CO. Those compounds that are able to serve an analogous role in the biorefinery will be of high importance
9.	*Commercial production of the compound from renewable carbon is well established.* The potential utility of a given compound is improved if its manufacturing process is already recognized within the industry

1.3 REVISITING THE CLASSIFICATION SYSTEM—GOALS AND SCOPES OF BIOREFINERIES

Biorefineries hold the promise of partly substituting for petroleum refineries. Besides this goal, the possibility of valorizing the whole biomass in a sustainable way is another goal. Existing biorefineries such as the wet mill of corn are already examples that can be extended to the valorization of straw. Algal biorefineries also have the potential to valorize the different parts of the biomass into energy and high value-added products. Other goals are reduction of GHGs, contributing to the slow down of the depletion of fossil resources.

With respect to their goals, the scope of biorefineries must include feedstock as an integral part. That means conceptualizing biorefineries not only as conversion plants but also as systems from the production of the feedstock to the substitution for a reference product system. The justification of this larger scope is the high heterogeneity of the feedstock in the case of biorefineries, compared to petroleum refineries, and the high dependency of the characteristics of the biorefineries' products on those of the feedstocks used. Such high dependency means that production, design, and engineering of the primary treatments of the feedstocks must be dependent on the characteristics of the needed products.

TABLE 1.4 Performance of the Promising Biochemicals [64]

Compound	1. Extensive Recent Literature	2. Multiple Product Applicability	3. Direct Substitute	4. High Volume Product	5. Platform Potential	6. Industrial Scaleup	7. Existing Commercial Product	8. Primary Building Block	9. Commercial Biobased Product
Ethanol	+ + +	+ + +	+ + +	+ + +	+ + +	+ + +	+ + +	+ + +	+ + +
Furfural	+ + +	+ +	+	+ +	+	+	+ + +	+ +	+ + +
HMF	+ + +	+ +	+	+	+ +	+	+	+ +	+
FDCA	+ + +	+	+	+ + +	+ +	+	+	+	+
Glycerol/ derivatives	+ + +	+ + +	+ + +	+ + +	+ + +	+ + +	+ + +	+ + +	+ + +
Isoprene	+ + +	+ +	+ + +	+ + +	+	+ + +	+ + +	+	+
Biohydrocarbons	+ + +	+ +	+ + +	+	+	+	+	+ +	+
Lactic acid	+ + +	+ + +	+	+ + +	+ +	+	+ +	+	+
Succinic acid	+ + +	+ + +	+	+	+ + +	+ + +	+	+	+
HPA	+ + +	+	+ + +	+ + +	+ +	+	+	+	+
Levulinic acid	+ + +	+ +	+ + +	+ +	+ + +	+ + +	+	+ + +	+
Sorbitol	+ + +	+ + +	+ + +	+ + +	+ + +	+ + +	+ + +	+ + +	+ + +
Xylitol	+ + +	+ + +	+	+	+ + +	+	+ +	+ + +	+ +

+ + +, good performance against criterion; + +, emerging performance against criterion; +, lower performance against criterion.

1.4 INCLUSION OF SUSTAINABILITY IN THE CLASSIFICATION SYSTEM

With consideration to the goals of biorefineries, it is important to include sustainability criteria in their classification system. Several criteria can be envisaged. Chapter 10 reviews some of them. Sustainability criteria adopted by the Bioenergy Technologies Office of the US Department of Energy (DOE) [65] for the assessment of the multiyear program plan of biorefineries are used to illustrate that issue.

1.4.1 Sustainability Criteria

1.4.1.1 Environmental Sustainability

Environmental concerns may be related to local or global issues. Examples of global issues are GHG emissions or stratospheric ozone depletion, whereas local environmental disturbances may include pollution of air, soil, and water. Loss of biodiversity is between these two categories (i.e., global and local). The criteria of the DOE include the following:

- Reduction of GHG emissions
- Maintaining or improving soil quality
- Maintaining or improving water quality, avoiding negative impact on water resources, improving efficient use of water
- Maintaining or improving air quality
- Conserving plant and animal diversity and protecting habitat and ecological systems
- Enhancing beneficial land use management and maintaining or improving land productivity

1.4.1.2 Economic Sustainability

The economic sustainability may also include several criteria. The following indicators can be selected:

- Economic viability of the whole biorefinery
- Avoiding economic cross subsidy between the products value chain

1.4.1.3 Social Sustainability

Social sustainability criteria are those that are most difficult to define. They are mainly qualitative. The following indicators are given as an example:

- Social acceptability: minimizing risk, maximizing transparency, and ensuring effective stakeholder participation
- Social well-being: maintaining or improving prosperity, safety, health and food security
- Energy security and external trade: Reducing dependence on foreign risky energy import, access to affordable energy, demonstrating a positive net energy balance relative to fossil fuels, improving the balance of trade between imports and exports for energy-related materials

- Resource conservation: Minimizing use of nonrenewable resources relative to renewable resources and enhancing the energy return on investment
- Rural development and workforce training: Creating job opportunities, enhancing rural livelihoods, and developing a skilled bioenergy workforce

1.4.2 Defining Sustainability Classes Using a Logic Based Model

For a given biorefinery, its sustainability performance can be estimated against various criteria. Then it would be possible to assess its global sustainability using any relevant aggregation method. As an example the Logic based model developed by Gnansounou [66] is used to define sustainability classes. The model consists in designing a rule-based built on values of three types of hierarchical indicators, namely determinants (e.g., environmental sustainability), general indicators (e.g., disturbances of the local environment), and specific indicators (e.g., air quality). Determinants and general indicators take only qualitative values whereas specific indicators may also be assessed using quantitative values. The values of the general indicators are used to define those of the determinants and combinations of values of specific indicators determine general indicators. Finally, combinations of the values of determinants are classified into three hierarchical groups corresponding to different sustainability performances. Table 1.5 shows the hierarchy of the indicators.

1.4.2.1 Modeling Economic Sustainability of Biorefineries

Without any loss of generality, it is assumed that the biorefineries to be assessed generate three coproducts. General indicator D1.1, economic viability of the whole biorefinery, is evaluated, taking into account the following components: the feedstock price composed of the farm gate price (D1.1.1) and the logistic cost (D1.1.2); and the conversion cost (D1.1.3). The latter can be evaluated through a benchmark with comparable biorefineries. D1.1.2 mainly depends on the distance between the biorefinery plant and the location of biomass resources. Finally, the farm gate price of the feedstock depends on the global availability of the biomass and the competition between the uses. D1.1.1 is subject to a high uncertainty during the lifetime of the biorefinery. General indicator D1.2, economic cross subsidy between the products' value chains, is meant to check that some beneficial coproducts will not subsidize other nonbeneficial ones. Indeed, in case of subsidies, the investors would prefer to remove the noneconomical coproducts. Hence, the principle of economic sustainability of the biorefinery used in this rule-based approach means that the economic value of the whole biorefinery is higher than or equal to the sum of that of each coproduct.

1.4.2.2 Modeling Social Sustainability of Biorefineries

The determinant D2, social sustainability, is modeled using four general indicators: social acceptability (D2.1), social well-being (D2.2), energy security (D2.3), and resource conservation (D2.4). D2.1 is explained by using combinations of values of three specific indicators, namely social risk (D2.1.1), transparency (D2.1.2), and stakeholders participation. An example of social risk is related to the changes in land ownership and rights of the poor people. Transparency is concerned with social accountability and information sharing with the potential impacted or concerned persons and institutions. Stakeholders participation is

TABLE 1.5 Determinants, General Indicators, and Specific Indicators

Determinants and General Indicators	Specific Indicators		
	1	**2**	**3**
D1 Economic sustainability			
D1.1 Economic viability of the whole biorefinery	D1.1.1 Economic affordability of the farm gate price of the feedstock	D1.1.2 Economic affordability of the feedstock logics cost	D1.1.3 Economic viability of the conversion cost
D1.2 Economic cross subsidy between the products' value chains	D1.2.1 Economic viability of the co-product 1	D1.2.2 Economic viability of the co-product 2	D1.2.3 Economic viability of the co-product 3
D2 Social sustainability			
D2.1 Social acceptability	D2.1.1 Social risk	D2.1.2 Transparency	D2.1.3 Stakeholders participation
D2.2 Social well-being	D2.2.1 Social prosperity	D2.2.2 Safety and health	D2.2.3 Food security
D2.3 Energy security	D2.3.1 Energy dependency	D2.3.2 Energy affordability	D2.3.3 Net energy balance
D2.4 Resource conservation	D2.4.1 Conservation of fossil fuels	D2.4.2 Water conservation	D2.4.3 Landscape and cultural heritage
D2.5 Rural development and workforce training	D2.5.1 Job opportunities	D2.5.2 Rural livelihoods	D2.5.3 Development of skilled workforce
D3 Environment Sustainability			
D3.1 Land use	D3.1.1 Land use management	D3.1.2 Land productivity	–
D3.2 Disturbances of local environment	D3.2.1 Soil quality	D3.2.2 Air quality	D3.2.3 Water quality
D3.3 Biodiversity	D3.3.1 Plant diversity	D3.3.2 Animal diversity	D3.3.3 Habitat and ecological systems
D3.4 Disturbances of global environment	D3.4.1 Reduction of GHG emissions	D3.4.2 Net nonrenewable energy use	–

related to the governance of the biorefinery system, which should get the different actors in the decision process involved. General indicator D2.2, social well-being, considers three dimensions: social prosperity (D2.2.1) is related to the increase of revenues induced by the biorefinery and its social redistribution; safety and health are about potential hazards to the population; food security (D2.2.3) is the main social dimension related to biorefineries/biofuels discussed in the literature. Biorefineries are supposed to alleviate the potential competition of first-generation biofuels with food production, by developing complementary schemes between feed/food, energy and biomaterials. This is achieved by using nonedible feedstocks such as agricultural residues. Energy security (D2.3) is explained through three specific indicators: energy dependency (D2.3.1), energy affordability to poor people (D2.3.1) and net energy balance (D2.3.3). The latter measures the balance between the energy needs and generation by the biorefinery. A positive net balance means that the biorefinery generates more energy than it consumes, which is favorable to the availability of energy supply. Resource conservation (D2.4) is related to rational use of fossil fuels (D2.4.1), water resources (D2.4.2), and protection of the landscape and cultural heritage (D2.4.3). General indicator D2.5, rural development and workforce, is explained through the human resources in rural areas: job opportunity (D2.5.1), development of skilled workforce in rural areas (D2.5.2), and purchasing power in rural areas (D2.5.2).

1.4.2.3 *Modeling Environmental Sustainability of Biorefineries*

Environmental sustainability (D3) is explained through four general indicators: land use (D3.1), disturbances of local environment (D3.2), biodiversity (D3.3) and disturbances of global environment (D3.4). Land use (D3.1) is related to the territorial policy including land planning, agricultural practices and their impacts on land fertility (D3.1.1) and land productivity (D3.1.2) that affect the demand for land. The disturbances of the local environment (D3.2) and those of the global environment (D3.4) as well as biodiversity (D3.3) are largely covered in Chapter 2.

1.4.3 Rule Base

A set of rules is used to define each value of a given indicator as a function of the appropriate combination of dependent lower-level indicators. Each indicator (specific, general indicators, determinant) has three exclusive values (1: low performance; 2: medium performance; 3: high performance). Combinations of values of specific indicators are used to define the values of the general indicators; the same is true between the general indicators and the determinants and between the determinants and the values of the global sustainability. Table 1.6 shows the rules that define the values of global sustainability in relation to the values of the three determinants.

According to the rules in Table 1.6, to be rated high in global sustainability, a biorefinery must rate high in two determinants and at least medium in the third.

1.4.4 Illustration of Sustainability Potentials

As an illustration, Table 1.7 shows the rating of selected biorefineries discussed in this chapter according to their sustainability potentials.

Only Case 1 is rated as a high sustainability class. Despite rating medium for economic sustainability, it performs well against social and environmental criteria. The weak point is the economic performance. Second-generation bioethanol remains uncompetitive compared to the first generation; furthermore, a wide-scale vetiver plantation is required, which implies adoption by the farmers of a new culture in degraded lands. Finally, using only the leaves would be in competition with the use of the roots for producing essential oil. The incentives given would not be enough to fill the gap. The same weakness is noted for Case 2: the production of furfural would not improve economic weakness enough. Moreover, as furfural is already produced from biomass, the reduction of GHG from this new production line is not high, reducing the environmental performance. However, Case 2 performs well for social

TABLE 1.6 Rules Using the Qualitative Values of the Determinants to Define the Qualitative Values of the Global Sustainability

No. of the Rule	D Global Sustainability	D1 Economic Sustainability	D2 Social Sustainability	D3 Environmental Sustainability
1	1	1	1	1
2	1	1	2	1
3	1	1	1	2
4	1	2	1	1
5	1	1	2	2
6	1	2	1	2
7	1	2	2	1
8	1	3	1	1
9	1	1	3	1
10	1	1	1	3
11	2	2	2	2
12	2	3	2	2
13	2	2	3	2
14	2	2	2	3
15	2	3	2	1
16	2	2	3	1
17	2	1	3	2
18	2	3	1	2
19	2	1	2	3
20	2	2	1	3
21	2	3	3	1
22	2	3	1	3
23	2	1	3	3
24	3	3	3	3
25	3	3	2	3
26	3	2	3	3
27	3	3	3	2

1, low performance; 2, medium performance; 3, high performance.

TABLE 1.7 Sustainability Class of Selected Biorefineries

Determinants. Strengths and Weaknesses	Case 1: A C5/C6 Plant Producing Bioethanol From Vetiver Leaves Using a Biochemical Route	Case 2: A C5/C6 Plant Producing Bioethanol and Furfural From Vetiver Leaves Using a Biochemical Route	Case 3: An Oil/C5/C6 Biorefinery Producing Biodiesel, Glycerin, Protein From Microalgae Using a Biochemical Route	Case 4: An Oil/C5/C6 Biorefinery Producing Biodiesel, Glycerin, Protein and Succinic Acid From Microalgae Using a Biochemical Route	Case 5: A Synthesis Gas Biorefinery Producing Biomethane and Fertilizer From Oil Palm EFB Using Hydrothermal Gasification Route
D1: Economic sustainability	Medium	Medium	Low	Low	Low
D2: Social sustainability	High	High	Medium	Medium	High
D3: Environmental sustainability	High	Medium	High	High	High
D: Global sustainability	High	Medium	Medium	Medium	Medium
Strengths	*Social* acceptability *Social* well-being *Energy* security *Rural* development *Land use* *Local* environment *Resource* conservation *Biodiversity* *Global* environment	*Social* acceptability *Social* well-being *Rural* development *Land use* *Local* environment *Resource* conservation *Biodiversity*	*Social* acceptability *Social* well-being *Land use* *Local* environment *Resource* conservation *Biodiversity* *Global* environment	*Social* acceptability *Social* well-being *Land use* *Local* environment *Resource* conservation *Biodiversity* *Global* environment	*Social* acceptability *Social* well-being *Rural* development *Land use* *Local* environment *Resource* conservation *Biodiversity* *Global* environment
Weaknesses	*Economic* viability	*Economic* viability *Global* environment	*Economic* viability	*Economic* viability	*Economic* viability

sustainability due to its contribution to reduce feed and chemicals competition by diversifying the feedstock used for furfural production. Since the use of microalgae for biodiesel production is not mature enough, Cases 3 and 4 perform badly in economic sustainability; furthermore, their energy balances are significantly positive and the creation of new jobs is not considered as significant compared to the case of agriculture, which leads to a medium rating for social sustainability. Finally, Case 5 performs well for social and environmental sustainability but badly for the economic one due to the low maturity level of hydrothermal gasification. It is worth noting that the class of sustainability that is attributed to a given biorefinery can change during its development if the weak points are improved. Hence, including sustainability in the classification of biorefineries should not be interpreted as a way to definitely discriminate a few pathways, but rather as an incentive for improving, whenever possible, performances against the three pillars of sustainability.

1.5 INCLUSION OF FLEXIBILITY

The development of biorefineries is subject to several uncertainties stemming from various factors such as feedstock procurement, demand, and prices of the products. Flexibility is a means for coping with these uncertainties. It is worth distinguishing between strategic flexibility and operational flexibility, even though the extent of the latter depends on the design choices that result from the former.

1.5.1 Strategic Flexibility

Strategic flexibility is related to the design and stepping in the development of the biorefinery. Chapter 4 of this book gives a few examples of strategic flexibility, consisting of expanding bioethanol production by integrating a second-generation plant with a first-generation one. Both plants mutualize the fermentation reactor, distillation and dehydration facilities, and the utilities; furthermore, the second-generation plant adds innovative facilities such as pretreatment and enzymatic hydrolysis, which improves the value of the whole integrated plant. Different schemes are discussed in Chapter 4: integration from a first-generation ethanol distillery (1G-ED) and integration from a sugar mill producing first-generation ethanol (SM). The integrated biorefinery can produce only fuels (OF), ethanol and surplus electricity—ED-OF and SM-OF—or ethanol and syrup for animal feed (fuel and feed, FF)—ED-FF and SM-FF. These integration schemes also show an interesting feature related to the feedstock. The second-generation bioethanol uses bagasse that is a residue of the first-generation plant and the cogeneration is fueled by bagasse complemented by green harvesting residues (GHRs) from the sugarcane field. Part of the GHRs are left on the field for soil regeneration. One advantage of the 1G-2G integration is the containment of the expansion of the sugarcane culture by relying on residues for the increase of bioethanol production, developing a multifeedstock plant in addition to process mutualization. At the time for investment renewal gives the best opportunity to invest in such an expansion. At that time, the investor has to choose between expansion of first-generation bioethanol and integration between first generation and second generation. The latter option even more flexible would be chosen only if the difference of cost between the second generation and the first generation is compensated for by the mutualizing economy, environmental and social positive externalities, and capability reward.

1.5.1.1 Cost Reductions Due to Mutualization

The economies from mutualization of facilities between plants are from three origins: the scale economy, the thermodynamic process integration, and the simultaneousness factor. The scale economy is related to the reduction of the specific investment cost due to a higher scale. In the case of integration of second-generation (2G) with first-generation (1G) bioethanol, the economy of scale is linked to the comparison of two separate standalone 1G plants with an integrated plant (1G-2G). In the former case, the scale of the new facilities (fermenter, distillation, and dehydration) would be lower, leading to higher investment. Furthermore, the production of sugarcane would increase, which implies using more land and inputs. In the case of integrated schemes, the economies related to thermodynamic integration of processes (water, energy) would also be important dependent on the state of the technology used in the 1G plant; thermodynamics integration of processes can be modeled using the PINCH method [67].

The cost reduction related to the simultaneousness factor arises when the plants that share the same facilities have their peaks at different times, leading to a capacity economy. That is not the case for the integrated 1G-2G plant since the streams of both subplants are gathered before entering the shared facilities. Another example for illustrating the simultaneousness factor is that of a biorefinery cluster based on an industrial ecology principle. Such clusters emerge from "ecosystems" of factories located on the same site, which provides the opportunity to share utilities and other facilities of common scope. The Bazancourt-Pomacle Biorefinery [68] illustrates that case. Located in France, that cluster of industries includes starch and glucose processing, a 1G ethanol plant, a sugar factory and dehydration plant, the pilot plant of the French FUTUROL second-generation project, along with several research centers. In this context, where the facilities are distinct even on the same site, the load curves of the shared facilities are not well correlated from one factory to the other, leading to economy of capacities.

1.5.1.2 Positive Environmental and Social Externalities

Several positive environmental externalities result from 1G-2G integrated biorefineries. In comparison with expansion of 1G bioethanol, reduction of land use and avoiding supplementary fertilizers, herbicides, and pesticides are a few examples of positive environmental externalities. In the case of the conventional practice of the 1G value chain, part of the residues is burnt on the field, releasing particles, carbon monoxide and dioxide, and other regulated pollutants; these releases contribute to air pollution. Finally, the collection of GHRs can induce new jobs in rural areas and well-skilled human resources in the factories.

1.5.1.3 Capability Reward

In the context of investment in 2G ethanol and integration with 1G, the capability reward can be defined as the net premium for further expanding and adding biobased product lines (lactic acid, succinic acid, biocomposites, etc.) to the ethanol production. That net premium can be estimated by using real option theory [69]. Private investors would not enter 2G bioethanol if that net premium is too high; the term *net* means the premium reduced by all incentives.

1.5.2 Operational Flexibility

Once the plant is designed, the optimal point of operation is generally fixed. However, that optimum depends on uncertain variables such as quantity and prices of inputs and coproducts. Operational flexibility refers to the ability of the plant to operate with a certain deviation from the optimal conditions without technical damages. Swaney and Grossmann [70] developed a flexibility index in chemical design. This index provides the range in which the parameters must locate to make the operation technically feasible. Kasivisvanathan et al. [71] proposed an operational flexibility model using fuzzy optimization. In a framework of biorefineries, the aim would be to adapt the portfolio of products to the variation of quantity and prices of products and inputs in order to maximize the benefits subject to constraints on capacity and technical compatibility between the states of the operational parameters. The value of strategic and operational flexibility is high in an uncertain environment.

1.6 THE RATIONALE OF PUBLIC AND PRIVATE INCENTIVES: THE ROLE OF CLASSIFICATION

Biorefineries development in general contributes to achieving the environmental and socio-economic objectives of public policy by fostering a biobased economic growth and anticipating depletion of fossil resources. However, the importance of this contribution may change significantly from one biorefinery to another. At the extremum, the development of some biorefineries must be avoided due to their potential harms to the environment and the people and therefore would not deserve receiving any public or private support. Even for a sustainable biorefinery, questioning at what extent public support can appropriately enhance the development of a biorefinery remains a hotspot theme of public choice. Depending on the state in the cycle of research-development-demonstration-deployment, the need for support varies greatly, as well as the role of public/private incentives and type of instrument. This state is qualitatively measured by the technology readiness level (TRL) that varies from 1 (basic research) to 9 (system proven and ready for full commercial deployment). Table 1.8 shows a few examples of classification of selected biorefinery platforms according to their TRL.

However, TRLs are not the only criterion for guiding incentives and promotion of R&DDD on biorefineries. As stated in Section 1.4, sustainability is another important criterion and, above all, potential effects on environment and the society are key factors for gaining a sustained policy acceptance. This is why life cycle assessment (LCA) as a method for evaluating environmental performance has come to the forefront of assessment methodologies used in designing and analyzing biorefinery concepts and projects at the feasibility, basic research, pilot, demonstration and deployment levels. LCA also plays an important role in discriminating between options of biorefineries within energy and environmental policies. As an example, the requirements on GHG emission reduction for eligible biofuel pathways in the European renewable energy directive necessitate using LCA for checking their compliance.

TABLE 1.8 Technology Readiness Level (TRL) of Selected Biorefineries

Determinants. Strengths and Weaknesses	Case 1: A C5/C6 Plant Producing Bioethanol From Vetiver Leaves Using a Biochemical Route	Case 2: A C5/C6 Plant Producing Bioethanol and Furfural From Vetiver Leaves Using a Biochemical Route	Case 3: An Oil Biorefinery Producing Biodiesel, Glycerin, Protein From Microalgae Using a Biochemical Route	Case 4: An Oil/C5/C6 Biorefinery Producing Biodiesel, Glycerin, Protein and Succinic Acid From Microalgae Using a Biochemical Route	Case 5: A Synthesis Gas Biorefinery Producing Biomethane and Fertilizer From Oil Palm EFB Using Hydrothermal Gasification Route
TRL	7	7	7	6	5
Justification	Vetiver as a feedstock for bioethanol: TRL=8 C5/C6 platform: TRL=7 Fermentation of C5: TRL=7	Conversion of C5 from vetiver into furfural: TRL=7 • Fractionation of vetiver leaves into C5 and C6 sugars • Purification	Extraction of algae oil: TRL=7	Combined algae oil and C5/C6 platforms: TRL=5 Succinic acid: TRL=8	Hydrothermal gasification: TRL=5

1.7 CONCLUSIONS AND PERSPECTIVES

Under the term *biorefinery* there is a large diversity of concepts, facilities, and systems. A biorefinery can have diverse realities. It could be an existing facility, a project, or a prospective or potential cluster between existing plants. This chapter elaborates on the classification proposed by IEA T42 and highlights the importance of including the feedstock production and its primary treatments into the classification system. Sustainability performance should be an important characteristic for qualifying a biorefinery and benchmarking different biorefineries. Concerning the evaluation of environmental sustainability, Chapter 2 presents the fundamentals of LCA, which is the main topic of the book; a specific approach is dedicated to the case of biorefineries. For economic sustainability, an appropriate techno-economic methodology has been developed that is used in connection with the LCA in Chapter 4. Social sustainability is considered qualitatively. Finally, in order to deal with the inherent risk related to these biorefineries, flexibility is a key feature; a possible strategy is expansion from existing plants. In that respect, Chapter 4 presents the case of integrated biorefineries between first and second generation. Comparison of biorefineries is tricky as they are at various maturity levels. As far as a benchmark is concerned, TRLs must be considered in the classification in order to take into account the closeness to market. That factor is difficult to evaluate and is somewhat qualitative since it involves the feedstock, the platforms, the processes, and the products. The perspectives of biorefineries related to assessment are to improve the data as well as the methods of impact assessment that still suffers from insufficient consensus in the scientist community. Efforts devoted to classification of biorefineries will contribute to the improvement of the assessment methodology by contextualizing the study cases more efficiently and enhancing the relevancy of the modeling choices.

References

[1] de Jong E, van Ree R. Biorefineries: adding value to the sustainable utilization of biomass. IEA Bioenergy T42, 2009:01; 2009.

[2] NREL. Biomass research—what is biorefinery? http://www.nrel.go/biomass/biorefinery.

[3] de Jong E, Jungmeier G. Biorefinery concepts in comparison to petrochemical refineries. In: Pandey A, editor. Industrial biorefineries and white biotechnology. Elsevier, Walthum, MA; 2015.

[4] European Union. Directive 2009/EC of the European Parliament and the Council on 23 April 2009 on the promotion of the use of energy from renewable sources. Off J Eur Union; 05.06.2009.

[5] Luguel Ch, Annelink B, Burel C, Gosse G, editors. Star-Colibri. Joint European biorefinery vision for 2030. www.star-colibri.eu.

[6] OECD-FAO. Agricultural outlook 2015.

[7] Brandt A, Gräsvik J, Hallett JP, Welton T. Deconstruction of lignocellulosic biomass with ionic liquids. Green Chem 2013;15:550–83.

[8] U.S. Department of Energy Genome Programs image gallery. http://genomics.energy.gov.

[9] Arantes V, Saddler J. Access to cellulose limits the efficiency of enzymatic hydrolysis. The role of amorphogenesis. Biotechnol Biofuels 2011;3:4.

[10] Monti A, editor. Switchgrass: a valuable biomass crop for energy. London: Springer-Verlag; 2012.

[11] Carle J, Holmgren P. Wood from planted forests. A global outlook 2005–2030. For Prod J 2008;58(12):6–18.

[12] Smil V. Crop residues: agriculture's largest harvest. BioScience 1999;49(4):299–308.

[13] Brosse N, Dufour A, Meng X, et al. Miscanthus: a fast-growing crop for biofuels and chemicals production. Biofuels Bioprod Biorefin 2012;6:580–98.

[14] United States Department of Energy, Biomass Feedstock Composition and Property Database, Energy Efficiency and Renewable Energy, Biomass Program. http://www.afdc.energy.gov/biomass/progs/search1.cgi.

[15] Ogden CA, Ileleji KE, Johnson KD, Wang Q. In-field direct combustion fuel property changes of switchgrass harvested from summer to fall. Fuel Process Technol 2010;91:266–71.

[16] Kim M, Day DF. Composition of sugar cane, energy cane, and sweet sorghum suitable for ethanol production at Louisiana sugar mills. J Ind Microbiol Biotechnol 2011;38:803–7.

[17] Shah A, Darr MJ, Webster K, Hoffman C. Outdoor storage characteristics of single pass large square corn stover bales in Iowa. Energies 2011;4:1687–95.

[18] Jeschke M, Heggenstaller A. Sustainable corn stover harvest for biofuel production. Crop Insights 2012;22:1–6.

[19] Zhang Y, Ghaly AE, Li B. Physical properties of wheat straw varieties cultivated under different climatic and soil conditions in three continents. Am J Eng Appl Sci 2012;5:98–106.

[20] Tarkalson DD, Brown B, Kok H, Bjorneberg DL. Impact of removing straw from wheat and barley fields: a literature review. Better Crops 2009;93:17–9.

[21] Stolarski MJ, Szczukowski S, Tworkowski J, Klasa A. Yield, energy parameters and chemical composition of short-rotation willow biomass. Ind Crops Prod 2013;46:60–5.

[22] Hinchee M, Rottmann W, Mullinax L. Short-rotation woody crops for bioenergy and biofuels applications. In Vitro Cell Dev Biol Plant 2009;45:619–29.

[23] Laser M, Lynd L. Introduction to cellulosic energy crops. In: Karlen DL, editor. Cellulosic energy cropping systems. Stevens CV, editor. Wiley series in renewable resources. Chichester: John Wiley & Sons, Ltd.; 2014.

[24] Panopoulos KD, Vamvuka D. Experimental evaluation of thermochemical use of two promising biomass fuels, In: Proceedings of the European combustion meeting, Vienna, Austria; 2009. p. 1–6.

[25] Wang S, Littell RC, Rockwood DL. Variation in density and moisture content of wood and bark among twenty *Eucalyptus grandis* progenies. Wood Sci Technol 1984;18:97–100.

[26] Spinelli R, Nati C, Magagnotti N. Using modified foragers to harvest short-rotation poplar plantations. Biomass Bioenergy 2009;33:817–21.

[27] Bridgeman TG, Jones JM, Shield I, Williams PT. Torrefaction of reed canary grass, wheat straw and willow to enhance solid fuel qualities and combustion properties. Fuel 2008;87:844–56.

[28] Mosseler A, Zsuffa L, Stoehr MU, Kenney WA. Variation in biomass production, moisture content, and specific gravity in some North American willows (*Salix* L.). Can J For Res 1988;18:1535–40.

[29] Patterson DW, Doruska PF. Effect of seasonality on bulk density, moisture content, and specific gravity of loblolly pine tree length pulpwood logs in southern Arkansas. For Prod J 2005;55:204–14.

[30] Chiesa S. Bioethanol production from oil palm empty bunches: a comparison among different pretreatment options [Dissertation no. 5544]. Lausanne: EPFL; 2012.

[31] Mosier N, Wyman Ch, Dale B, et al. Features of promising technologies for pretreatment of lignocellulosic biomass. Bioresour Technol 2005;96:673–86.

[32] Kumar P, Barrett DM, Delwiche MJ, Stroeve P. Methods for pretreatment of lignocellulosic biomass for efficient hydrolysis and biofuel production. Ind Eng Chem Res 2009;48:3713–29.

[33] Hendriks ATWM, Zeeman G. Pretreatments to enhance the digestibility of lignocellulosic biomass. Bioresour Technol 2009;100:10–8.

[34] Medina JDC, Woiciechowski A, Filho ZA, Noseda MD, Kaur BS, Soccol CR. Lignin preparation from oil palm empty fruit bunches by sequential acid/alkaline treatment—a biorefinery approach. Bioresour Technol 2015;194:172–8.

[35] Sun Y, Cheng J. Hydrolysis of lignocellulosic materials for ethanol production. A review. Bioresour Technol 2002;83:1–11.

[36] Toledano A, García A, Mondragon I, Labidi J. Lignin separation and fractionation by ultrafiltration. Sep Purif Technol 2010;71:38–43.

[37] Creighton RHJ, McCarthy JL, Hibbert H. Studies on lignin and related compounds. LIV. Alkaline nitrobenzene oxidation of plant materials and application to taxonomic classification. J Am Chem Soc 1941;66:32–7.

[38] Creighton RHJ, Gibbs RD, Hibbert H. Studies on lignin and related compounds. LXXV. Alkaline nitrobenzene oxidation of plant materials and application to taxonomic classification. J Am Chem Soc 1944;66:32–7.

[39] Creighton RHJ, Hibbert H. Studies on lignin and related compounds. LXXVI. Alkaline nitrobenzene oxidation of corn stalks. Isolation of p-hydroxybenzaldehyde. J Am Chem Soc 1944;66:37–8.

[40] Faix O. Classification of lignins from different botanical origins by FT-IR spectroscopy. Holzforschung 1991;45 (Suppl.):21–7.

[41] García A, Toledano A, Serrano L, et al. Characterization of lignins obtained by selective precipitation. Sep Purif Technol 2009;68:193–8.

[42] Passoni V, Scarisa C, Levi M, Turri S, Griffini G. Fractionation of industrial softwood Kraft lignin: solvent selection as a tool for tailored material properties. ACS Sustain Chem Eng 2016;4:2232–42.

[43] Lange H, Decina S, Crestini C. Oxidative upgrade of lignin—recent routes reviewed. Eur Polym J 2013;49:1151–73.

[44] Kumar A, Jones DD, Hanna MA. Thermochemical biomass gasification: a review of current status of technology. Energies 2009;2:556–81.

[45] Gani A, Naruse I. Effect of cellulose and lignin content on pyrolysis and combustion characteristics for several types of biomass. Renew Energy 2007;32:649–61.

[46] Hosoya T, Kawamoto H, Saka S. Cellulose-hemicellulose and cellulose-lignin interactions in wood pyrolysis at gasification temperature. J Anal Appl Pyrol 2007;80:118–25.

[47] Qin Y, Campen A, Wiltowski T, Feng J, Li W. The influence of different chemical compositions in biomass gasification tar formation. Biomass Bioenergy 2015;83:77–84.

[48] E4tech. Review of technologies for gasification of biomass and wastes. Final report NNFC project 09/008; 2009.

[49] Ciferno JP, Marano JJ. Benchmarking biomass gasification technologies for fuels, chemicals and hydrogen production. Washington, DC: U.S. Department of Energy, National Energy Technology Laboratory; 2002.

[50] Devi L, Ptasinski KL, Janssen FJJG, et al. Catalytic decomposition of biomass tars: use of domolite and untreated olivine. Renew Energy 2005;30:565–87.

[51] Horvat A, Kwapinska M, Xue G, et al. Detailed measurement uncertainty analysis of solid-phase adsorption—total gas chromatography (GC)—detectable tar for biomass gasification. Energy Fuels 2016;30:2187–97.

[52] IEA T33. Fact sheet gas cleaning and tars. http://www.ieatask33.org/app/webroot/files/file/publications/Fact_sheets/IEA_Gas_cleaning_and_tars.

[53] Milne TA, Abatzoglou N, Evans RJ. Biomass gasifier tars: their nature, formation and conversion. Golden, CO: NREL, U.S. Department of Energy; 1998.

[54] Morf P, Hasler Ph, Nussbaumer T. Mechanisms and kinetics of homogeneous secondary reactions of tar from continuous pyrolysis of wood chips. Fuel 2002;81:843–53.
[55] Hofbauer H, Rauch R, Ripfel-Nitsche K. Gas cleaning for synthesis applications. Work package 2E "gas treatment" Project ThermalNet funded by the European Commission, Intelligent Energy, Vienna: Vienna University of Technology; 2007. October.
[56] Yeh KL, Chang JS. Nitrogen starvation strategies and photobioreactor design for enhancing lipid production of newly isolated microalga *Chlorella vulgaris* sp-31: implications for biofuels. Biotechnol J 2011;6:1358–66.
[57] Torri IDV, Paasikallio V, Faccini CS, et al. Bio-oil production of softwood and hardwood forest industry residues through fast and intermediate pyrolysis and its chromatographic characterization. Bioresour Technol 2016;200:680–90.
[58] Mohan D, Pittman ChU, Steeke H. Pyrolysis of wood/biomass for bio-oil: a critical review. Energy Fuels 2006;20:848–89.
[59] Pukorz J, Scott DS, Radlien D. Composition of oils obtained by fast pyrolysis of different woods. In: Pyrolysis oils from biomass: producing, analyzing and upgrading. Washington, DC: American Chemical Society; 1988. p. 167–78.
[60] Praveenkumar R, Shameera K, Mahalakshmi G, et al. Influence of nutrient deprivations on lipid accumulation in a dominant indigenous microalga *Chlorella* sp. BUM11008: evaluation for biodiesel production. Biomass Bioenergy 2012;37:60–6.
[61] Sun Z, Zhou Z-G, Gerken H, Chen F, Liu J. Screening and characterization of oleaginous *Chlorella* strains and exploration of photoautotrophic *Chlorella protothecoides* for oil production. Bioresour Technol 2015;184:53–62.
[62] Lin T-S, Wu J-Y. Effect of carbon sources on growth and lipid accumulation of newly isolated microalgae cultured under mixotrophic condition. Bioresour Technol 2015;184:100–7.
[63] Werpy T, Petersen G, editors. Top value added chemicals from biomass, vol. I. PNNL and NREL, US Department of Energy; 2004. Washington, DC, August.
[64] Bozell JJ, Petersen GR. Technology development for the production of biobased products from biorefinery carbohydrates—the US Department of Energy's "Top 10" revisited. Green Chem 2010;12:539–54.
[65] U.S. Department of Energy, Bioenergy Technology Office. Multi-year program plan; 2016, March, http://www.energy.gov/sites/prod/files/2016/03/f30/mypp_beto_march2016_2.pdf.
[66] Gnansounou E. Assessing the sustainability of biofuels: a logic-based model. Energy 2011;36(4):2089–96.
[67] Martinez-Hernandez E, Sadhukhan J, Campbell GM. Integration of bioethanol as an in-process material in biorefineries using mass pinch analysis. Appl Energy 2013;104:517–26.
[68] Schieb P-A, Lescieux-Katir H, Thénot M, Clément-Larosière B. Biorefinery 2030—future prospects for bioeconomy. Heidelberg: Springer-Verlag; 2015.
[69] Trigeorgis L. Real options—managerial flexibility and strategy in resource allocation. Cambridge, MA/London/England: The MIT Press; 2000.
[70] Swaney RE, Grossmann IE. An index for operational flexibility in chemical process design. Part 1. Pittsburgh, PA: Carnegie Institute of technology, Department of Chemical Engineering, Carnegie Mellon University; 1983.
[71] Kasivisvanathan H, Ng DKS, Poplewski G, Tan RR. Flexibility optimization for palm oil-based integrated biorefinery with demand uncertainties. Ind Eng Chem Res 2016;55:4035–44.

CHAPTER

2

Fundamentals of Life Cycle Assessment and Specificity of Biorefineries

E. Gnansounou

Bioenergy and Energy Planning Research Group (BPE), Ecole Polytechnique Fédérale de Lausanne (EPFL), Lausanne, Switzerland

2.1 LIFE CYCLE ASSESSMENT: FROM INFANCY TO A STANDARDIZED METHODOLOGY

Most of the primary raw materials consumed worldwide are based on fossil sources. That context is deemed to be a cause of threats such as resource depletion, greenhouse gas (GHG) emissions, and local pollution. The early hopes that biomaterials and biofuels could be the solution were followed by disenchantment when a few scientists asked questions about nonrenewable sources embodied in biomaterials, as well as issues such as land use changes. The performance of public policies that support biofuels were questioned and the need increased more and more for rigorous tools that could help in assessing advantages versus burdens on the environment. The issues of waste management and end-of-life of products are other concerns. During the last three decades, the awareness that performance of a product or a system must be evaluated along its whole life cycle has been rapidly increasing.

Post-World War II consumerism in industrialized countries played a major background role in worries about technology's environmental footprint. A value system based on anthropocentrism was dominant. In the 1960s several social movements alerted the public decision makers to the limits of the planet and the necessity of a rational use of nonrenewable natural resources. One of the representative studies in that sense was the Meadows report mandated by the Club of Rome [1]. That report concluded that there was a high risk of negative

http://dx.doi.org/10.1016/B978-0-444-63585-3.00002-4

retroactions from the impacted environment on living organisms, and it recommended strong measures such as birth control and severe fiscal disincentives against misuses of natural resources and pollution. In parallel, the paradigm of nature conservation emerged from environmentalist groups. At a more prosaic level, accounting methods started taking into account natural resources embodied into products and emissions throughout industrial processes. With the natural resources that were considered, energy was prioritized. Mass and energy balances became the backbone of these accounting methods.

The origin of the life cycle assessment (LCA) dated in the US from that period and was attributed to Harry Teasley in 1969 in the framework of a study sponsored by Coca-Cola and undertaken by the US Midwest Research Institute. That study compared several containers of beverages in order to find those that resulted in less natural resource consumption and fewer environmental burdens [2]. The methodology used was referred to as resource and environmental profile analysis (REPA) and consisted of the following steps:

(1) expression of needs—identification of the materials and products to be studied,
(2) detailed definition of the system flow diagram (FD),
(3) energy and materials balances,
(4) evaluation of consumption of natural resources, and
(5) evaluation of environmental emissions.

Data needed to implement the methodology included the following: raw materials, water usage, energy and material resources, process energy and transportation energy, emission in air, waterborne wastes, and solid wastes [2]. Since the early 1970s, a number of REPAs have been issued on several areas including beverages, food products, chemicals, plastic, and other industrial products. In the same period and for similar reasons, comparable methodologies were developed in Europe and all over the world under the term Eco Balance. REPA as well as Eco Balance studies were focused on data inventories of natural resource consumption and releases of substances into the natural environment. The concerns were mainly about industrial products, packaging and solid wastes [3–5]. Furthermore, the studies were based on a variety of methods without the same theoretical backgrounds and procedures.

As far as assessment of environmental performance is concerned, the lack of harmonization between the methods does not allow fair comparison between products, processes and systems. One of the goals of the standardization that started in the early 1990s was to set up common guidelines for practitioners and researchers under the umbrella of the European and North American Chapters of the Society of Environmental Toxicology and Chemistry (SETAC) followed by the International Organization for Standardization (ISO). In 1993 guidelines for LCA were established by SETAC [6]. Since the end of the 1990s, ISO has been making a considerable effort to harmonize LCA procedures, resulting in the so-called ISO 14040 series, which includes standards, technical reports (TRs) and technical specifications (TSs):

- ISO 14040: International standard presenting the principles and framework of LCA [7].
- ISO 14044: International standard providing requirements and guidelines for LCA [8].
- ISO/TR 14047: TR related to life cycle impact assessment (LCIA) [9].
- ISO/TS 14048: TSs related to data documentation format [10].
- ISO/TR 14049: Informative document with an example related to goal and scope definition and inventory analysis [11].

Other TSs or TRs have included those like ISO 14067, related to the carbon footprint of products [12]. Standardization work related to LCA as well as that related to connected activities are continuing. New TSs and TRs have been recently issued, or will soon be issued, [13] such as:

- ISO TS 14071: Critical review processes and reviewer competencies [14].
- ISO TS 14072: Additional requirements and guidelines for organizations [15].
- ISO/AWI TR 14073: Water footprint—Illustrative examples of how to apply ISO 14046 (water footprint) [16].

According to ISO standards, the LCA is a methodology that addresses "the environmental aspects and potential environmental impacts throughout a product's life cycle including raw materials acquisition, production, use, end-of-life treatment recycling and final disposal" [8].

The methodology of LCA involves the following four stages (Fig. 2.1):

- the goal and scope definition,
- the life cycle inventory (LCI),
- the LCIA, and
- the interpretation.

LCA is a holistic methodology and its background is general systems theory (GST) [17]. A product is not only considered intrinsically according to its substance, it is a continuously transforming product perceived at a given time during an ongoing process, from the extraction of its constitutive materials to the end of its life throughout the logistics of the materials, the production, the logistics and use of the product, the rehabilitation of part of the product, and the recycling, disposal or elimination at the end of its life. According to that theoretical elaboration, the LCA methodology considers a product system in which that trajectory (from cradle-to-grave) is embodied along with the activities, the resources consumed, and the release

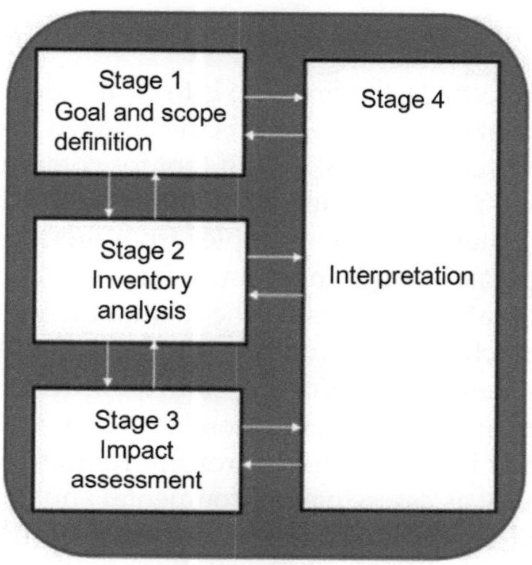

FIG. 2.1 Life cycle assessment framework according to ISO 1040:2006.

of substances involved during the whole lifetime. The product system is at any time in interactions with its natural, social and economic environments.

The application of the systems approach to the so-described product system includes the following tasks:

(1) decomposition of the system into relevant segments,
(2) inventory of the inputs and outputs for each segment,
(3) investigation of the interactions of each segment with its environment, and
(4) analysis of the interactions between the different segments.

The standardized LCA restricts the product system to its technological component that is expressed as a process, the description of which depends on the goal and the scope of the study (relevance criterion). The environments are also limited to the natural environment. By doing this, the standardized LCA ignores the actors' component as well as organizational issues of the product system, the economic dimension and several other environments (social, cultural, political, time-space dynamics, etc.) involved. However, that reductionism is consistent with the main objective of the conventional LCA, which originally aimed to analyze the impacts of a given industrial product on the natural environment along its lifetime from the raw materials procurement to the end of its life.

Nevertheless, the GST bears a theoretical framework for a more complete LCA, which would include environmental LCA (the conventional and normative one), economic LCA (life cycle costing), social and organizational LCA within an integrated life cycle sustainability assessment (LCSA). Such a complete scheme was initiated by UNEP/SETAC in 2011 [18].

This chapter mainly discusses the normative LCA that is standardized throughout the ISO 14040 series and surrounding standards. LCSA is outlined as well.

2.2 DEFINITION OF THE GOAL AND SCOPE

2.2.1 The Goal

When carrying out an LCA study, the first step is to clearly define the goal. The motivation and purpose of the study must be stated as well as the targeted people and the intended uses of the results. LCA is often used as a background tool for communication. It is important to clarify the targeted audience in order to design the scope and then the boundary of the system. This is illustrated hereafter by a few examples from literature.

In a study comparing small- and medium-to-large-scale production of bioethanol, Bernesson et al. [19] stated that the purpose of the related LCA was to analyze the influence of the scale of the fuel production system on the reduction of environmental burdens. Then the authors explained how their LCA was designed in accordance with the goal of the study. Guinée and Heijungs [20] compared different scenarios of allocation in fossil fuels chains with the purpose of analyzing their influence on LCA results. Kamidelivand and Gnansounou [21] defined the goal of their study as assessing environmental and land use change impacts that are related to products and coproducts from palm fresh fruit bunch (FFB) in a conceptual biorefinery that would be located in North Brazil. Panichelli et al. [22] studied the LCA of soybean-based biodiesel with the goal of evaluating the impact on the environment of its production in Argentina for export in Switzerland. Cherubini and Jungmeier [23] studied an LCA of a biorefinery concept producing bioethanol, other bioenergy and chemicals from grass, where the goal was implicitly a comparison with a fossil-based reference system. The goal of the study of Madival et al. [24] was to compare polylactic acid (PLA), polyethylene terephthalate (PET), and polystyrene (PS) clamshell containers based on their environmental profiles. Finally, Coltro et al. [25] presented the goal of their study as to undertake "the LCI of green coffee produced in Brazil for the reference crops 2001/02 and 2002/03 in order to obtain detailed production inventory data, as well as to quantify the potential environmental impacts of this tillage in order to realize how to reduce those impacts and increase environmental sustainability of this product."

More generally, the goal of an LCA may concern the environmental profile of a product in order to increase knowledge on its performance. It may also be a comparative LCA, in which case the goal is to compare several products that provide the same service in order to find out which is the best or to know more about their respective strengths and weaknesses.

2.2.2 The Scope

The goal of a study determines the scope. For a few of the papers cited previously, the goal as well as the scope limited to system boundary and functional unit are presented in Table 2.1.

According to ISO 14044, the description of the scope must include several issues. The main topics are discussed in the next subsections.

TABLE 2.1 Illustration of Goal and Scope Definition

Article	Goal	Extent and Categories of Impacts	Boundary of the System	Functional Unit
Bernesson et al. [19]	Analysis of the scale influence of fuel production system on the reduction of environmental burdens	Limited to air emissions and energy requirements	Well-to-tank	1 MJ
Guinée and Heijungs [20]	Evaluation of the influence of different allocation scenarios of petroleum on the results of a comparative LCA with ethanol	Evaluation of 11 impact categories	Well-to-wheels	1 km
Kamidelivand and Gnansounou [21]	Assessment of environmental and land use change impacts related to products and coproducts from the palm Fresh Fruit Bunch (FFB)	Evaluation of four impact categories	Well-to-wheels	100 km
Panichelli et al. [22]	Evaluation of the impacts on environment of soybean-based biodiesel production in Argentina for export in Switzerland	Comparison with 11 pathways and with fossil diesel. Seven impact categories along with land use changes	Well-to-wheels	1 km
Cherubini and Jungmeier [23]	Comparison of a biorefinery concept from grass with a fossil-based reference system.	Eleven impact categories	Well-to-wheels	1 km

2.2.2.1 *Function of the Product System and Functional Unit*

The service provided by the product system must be clearly stated. A motor fuel is for a mobility service or a stationary motor for generating mechanical energy. Depending on the goal of the study, the functional unit could be 1 km or 1 MJ of energy. For instance, when comparing two pathways of biodiesel production where the products have the same specifications, the functional unit may be 1 MJ of energy, 1 L or 1 ton at the gate-out of the biodiesel plant, as the pathway from that point is the same for both product systems. That is not the same when comparing biodiesel with ethanol fuels, because these two fuels will be used in vehicles with different combustion efficiencies; the functional unit here is a mobility indicator such as a multiple of kilometers.

2.2.2.2 *Description of the Product System*

The FD of the system to be assessed must be described showing the relevant units in the system as well as their inputs and outputs. The level of required detail depends on the goal and scope of the study; as examples in the next chapters, several simplified FDs describe the inputs as well as the products of the biorefineries that have been studied. An effective drawing of the FD is required in order to prepare the collection of the relevant data. Before drawing the FD, several questions need to be answered, such as:

(1) What are the main block diagrams that should be considered and why?
(2) For each block diagram, what is the unit process, the smallest element that should be considered?
(3) What are the elementary flows entering or leaving the product system?
(4) What are the reference flows that should be considered, taking into account the services required by the functions of the product system?
(5) What intermediate flows should be considered and which are to be neglected due to their insignificant contribution to the performances to be estimated?

Since responding to these questions is not straightforward, the definition of the product system and drawing of the FD of the system may be an iterative procedure based on the interpretations of the results of the LCI and LCIAs.

2.2.2.3 *The Boundary of the System*

Depending on the goal of the study, a complete (cradle-to-grave) or a partial LCA can be undertaken. The main units that are not considered (cut-offs) in the FD of the system must be stated and the main reasons for the choice to exclude them must be elucidated. That also applies to the main inputs and outputs that are excluded from the system. The criteria used to decide to consider items within or out of the system may depend on their contributions to mass, energy or environmental impacts.

2.2.2.4 *Procedures of Allocation*

Processes often generate several products that share the upstream resources and releases. Allocating inputs and outputs between the different coproducts is not straightforward. In general, there is no significant physical relationship between the coproducts on one side and the inputs and outputs on the other side. In the early age of LCA one product was focused on by the study and the coproducts were relegated to the status of byproducts. Implicitly a byproduct is a coproduct with a minor economic value. All the consumed resources and the releases were then attributed to the product being focused on, possibly resulting in too high an estimated environmental impact. However, in several cases the economic values of the coproducts are significant. The resources and releases can be allocated to them using economic values of the products, mass, energy content or carbon content. As different allocation procedures may produce various results, allocation is the Achilles heel of LCA. That is why in ISO 14044 it is recommended, wherever possible, to avoid allocation by using the following two-step method:

- To partition the unit process into as many subunits as required and attribute to each of them its specific inputs and outputs.
- To expand the product system in order to take into account the functions of its coproducts.

However, in most cases following this recommendation is not tractable. Multifunctional systems can be grouped into several categories. Frischknecht [26] proposed a typology based on three criteria:

- combined or joint production of products,
- successive or simultaneous production of products, and
- the number of decision makers involved (one or several).

The first criterion is more sensitive with regard to the allocation issue. In a combined system, the ratio between the amounts of the coproducts may change depending on the operation strategies. For instance, in a flexible sugar mill/bioethanol distillery the ratio between the amount of crystallized sugar and bioethanol produced may change depending on the relative market prices of both products. For simplicity it is assumed that bioethanol is produced using a portion of sugar juice. In reality molasses is used and it would be complemented by sugarcane juice. According to that simplification, if the price of ethanol is high, a more important share of the sugar juice would be sent to the distillery for ethanol production. As the common resource of both products is sugar juice that is supposed to be homogeneously divisible, the production system can be split into two subprocesses, each associated with the corresponding partitioned amount of identical product, here sugar juice. Considering the crystallized sugar as the coproduct that determines the amount of sugarcane (coproduct A), process A must be expanded to consider the ethanol production process (coproduct B). Such an expansion is tractable as both processes are physically based on the same intermediate feed—that is, sugarcane juice. The case where ethanol is produced using molasses is not so different. Both processes are successive (criterion 2 of Frischknecht's typology).

The case of sugar juice and bagasse can illustrate a joint production scheme. The ratio between the total amounts of juice and bagasse is fixed for a given sample of sugarcane, which is the general case in joint production. Sugarcane juice and bagasse are separable. However, the two streams after extraction have different physical and chemical properties—that is, a separation into complementary coproducts. Let us consider that the bagasse is used as a combustible to replace gasoil. The substitution approach recommended by ISO 14044 would consist of the following two steps:

- Attribute all the impact to the sugarcane juice considered as the focused product.
- Then credit sugarcane juice with the impact of gasoil procurement, considering an amount of energy that is equivalent to that which the bagasse would provide.

Weidema [27] claimed that it is always recommended to avoid allocation between coproducts, by using the substitution approach. However, even though that statement is theoretically undoubtable, the result of the substitution method depends on the displaced process considered in the substitution method (gasoil in the example). Furthermore, the displaced process may in turn involve a multifunctional system, as is the case for gasoil which is one of the several coproducts of a petroleum refinery. If the displaced product was natural gas or biomass, the result of the substitution approach would be far different, which raises the contextual question of how to choose the relevant displaced product.

In practice, all situations exist in the literature; different allocation procedures are used as well as substitution methods, implying discrepancies among the results. Instead of making a given procedure mandatory, it would be more scientifically sound to consider that there is no unique solution to the problem of partitioning common resource consumption and releases between joint coproducts. The validity of the methods used would be assessed based on their relevance dependent on the contexts of the LCA.

2.2.3 Specificity of Biorefineries With Regard to the Goal and Scope

2.2.3.1 The Context of Biorefineries

LCA is a general methodology that is applicable to every product and system in various contexts. That can explain the variety and the differences among the results. Therefore, maximizing the value of LCA implies clarifying the context of its application. The contextualization of the assessment should be included in the goal and scope stage. The renewal of biorefinery concepts can be explained by the perspectives of long-term depletion of fossil fuels, the threatened climate change and their numerous potential irreversible damages. Policy measures are taken by decision makers and need to be assessed from a sustainability point of view. The idea of bioproducts is the use of biomass to produce a matrix of valuable products including bioenergy that would replace equivalent fossil-based materials. Policy measures are presently focused on fossil fuels and their renewable substitutes. They mainly consist of promoting efficient use of depleting fuels, substitution with renewable energy, and GHG emission reduction. Biorefineries may comply with policy constraints if both refineries and constraints are sustainable. It is worth noting that since policy measures do not involve the whole biorefinery, there is no required framework for assessing the sustainability of biorefineries. With regard to LCA, biorefineries are multifunctional systems that may involve recycling processes. For such systems, and with consideration to the decision background, the problematics of LCA are related to comparison of performances between the system under study and a reference system. As for a multifunctional and mostly joint-multiproducts case, biorefineries assessment leads to the issue of allocation versus substitution. That issue is caused by the necessity of changing from the system as a whole to its products where the compliance to policy measures has to be investigated. The case of GHG emission reduction is considered in this section. Such policy measures are, for instance, induced by the European Renewable Energy Directive (RED) [28] where for the transport sector, the eligible renewable fuels must show a reduction of GHG emissions not lower than 35% for installations in operation before the entry into service of the directive. The percentage is related to comparison with the equivalent fossil fuel and is set to 50% for installations started in Jan. 2017 and

60% in Jan. 2018 onwards. As an illustration, if a biorefinery that will be installed in 2020 somewhere in the European Union is deemed to produce jointly bioethanol and a surplus of electricity, a comparative LCA must be conducted to answer the following question: will the reduction of GHG emission from the ethanol produced by the biorefinery be higher than the one from gasoline to the extent of 60% or more? It is stated in this section that the answer will depend on a precise elucidation of the context that is addressed. If there is no constraint on the GHG emission reduction of the surplus electricity, all the emission of GHG of the biorefinery will be attributed to the bioethanol which will be credited at the same time with the emission of GHG of the electricity reference process in the reference system, avoiding allocation. That is the substitution approach or expansion of the system promoted by Weidema [27]. What is the case when both products are to face compliance to policy thresholds on GHG emission reductions?

2.2.3.2 Beyond Allocation and System Expansion: A Claiming-Based Approach

Let us assume a biorefinery the GHG emission performance of which is X_T, the performances x_i of the joint coproducts ($I=1,\ldots,n$) being unknown. Let us assume that there exists a reference system where the performance y_i ($I=1,\ldots,n$) of the reference process of each coproduct i is known. The reference system is supposed to be separable, i.e., each reference process can be assessed separately from the other processes of the reference system. Then the global GHG emission of the reference system is:

$$Y_T = \sum_i y_i \tag{2.1}$$

The question that is asked in this section is as follows:

How can we estimate the performance x_i of each coproduct i?

ISO 14040 recommends whenever it is possible to avoid allocation and use the substitution method. However, in the case of biorefineries, the use of the substitution method would result in the following (Fig. 2.2):

$$x_i = X_T - Y_T + y_i \quad i=1,2,\ldots,n \tag{2.2}$$

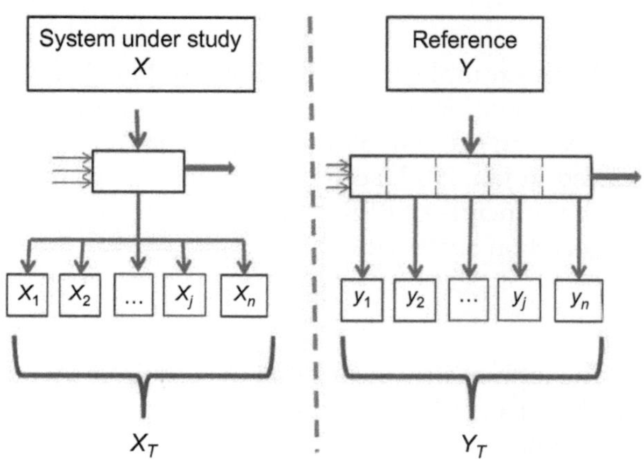

FIG. 2.2 System under study claiming for carbon sustainability.

$$\sum_i x_i = nX_T - nY_T + \sum_i y_i \tag{2.3}$$

$$(n-1)Y_T = (n-1)X_T$$

Then: $X_T = Y_T$.

Since in the case of a biorefinery, each coproduct must be considered in the same way, the use of the substitution method is not possible except when the reference and the studied system have the same GHG emission performance. In that case, the GHG emission performance of each coproduct equals the one of its reference product.

In the present section, a particular allocation method is proposed for cases where $X_T \neq Y_T$. It is termed: Allocation according to the principle of relative environmental requirement (RER).

Principle of RER

The following formula expresses the relations between the system under study and the reference system:

$$X_T \leq (1-a)Y_T \tag{2.4}$$

$$X_T = (1-c)Y_T \tag{2.5}$$

$$x_i \leq (1-a_i)y_i, \quad i=1,2,\ldots,n \tag{2.6}$$

$$y_i = b_i Y_T, \quad i=1,2,\ldots,n \tag{2.7}$$

- a and a_i are the required improvement of the GHG emission performance of the system under study (respectively the coproduct i) with respect to the reference system (respectively the reference process of the product i); c is the actual improvement of the system under study with respect to the reference system.
- b_i is the share of the GHG emission of the reference process i in the global GHG emission of the reference system Y_T.
- c is the global improvement of the biorefinery with regard to the reference system.

By definition:

$$a = \sum_{i=1}^{n} a_i b_i \tag{2.8}$$

As an example, if x_i is the share of bioethanol in the GHG emission of the system under study, y_i will be the share of gasoline in the GHG emission of the reference system and a_i is 35%, respectively 50% reduction of GHG emission in 2013 respectively for installation started in Jan. 2017, according to the European Directive on renewable Energy (RED).

The principle of RER proposed in this section states that a biorefinery is considered to be fully carbon sustainable if:

$$c \geq \sum_i (a_i b_i) \tag{2.9}$$

The term *relative* means that the requirement on the GHG emission performance of the coproduct i (say bioethanol) is defined with respect (relative) to a reference process (say gasoline).

Let us define a carbon sustainability index s as follows:

$$s = \frac{c}{a} \tag{2.10}$$

The allocation method according to the principle of RER is:

- If $a_i = 0, i = 1, 2, \ldots, n$, there is no environmental requirement on the biorefinery. In that case:

$$x_i = b_i X_T, \quad i = 1,2,\ldots,n \tag{2.11}$$

- If $a_i \neq 0, i = 1, 2, m$ and $a_i = 0, i = m+1, m+2, \ldots, n$

  Let e and c_i be the numbers such that:

$$e = c - \sum a_i b_i \text{ and } c_i = (a_i + e), \quad i = 1,2,\ldots,n \tag{2.12}$$

If the biorefinery is fully carbon sustainable, then

$$x_i = (1 - c_i) y_i, \quad i = 1,2,\ldots,n \tag{2.13}$$

$$\frac{x_i}{X_T} = b_i \frac{1 - c_i}{1 - c}, \quad i = 1,\ldots,n \tag{2.14}$$

If the biorefinery is not fully carbon sustainable ($s < 1$), then the coproducts can be ordered in the priority of the GHG emission requirement. Let k be an integer such that:

$$\sum_{i=1}^{k} (a_i b_i) \leq c < \sum_{i=1}^{k+1} (a_i b_i) \tag{2.15}$$

Then,

$$x_i = (1 - a_i) y_i, \quad i = 1,2,\ldots,k$$

$$x_{k+1} = \left(1 - \frac{e_{k+1}}{b_{k+1}}\right) y_{k+1} \tag{2.16}$$

$$\frac{x_i}{X_T} = b_i \frac{1 - a_i}{1 - c}, \quad i = 1,\ldots,k$$

$$\frac{x_{k+1}}{X_T} = \frac{b_{k+1} - e_{k+1}}{1 - c} \tag{2.17}$$

where

$$e_{k+1} = c - \sum_{i=1}^{k} (a_i b_i) \tag{2.18}$$

2.2.3.3 *Illustration of the Claiming-Based Allocation*

Let us consider a biorefinery consisting of an ethanol distillery (ED). As is presented in Chapter 4, the plant uses sugarcane juice to produce first-generation bioethanol (1G); 60%

of the remaining fibrous residue, called bagasse, is used to produce second-generation bioethanol (2G), while the remaining bagasse along with 50% of the green harvesting residue (mixtures of leaves, tops, and logs) and the residual lignin-rich stream from the distillation are sent to the cogeneration unit that provides steam and electricity to fuel the whole integrated 1G/2G ED. The biorefinery provides only two coproducts, bioethanol and surplus electricity, in the following amounts: with 1 ton of sugarcane, 116.6 L of ethanol and 25 kWh of surplus electricity are produced. While the biogenic GHG absorbed by the crop during the growth phase is supposed to be released during the conversion into energy, the nonbiogenic sourced GHGs, mainly due to field and logistics operations, are considered to be the net GHG emission balance of the biorefinery. For 1 ton of sugarcane, the total net GHG emission balance amounts to $X_T = 108.5$ kgCO_2eq.

Question: What would be the allocation factors with respect to the claiming-based approach?

To answer this question, the context is elucidated as follows.

The biorefinery will be built in a country where the ethanol will be blended with gasoline and together they will form an E85 fuel that will be used in flexible vehicles. Calculations of fuel efficiency show that the volume of 116.6 L of ethanol that will be produced by processing 1 ton of sugarcane corresponds to 81.1 L of gasoline for an equivalent mobility service. The GHG emission of gasoline is 4.7 kgCO_2eq/L, resulting in:

$$y_1 = 381.17\,\text{kgCO}_2\text{eq}$$

The GHG emission of the electricity mix of the country is estimated to be 0.21 kgCO_2eq/kWh, leading to:

$$y_2 = 5.32\,\text{kgCO}_2\text{eq}$$

Then the total GHG emission of the reference system is:

$$Y_T = y_1 + y_2 = 386.49\,\text{kgCO}_2\text{eq}$$

The actual improvement in GHG emission provided by the biorefinery in comparison to the reference system is given by:

$$c = 1 - \frac{X_T}{Y_T}$$

$$c = 71.93\%$$

Case 0: Applying economic-based allocation at different points of the system results in global allocation factors of the net GHG balance of 99.4% for the bioethanol and 0.6% for the electricity.

Case 1: Let us consider that the required improvement for biofuels and electricity with regard to GHG emissions are

$$a_1 = 60\% \text{ and } a_2 = 50\%$$

Considering that the shares of gasoline and electricity in the GHG emission of the reference system are:

$$b_1 = \frac{y_1}{Y_T} = 98.6\% \text{ and } b_2 = 1.4\%$$

The required minimum GHG emission reduction for the carbon sustainability of the biorefinery is:

$$a = a_1 b_1 + a_2 b_2 = 59.86\%$$

Since $c > a$ then the biorefinery is considered to be fully carbon sustainable.

Applying Eqs. (2.12), (2.13) gives the following results:

$$x_1 = 106.46\,\text{kgCO}_2\text{eq}$$

$$x_2 = 2.02\,\text{kgCO}_2\text{eq}$$

Applying Eq. (2.14) leads to allocation factors of 98.1% for ethanol and 1.9% for electricity.

Case 2: The emission reduction requirement for the ethanol remains at least 60% and there is no carbon policy requirement on the GHG emission of electricity.

Applying Eqs. (2.12), (2.13) with $a_2 = 0$ gives:

$$x_1 = 103.8\,\text{kgCO}_2\text{eq}$$

$$x_2 = 4.7\,\text{kgCO}_2\text{eq}$$

The resulting allocation factors are 95.7% for bioethanol and 4.3% for electricity.

Case 3: In this case, bioethanol is considered the focused product. Using the system expansion gives the following results:

$$x_1 = X_T - y_2 = 103.18\,\text{kgCO}_2\text{eq}$$

and then

$$x_2 = X_T - x_1 = y_2 = 5.32\,\text{kgCO}_2\text{eq}$$

The resulting allocation factors are 95% for ethanol and 5% for electricity.

Case 4: Finally, being more stringent on carbon policy with 72% GHG emission reduction for bioethanol and 80% for electricity would lead to the following results:

$$a = 72.11$$

As $c < a$, the biorefinery is not fully sustainable. Let us assume that the bioethanol is prioritized on the surplus electricity; then applying Eqs. (2.16)–(2.18) results in the following:

$$x_1 = 106.7\,\text{kgCO}_2\text{eq}$$

$$e_2 = c - a_1 b_1 = 71.93 - 70.99\% = 0.94\%$$

$$x_2 = 1.7\,\text{kgCO}_2\text{eq}$$

TABLE 2.2 Allocation Factors in Different Contexts

	Case 0—Economic Allocation	Case 1—Both Coproducts are Policy Restrained	Case 2—Only Bioethanol is Policy Restrained	Case 3—Substitution Approach	Case 4—Stringent Carbon Policy
a_1—Required improvement of GHG emission of bioethanol (%)	–	60	60	–	72
a_2—Required improvement of GHG emission of electricity (%)	–	50	0	–	80
Allocation factor of GHG emission to ethanol (%)	99.4	98.1	95.7	95.0	98.4
Allocation factor of GHG emission to electricity (%)	0.6	1.9	4.3	5.0	1.6

The improvement of GHG emission of electricity is only 67%, failing to comply with the policy requirement of 80%. The allocation factors are 98.4% for bioethanol and 1.6% for electricity surplus.

The main results are summarized in Table 2.2.

2.2.4 Choices of Impact Categories

The types of impacts that are considered depend on the goal and scope of the LCA. With regard to biorefineries, depletion of nonrenewable resources and global warming are the impacts that would at least be considered. More details are given in Section 2.4.

The data quality required depends on the objectives of the LCA. In the case of LCAs of existing systems, most of the data must be collected in situ. That is not the case for a prospective LCA, where the system under study may be futuristic and the data result from simulations of conceptual designs of the system. Other data related to the inputs of the system are taken from the literature or from inventory databases. According to ISO 14044 data quality requirements should address issues such as time-related concerns, geographical issues, technology coverages, precision, completeness, representativeness, consistency, reproducibility, sources of the data, and uncertainty.

2.3 LIFE CYCLE INVENTORY

2.3.1 Aim of Life Cycle Inventory

The product system is described in the goal and scope stage. The LCI aims to collect and quantify the input and output data at each unit process or block process level, validate these data, and analyze the data flows between the units throughout the whole system. According to ISO 14044 the LCI includes the following six steps:

(1) preparation for collection of data,
(2) collection of data,
(3) validation of data,
(4) relating data to unit process,
(5) relating data to the functional unit, and
(6) aggregation of data.

That procedure is not linear. Several feedbacks and iterations are possible, including revising the goal and the scope stage as well as redefining the detail level of the system FD. As the amount of data to be collected is huge, the main risk is inconsistency. Indeed, multisources of data are often used with sometimes significant differences in the ways data have been produced. The time during which the different data are estimated may also be different, requiring adaptation and harmonization. The technology under study may not be mature and the data used may be related to proximate existing technologies. One important issue that is not often addressed is the scaling effect. Very often the data are collected at different scales and must be harmonized when relating them to the unit processes and to the functional unit, respectively. Assumptions of linearity are often made, meaning that the quantity of different inputs is proportional to the scale of the output of the unit process, which can induce biases when assessing real systems. With regard to that issue, distinction is also made between attributional and consequential LCAs.

2.3.2 Attributional LCI

In an attributional approach to LCI (ALCI), the results in terms of releases of substances or consumption of resources are related to a functional unit. The unit processes are linked or portioned according to the modeling of the product system. The releases or consumptions are linearly scaled to get the ones of the whole system under study (geographic system, company production, etc.). As an example, let us consider the case of a country producing yearly 10^9 L of bioethanol from sugarcane juice. This amount of bioethanol is blended to X liters of gasoline for use in vehicles. Let us consider that the GHG balance attributed to 1 L of produced ethanol and gasoline are, respectively, Ge and Gg; then the total GHG balance due to ethanol (respectively) gasoline production will be $10^9 \times \mathrm{Ge}$ and $X \times \mathrm{Gg}$, respectively, leading to a total GHG balance of: $10^9 \times \mathrm{Ge} + X \times \mathrm{Gg}$.

This simple calculation is possible because it is implicitly admitted that X is known, which is relevant in a retrospective case. In a prospective case, X depends on how the produced ethanol will be blended with gasoline and therefore on the amount of gasoline that will be replaced.

In an attributional approach to LCI, the effect of ethanol production depends only on the ethanol system and not on the gasoline system. Even though correlation exists between the ethanol and gasoline systems, this correlation is accounted for throughout the total amounts of produced ethanol and gasoline. In the literature, an attributional approach to LCI is often referred to as retrospective, while Klöpffer and Grahl [13] rightly observed that ALCI is often used for assessing environmental performances of new products, meaning that the association between ALCI and retrospective is somewhat abusive. Allocation is often used in ALCI.

2.3.3 Consequential LCI

The consequential approach to LCI (CLCI) aims to assess the environmental system-wide consequences in the level of a product. The change in the amount of given products may influence other products. The estimates of the consequences cannot be derived by linear extrapolation due to the magnitude effect of the change. Let us consider an example of an introduction of 10^9 L of ethanol produced in a country from sugarcane juice. As in the illustrative example for ALCA, the produced ethanol is blended with gasoline. The question now is: what is the change in the GHG balance due to the introduction of this amount of bioethanol? It is worth noting that the question is of a prospective nature. Furthermore, depending on the magnitude of change, different questions can arise, such as: What infrastructure is required to produce, store, blend, transport and distribute the fuel? What is the GHG balance related to these activities?

The high amount of ethanol will have significant impact on the supply chain of gasoline and on the vehicle fleets themselves, depending on the types of blends that are envisioned (E10, E25, E85, etc.). What would be the changes in GHG balance related to these changes? The types of questions to be solved depend on the goal and the scope of the LCA that determines the boundaries of the system to consider. However, compared to the ALCI approach where only the ethanol system was considered, in the case of CLCI, expansion of the system is required to account for most of the effects of the changes. Since the type of required expansion and its significance depend nonlinearly on the magnitude of changes, linear extrapolates of the case of an arbitrary functional unit are not valuable anymore. In the case of a CLCI, the chosen functional unit should reflect the magnitude of the changes.

However, in a number of studies, the magnitude effects are not considered because the effects related to the build-up of new infrastructures and underutilization of existing infrastructure of the replaced technologies are minored. The estimation of the effects of the changes is based on the comparison of the performances of the system under study with the ones of a reference system (a system without the changes).

2.3.4 Average Versus Marginal or Incremental Data

Depending on the context, the nature of the data required for the LCI may change. In the literature there is a debate whether the data must be average, incremental or marginal [29]. In order to illustrate this issue, let us consider the LCI of the electrical power generation in a country. The following three questions are to be discussed: (A) What is the GHG balance of 1 kWh produced in the country? (B) What would be the change in the GHG balance with regard to the increase of one additional kilowatt-hour produced in the country? (C) What would be the change in the GHG balance with regard to the introduction of 1 GW of wind energy in the electrical power generating system in the long term?

With regard to question A, the aim is to estimate the average GHG emission factor of the electricity generation in the country. The LCI must work with average data, meaning that the electricity generation mix in the country must be expressed for an average generated kilowatt-hour. In relation to question B, the effect of an increase by 1 kWh will depend on the time of occurrence. If the increase occurs during off-peak hours when there is an underutilization of even off-peak plants, the change in the GHG balance will be related to the additional operation cost of these plants. That means short-term marginal data. However, the increase in the electricity generation by 1 kWh may occur at the peak hour when all the capacity of the electrical generating system is already in operation. In that case, a new peaking plant would be required, which means marginal increase of both capacity and operation. Long-term marginal data will apply. Finally, in the case of question C, the LCI requires incremental data. The change is related to a high amount of additional capacity and the change in the GHG balance will depend on the nature and amount of generation of that additional capacity as well as to the change in the electrical generation of the existing capacity.

2.3.5 Computational LCI

Although LCA is oriented towards practitioners, there are a few computational frameworks, the major ones being elaborated by Heijungs and Suh [30]. In that reference a steady-state case is modeled using linear matrix representations of the processes and flows. Indeed, in an LCI the product system is a chain that includes for each unit process one input node and one output node. Each input can be in turn considered as associated to its embodied product system, which leads to a complex system. An LCI consists of gathering and analyzing the multidimensional data with the aim to carry out environmental balances. For the purpose of these analyses, it is convenient to partition the data between two categories: the technological conversion and the environmental issues. The formalization of Heijungs and Suh involves the following steps:

(1) Using the notion of linear space to define a multidimensional data point representing a unit process by a vector.
(2) Defining the product-process by a matrix, the columns of which represent distinct unit process.
(3) Partitioning the process matrix into two different parts. The first represents the "economic flows" also referred to as technology matrix and the second the flows from or to the environment, referred to as environmental or intervention matrix.
(4) Specifying the required performance of the system by expressing the flow objectivized by the system.
(5) Defining the inventory table, that is the set of environmental flows related to the objectivized flow.

Despite the rigor of that formalization, it is not fully used in the LCA software due to the complexity of actual cases that would make the implementation of such a theoretical model intractable. However, up-to-date databases adopt matrix formalization to present their procedures. Two of them are outlined in this article.

2.3.6 LCI Databases

2.3.6.1 Ecoinvent Inventory Database

In the early 2000s an initiative was made by the Swiss Federal Research Institutions and Administration aiming at building up and managing a consistent LCI database for supporting LCA works in the country. The initial project (Ecoinvent 2000) involved a few dozen products distributed into five groups: energy supply, materials and wastes, transport services, basic chemicals, and agricultural products and processes [31]. The new generation of Ecoinvent issued in 2015 includes over 11,500 inventory datasets on products or services such as: energy supply, chemicals, plastics and plastics production, production and processing of metals, transport and mobility, disposal, construction, agriculture, information and communication technology, electronics, biofuels and materials, engineering, paper industry, recycling processes, and water data [32]. Furthermore, it is required that the user choose between several system models, such as "Allocation, ecoinvent default" that uses an attributional approach, or "Substitution, consequential, long term" using system expansion [33]. These options improve the possibility for the users to link their datasets in a more consistent way compared to the previous versions.

2.3.6.2 GaBi LCI Database

The structure of this LCI database involves concepts such as *plan* and *process* (Fig. 2.3). The concept of the Main Plan System is the highest aggregate level of process chain description. At that level, the process is described through the concept of "Plan 1" which could be tuned at the conceptual Subsystem 1 level where components of Plan 1 can be described in more detail into Subplan 1. Finally, at the conceptual Subsystem 2 level, components of Plan 2 related to a process involved in Subsystem 1 can be described in an even more detailed way.

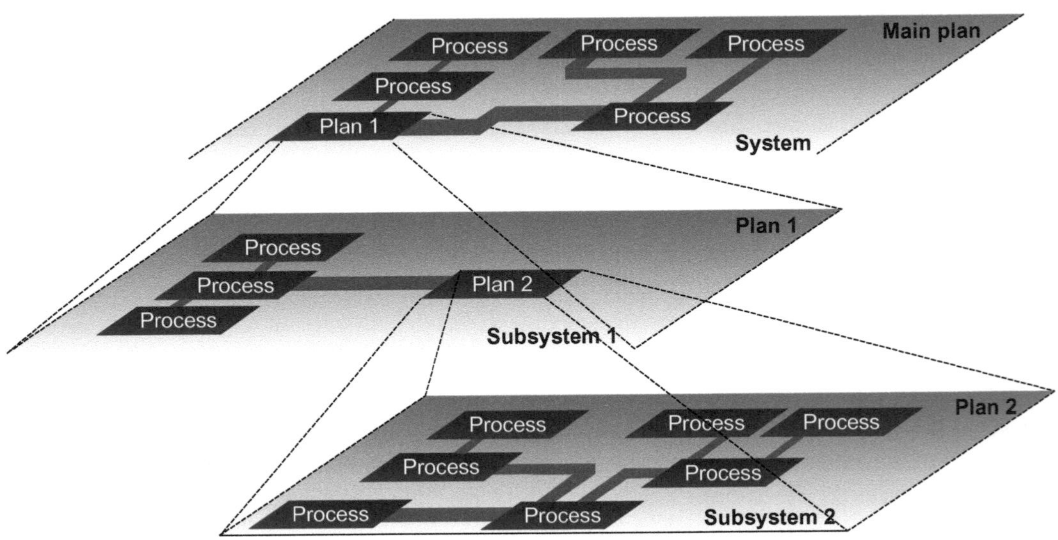

FIG. 2.3 Conceptual system and subsystem in GaBi LCI database [34].

The hierarchical structure provides a flexible conceptual model for representing the product system with any relevant detail. The master database involves about 10,000 generic Plan systems [34]. Aggregate models are provided by the GaBi database for generic systems such as refineries, power plants in the grid mix, country natural gas production within an interconnected natural gas mix. These models express the functional relations between a specific object (e.g., a power plant) and the interconnected system in which it operates, and finally to the functional unit (e.g., 1 kWh).

2.3.7 Data Inventory From Simulation

In the case of new products and systems where information in LCI databases is not available, process simulation is used, especially for the conversion step of the product system in order to provide material and energy balance consistent with thermodynamics principles and chemical reactions. Several tools can be used, such as Aspen Plus, Aspen HYSIS, CHEMCAD, and ProSimPlus. Process simulation is particularly useful in the case of biorefineries due to the impossibility of making the inventories of all schemes available in databases. In that particular case, process simulators are used for both the design of the biorefinery and for the assessment of mass and energy balance.

Process simulation describes processes in FDs using different theoretical models and computational techniques for the calculation of properties. The aim of this modeling is to approximate the real characteristics of a given process. The use of process simulation can be helpful in creating and evaluating new processes, as well as improving already-implemented processes. In process simulation a process is mathematically modeled and calculated dependent on thermophysical properties of basic components and mixtures.

The use of aspenONE in process simulation: AspenONE software optimizes chemical and other processes by adapting capacity, reducing costs, and improving energy efficiency. AspenONE Engineering is a series of programs that includes Aspen Plus and Aspen HYSYS process simulators, which provides the capability to understand and predict process behavior and optimize asset performance. Aspen Plus serves the needs of the chemical industry and provides additional modules for polymer and value chains modeling. The package also includes other programs, such as Aspen Simulation Workbook, which is helpful for implementing the simulation results in Microsoft Excel.

2.4 LIFE CYCLE IMPACT ASSESSMENT (LCIA)

2.4.1 Importance of the LCIA

The LCI provides a large amount of information relating to releases of various substances that can be harmful to the environment. However, the consequences on the environment are not estimated yet. As these impacts are the main reason for an LCA, their modeling and evaluation are of high importance. It is noteworthy to mention that the impacts of concern are not fully measurable but rather are potential. The role of modeling is therefore crucial. According to ISO 4044 LCIA activity involves three mandatory steps, namely the choice of the impact categories and the characterization models, the classification of the results of the LCI into these categories, and the characterization of each category by estimating impact indicators. The importance of the LCIA stems from the goals stated in the first stage of the LCA. The impacts that are estimated must clearly address the background issues related to the goals. Lack of clarification about these issues may make the interpretation stage just an imposed style.

2.4.2 Selection of the Impact Categories

Even though ISO 14044 gives general recommendations on how the selection procedure should be conducted, no mandatory list of impact categories is given. Each LCA analyst has to focus on relevant impacts according to the goal and scope of his study. Since the 1990s, a growing effort to gather a consensus among scientists on impact categories, characterization models and indicators has been undertaken, with the aim of increasing consistency and completeness and avoiding double counting. During recent years, this effort has been led within a United Nations Environment Programme (UNEP)/SETAC Life Cycle Initiative aiming to provide "global guidance on environmental LCIA indicators" [35,36].

The impact categories that are most used include the following:

(1) depletion of abiotic resources,
(2) depletion of biotic resources,
(3) climate change,
(4) stratospheric ozone depletion,
(5) human toxicity,
(6) ecotoxicity,
(7) freshwater aquatic ecotoxicity,
(8) marine ecotoxicity,
(9) terrestrial ecotoxicity,
(10) photooxidant formation,
(11) acidification,
(12) eutrophication,
(13) impact of ionizing radiation, and
(14) loss of biodiversity.

In ISO/TR 14047 the difference between the impact category, the category endpoints and the category midpoints is distinguished. As an example, for the impact category "Climate change" the category endpoints are consequences on coral reefs, natural vegetation, forest, crops, buildings, and human life expectancy, whereas category midpoints are infrared radiation, temperature, and sea level. In practice, the level at which the impact category should be defined is unclear. Once the inputs and outputs of the LCI are known and the category endpoints are chosen, ISO 14047 refers to intermediate variables between the LCI elements and the category endpoints as category midpoints. Sometimes the impact category is defined with reference to midpoints whereas the ISO 14044 makes the distinction between the impact category and the category midpoint. As an example, in the ILCD Handbook [37] "Climate change" is termed as midpoint impacts, whereas in ISO 14047 it is an impact category. Besides the issue of terminology, the category endpoint refers to a consequence such as damage, whereas the midpoint is a variable of the process producing a consequence from the releases accounted for with the LCI. For instance, the releases of GHG (LCI results) will potentially provoke an increase in the sea level (midpoint impact) leading to the deterioration of buildings (damage—endpoints). All that process from GHG emissions to the endpoint impacts pertains to the "climate change" impact category, which can be described by midpoint as well as endpoint indicators.

2.4.3 Selection of Characterization Models, Classification and Characterization

2.4.3.1 General Considerations on Characterization

Characterization is the quantitative evaluation of the impacts of each inventory emission and resource flow from the environment according to the underlying mechanisms. To allow characterization of the impacts, each input from the environment and release from the product system to the environment must be classified between the impact categories, which is the classification step. For each impact category, several characterization models exist and selecting a consistent collection of characterization models for the impact categories studied is not straightforward. The outputs of the selected models are characterization factors that apply to the inventory flows to estimate an aggregate category indicator. As an example, for the impact category "Climate Change," the LCI provides, per functional unit, emissions of different GHGs, such as CO_2, CH_4, N_2O, CFC, etc. The characterization model is chosen from those developed by the International Intergovernmental Panel on Climate Change (IPCC) [38]. The characterization factor is global warming potential (GWP) of each GHG expressed in $kgCO_2eq/kg$ gas, which after multiplying the related emission of gas and summing up over all GHGs results in an aggregate impact indicator in $kgCO_2eq$ per functional unit.

According to ISO 14044, the selection of characterization models should consider the following recommendations:

- The model should be accepted internationally, i.e., approved by an international body.
- The impact categories should represent the aggregated impacts of inputs and outputs of the product system on the category endpoint(s) through the category indicators.
- Value choices and assumptions made during the selection of impact categories, category indicators and characterization models should be minimized.
- The impact categories, category indicators and characterization models should avoid double counting unless required by the goal and scope definition.
- The characterization model for each category indicator should be scientifically and technically valid and based upon a distinct identifiable environmental mechanism and reproducible empirical observation.
- The extent to which the characterization model and the characterization factors are scientifically and technically valid should be identified.
- The category indicators should be environmentally relevant.

Since the early 2000s various LCIA methods have been proposed. They consist of a collection of characterization models of a number of category impacts. With the types and number of impact categories, the characterization models may be different from one method to the other. A few examples are CML 2002 [39], Eco-indicator 99 [40], and ReCiPe [41]. None of these models complies with all the requirements and recommendations of ISO 14044.

2.4.3.2 *Case of ReCiPe*

In other chapters of this book the ReCiPe LCIA method has been selected to estimate and analyze the impacts. The method was developed by LCA researchers in the Netherlands with the aim of harmonizing choices of characterization models and addressing category indicators at both midpoint and endpoint levels. Two groups of environmental mechanisms are defined (Fig. 2.4). The first one expresses the conversion of the LCI results into the category indicators at the midpoint level. The models of the second group link the category indicators at the midpoint level to the category impact indicators at the endpoint level.

In ReCiPe the same names are used for the impact categories at the midpoint level and for the related models in the first group of mechanisms.

In Table 2.3 the impact categories, category indicators and the names of characteristic factors are defined according to Goedkoop et al. [41].

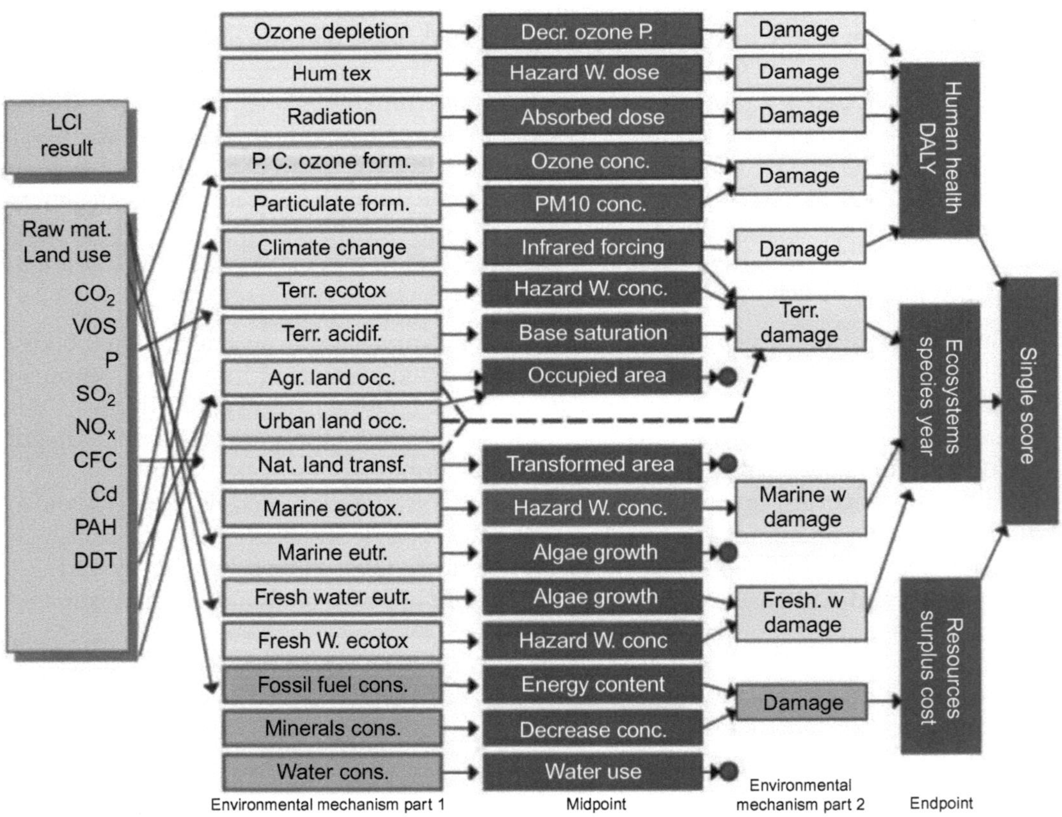

FIG. 2.4 Category indicators in ReCiPe [41].

TABLE 2.3 Impact Categories, Category Indicators and Characteristic Factors in ReCiPe Method

Impact Category	Indicator	Unit	Characterization Factor	Unit
Midpoint				
Climate change	Infrared radiative forcing	W × year/m^2	Global warming potential	kg (CO_2eq to air)
Ozone depletion	Stratospheric ozone concentration	ppt[a] × year	Ozone depletion potential	kg (CFC-11[b] to air)
Terrestrial acidification	Base saturation	year × m^2	Terrestrial acidification potential	kg (SO_2 to air)
Freshwater eutrophication	Phosphorus concentration	year × kg/m^3	Freshwater eutrophication potential	kg (P to freshwater)
Marine eutrophication	Nitrogen concentration	year × kg/m^3	Marine eutrophication potential	kg (N to freshwater)
Human toxicity	Hazard-weighted dose	–	Human toxicity potential	kg (14-DCB[c] to urban air)
Photochemical oxidant formation	Photochemical ozone concentration	kg	Photochemical oxidant formation potential	kg (NMVOC[d] to air)
Particulate matter formation	PM_{10} intake	kg	Particulate matter formation potential	kg (PM10 to air)
Terrestrial ecotoxicity	Hazard-weighted concentration	m^2 × year	Terrestrial ecotoxicity potential	kg (14-DCB to industrial soil)
Freshwater ecotoxicity	Hazard-weighted concentration	m^2 × year	Freshwater ecotoxicity potential	kg (14-DCB to freshwater
Marine ecotoxicity	Hazard-weighted concentration	m^2 × year	Marine ecotoxicity potential	kg (14-DCB to marine water)
Ionizing radiation	Absorbed dose	man × Sv	Ionizing radiation potential	kg (U^{235} to air)
Agricultural land occupation	Occupation	m^2 × year	Agricultural land occupation potential	m^2 × year (agricultural land)
Urban land occupation	Occupation	m^2 × year	Urban land occupation potential	m^2 × year (urban land)
Natural land transformation	Transformation	m^2	Natural land transformation potential	m^2 (natural land)
Water depletion	Amount of water	m^3	Water depletion potential	m^3 (water)
Mineral resource depletion	Grade decrease	kg^{-1}	Mineral resource depletion potential	kg (Fe)
Fossil resource depletion	Lower heating value	MJ	Fossil resource depletion potential	kg (crude oil in the ground)

Continued

TABLE 2.3 Impact Categories, Category Indicators and Characteristic Factors in ReCiPe Method—cont'd

Impact Category	Indicator	Unit	Characterization Factor	Unit
Endpoint				
Damage to human health	Disability-adjusted loss of life years DALY	year	–	–
Damage to ecosystem diversity	Loss of species during a year	year	–	–
Damage to resource availability	Increased cost	$	–	–

[a] *ppt: unit of equivalent chlorine.*
[b] *CFC-11: chlorofluorocarbon.*
[c] *14-DCB: 1,4 dichlorobenzene.*
[d] *NMVOC: nonmethane volatile organic carbon compound.*
Data from: Goedkoop MJ, Heijungs R, Huijbregts M, Schryver AD, Struijs J, van Zelm R. ReCiPe 2008. Report I: characterization. Ruimte en Milieu; May 2013.

Model of climate change in ReCiPe: For illustration, the model used to evaluate climate change indicators in ReCiPe is presented based on Schryver and Goedkoop [42]. A three-step model is used in ReCiPe in order to convert the GHG emissions into endpoint damage (Fig. 2.5).

Step 1: Model of radiative forcing used in ReCiPe

At this step the characterization factors (i.e., GWP) are estimated using the IPCC model. That model relates the absolute global warming potential for a timeframe T ($AGWP_{x,T}$) of a given GHGx to the product of its abundance ($x(t)$) [kg] and its radiative efficiency a_x [$Wm^{-2}\ kg^{-1}$]. The product is obviously summed along the timeframe T considered. The relative GWP of a given GHGx is estimated by dividing $AGWP_{x,T}$ by that of the GWP of a GHG r chosen as the reference $AGWP_{r,T}$.

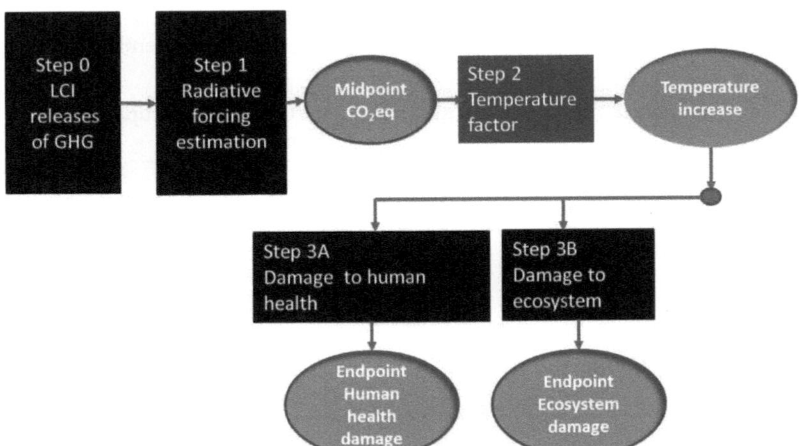

FIG. 2.5 Three-step climate change impact assessment.

$$GWP_{x,T} = \frac{AGWP_{x,T}}{AGWP_{r,T}} \quad (2.19)$$

$$AGWP_{i,T} = \int_0^T a_i \times i(t)dt \quad (2.20)$$

In Eq. (2.20) i denotes any GHG.

The reference GHG in Eq. (2.19) is CO_2. The dependence on T of $GWP_{x,T}$ of a given GHG x varies with regard to the lifetime of x. If the lifetime of CO_2 is close to that of x, $GWP_{x,T}$ will be insensitive to T. The lifetime of CO_2 is considered to be 150 years, according to IPCC 2001 [43].

The choice of T is subjective and depends on the value system. In ReCiPe the following value systems are considered:

- The hierarchist value system is consensual. A timeframe of 100 years is considered.
- The egalitarian attitude considers a very long-term perspective leading to a 500-year timeframe. That timeframe is considered to be the longest lifetime of the fate of the emitted GHG in the atmosphere.
- The individualist value system considers a short-term perspective leading to the choice of the 20-year timeframe.

The International Reference Life Cycle Data System recommends a timeframe of 100 years.

Step 2: Estimation of the temperature factor in ReCiPe

The temperature factor (TF) is used to link the emission of a given GHG to the increase of the average temperature of the earth. TF_{CO_2} is estimated as a product of the lifetime of CO_2 (LT_{CO_2}) and a factor c_{CO_2} that is the mitigation sensitivity of CO_2 expressed in [$°C\ kg^{-1}$].

$$TF_{CO_2} = LT_{CO_2} \times c_{CO_2} \quad (2.21)$$

Based on Meinshausen [44] it is found [42]:

$$c_{CO_2} = 7.09 \times 10^{-16} \left[°C\,kg^{-1}\right]$$

and then with $LT_{CO_2} = 150$ years it comes out that:

$$TF_{CO_2} = 1.064 \times 10^{-13} \left[°C \times year \times kg^{-1}\right]$$

De Schryver et al. [42] estimated the TF for the other GHG taking into account the lifetime of each GHG and using a flux-based approach.

Step 3A: Estimation of damage to human health in ReCiPe

The concept of "disability-adjusted life years" (DALY) is commonly used in LCA to assess damage to human health. DALY was introduced in the early 1990s in the World Health Organization (WHO) work on measuring the "burden of disease" [45]. This indicator emerged from several propositions aiming to design a composite index of the status of health that would integrate morbidity as well as mortality information. Therefore, it is composed of two elements: the number of years of life with a disability due to diseases or injuries (YLD), and the number of years of life lost due to premature death (YLL).

$$\mathrm{DALY} = \mathrm{YLD} + \mathrm{YLL} \tag{2.22}$$

The use of DALY, although largely widespread, is subject to criticism. For example, Anand and Hanson [46] argued that "the implications of age-weighting and discounting are found to be unacceptable." Indeed, DALY is dependent on age, disease, time and individual. When assessing the DALY for a geographic context, country for instance, how the different ages should be weighted is questionable. There are also several diseases that may be considered and these diseases are not necessarily independent of each other. The timeframe that is considered for estimating the impact is also a significant factor. Finally, variability of the individual response to disease may depend on biological and socioeconomic factors. The latter factor is often considered by assessing DALY for various regions. Murray et al. estimated DALYs for 291 diseases and injuries for the 1990–2010 period and 187 countries grouped into 21 regions [47]. They did not weight the ages and did not discount for time preference. The presentation of ReCiPe in this chapter is based on [41].

In ReCiPe, the following five health effects are considered with regard to the impacts of climate change:

(1) malnutrition,
(2) diarrhea,
(3) cardiovascular diseases,
(4) coastal and inland flooding, and
(5) malaria.

Six regions are considered as well.

The endpoint damage factor for human health due to climate change ($\mathrm{DF_{HH}}$ [DALY/(year °C)]) is calculated in ReCiPe using the following equation:

$$\mathrm{DF_{HH}} = \frac{\Delta\mathrm{IMPACT}}{\Delta T} = \frac{\Delta\mathrm{AttB_{tot}}}{\Delta T} \tag{2.23}$$

ΔT is the marginal difference in temperature rise between the baseline emission scenario chosen, such as for 1990 (in ReCiPe) and the emission scenario considered.

$\mathrm{AttB_{tot}}$ is the global attributable burden for all health effects [DALY/year]. It is estimated using Eq. (2.24):

$$\mathrm{AttB_{tot}} = \sum_r \sum_h \left((RR_{r,h} - 1) \times BoD_{r,h} \right) \tag{2.24}$$

where r and h stand for region and health effects, respectively.

RR is the relative risk factors depending on the chosen scenario of emissions and value choices (individualist, hierarchist, and egalitarian).

Finally, the endpoint characterization factor for climate change health damage ($\mathrm{CH_{HH}}$ [DALY/kgCO_2]) is calculated as a product of the TF estimated in step 2 and the damage factor from step 3A.

TABLE 2.4 Characterization Factors for Human Health

Value Choice	Adaptation Assumptions	Temperature Factor (°C×year/kgCO_2)	Damage Factor (kDALY/°C temp rise)	Characterization Factor (DALY/kgCO_2)
Individualist	Yes	1.064×10^{-13}	1.11×10^4	1.19×10^{-6}
Hierarchist	Yes	1.064×10^{-13}	1.31×10^4	1.40×10^{-6}
Egalitarian	No	1.064×10^{-13}	3.30×10^4	$3.51x\ 10^{-6}$

Data from: Goedkoop MJ, Heijungs R, Huijbregts M, Schryver AD, Struijs J, van Zelm R. ReCiPe 2008. Report I: characterization. Ruimte en Milieu; May 2013.

$$CF_{HH} = TF \times DF_{HH} \tag{2.25}$$

The endpoint TF, damage factor and characterization factor at endpoint are presented in Table 2.4 for each value choice.

It can be seen that, while the TF is obviously independent from value choice, the damage factor for the egalitarian value choice is three times that of the individualist value choice. This result is due to the assumption made in ReCiPe that for the egalitarian value choice there is no adaptation to climate change, leading to high risks for cardiovascular, coastal and inland flooding, and malnutrition, whereas for the individualist value choice there is a full adaptation to climate change.

Step 3B: Estimation of damage to ecosystem in ReCiPe

The endpoint characterization factor for damage to ecosystem diversification used in ReCiPe CF_{ES} is the product of the TF from step 2 and the damage factor DF_{ES} [1/°C] defined in step 3B as follows:

$$DF_{ES} = \frac{\Delta IMPACT}{\Delta T} \times SD_{terr} = \frac{\Delta PDF \times area \times SD_{terr}}{\Delta T} \tag{2.26}$$

with, SDterr is the density of species, ΔPDF is the marginal change in potential fraction of species that disappears, area is the total terrestrial area of the world.

The estimation of PDF is based on the paper of Thomas et al. [48] and assumptions related to dispersal of species are made depending on the value choices (see Table 2.5). For individualist and hierarchist value choices, dispersal is assumed. For egalitarian value no dispersal of species is considered, leading to a more than doubled damage factor.

TABLE 2.5 Characterization Factor for Damage to Ecosystem

Value Choice	Dispersal Assumptions	Temperature Factor (°C×year/kgCO_2)	Damage Factor (1/°C)	Characterization Factor (year/kgCO_2)
Individualist	Yes	1.064×10^{-13}	74.5×10^3	7.93×10^{-9}
Hierarchist	Yes	1.064×10^{-13}	74.5×10^3	7.93×10^{-9}
Egalitarian	No	1.064×10^{-13}	176×10^3	18.7×10^{-9}

Data from: Goedkoop MJ, Heijungs R, Huijbregts M, Schryver AD, Struijs J, van Zelm R. ReCiPe 2008. Report I: characterization. Ruimte en Milieu; May 2013.

2.4.4 Nonmandatory Elements of LCIA

Beside mandatory elements of LCIA, such as choice of impact category, characterization models, category indicators, classification and characterization, other elements such as normalization, grouping, weighting and data quality analysis are rather optional.

Normalization: For communication purposes the category indicators may be normalized with respect to reference values. Normalizing the results of the LCIA for different impact categories allows comparing at the same reference level the share of a given category indicator in the total impact of the category. When the reference level (e.g., per capita) is the same for two or more impact categories, their relative magnitude can be visualized without any possibility of interpretation since the impact categories are not weighted to reflect their relative importance.

Grouping: The category indicators may be sorted by groups such as global or local scales or ranked by levels such as high, medium and low according to the given value system. The optional grouping depends on the goals of the LCA.

Weighting: The various impact categories at midpoint or endpoint represent distinct dimensions of impact and may not be weighted if loss of information has to be avoided. Weighting means that substitution and compensation are admitted between distinct impact categories. As the weights would depend on the value choices of individuals, ISO 14044 recommends undertaking sensitivity analysis using different weights and providing the results with and without weighting.

Analysis of quality of the LCIA data: ISO 14044 recommends additional analyses to elucidate the significance of the LCIA results and their sensitivity to uncertain variables and value choices. Different methods can be used such as gravity, uncertainty, or sensitivity analyses.

2.5 INTERPRETATION

The interpretation stage closely interacts with the three other stages of the LCA. The issues linked to the goals and scope of the LCA must be discussed in view of the results of the LCI and the LCIA. The different results must be checked and analyzed with regard to their consistency, completeness, and sensitivity to relevant uncertain factors. A few examples of the interpretation are as follows: the influence on the results of the boundaries of the system, the detail level of the description of the product system, the functional unit, the type of allocation, assumptions made about missing data and value choices as well as temporal and spatial scopes; the importance of each step of the product system in the category indicator; the sensitivity to the choice of the reference system. When the LCA is undertaken in relation to an environmental policy, the extent to which a given target is complied with must be elicited.

2.6 IMPRECISION, UNCERTAINTIES AND MEANINGFULNESS IN LCA

2.6.1 General Considerations on Imprecision and Uncertainty

Imprecision and uncertainty have different meanings and their relationships are subject to various conceptions. French [49] included imprecision and ambiguity into uncertainty, whereas for Smets [50] imprecision and uncertainty are two distinct concepts. This latter way of defining the two concepts is particularly adopted by the researchers in approximate reasoning and is emphasized in this section. It can be illustrated by the following statements:

(1) The carbon performance per kilometer of a gasoline EURO4 vehicle is 0.35 $kgCO_2eq/km$.
(2) The carbon performance per kilometer of a gasoline EURO4 vehicle is about 0.35 $kgCO_2eq/km$.
(3) It is likely that the carbon performance of a gasoline EURO4 vehicle is 0.35 $kgCO_2eq$.
(4) It is likely that the carbon performance of a gasoline EURO4 vehicle is low.

- In the first statement there is no doubt that the performance is 0.35 $kgCO_2eq/km$, so there is no uncertainty in the statement. Furthermore, the performance is represented by a precise number (performance = 0.35).
- The second statement is certain; however, the performance is imprecise (performance = 0.35).
- The third statement is uncertain ("it is likely that"), but the performance is precise (performance = 0.35).
- Finally, the fourth statement is uncertain ("it is likely that") and the performance is imprecise and fuzzy (performance = low).

Imprecision expresses the lack of exactness of a measurement, a word, a sentence or a statement. Imprecision was initially conceptualized in the framework of the theory of errors in experimental measurements. Imprecision and inaccuracy were two forms of errors: inaccuracy is failure to get by measurement the true value of the attribute, while imprecision is the variability of the results of measurements from one experiment to the other depending on the imperfection of the instruments, the experimenter and the possible variations of the environment of measurements. Nowadays the meaning of imprecision is broader, especially in the fields of approximate reasoning and theory of fuzzy sets. Two ways of modeling imprecision are range of values and fuzzy numbers. Range of values can be discrete values; it is often continued values represented by intervals defined by their lower and upper boundaries. In an interval modeling of imprecision, each point of the interval has the same credibility with respect to the value that is represented. In the fuzzy set representation, there may be different grades of membership in the interval. Finally, intervals can be considered as a particular type of fuzzy numbers. Fig. 2.6 illustrates both interval and fuzzy number representations of imprecision.

Conversely, uncertainty should not relate to the precision of a measurement, a statement and a word but to the degree of confidence expressed. Uncertainty can be modeled using the theory of probability, the Dempster-Shafer theory of evidence, and the theory of possibility. If uncertainty results from random processes it can be measured by a probability function. When it is due to incomplete knowledge or ignorance, it is termed "epistemic uncertainty" and should be represented by imprecise probability or fuzzy measures such as belief or plausibility. Although distinct, there should be a connection between uncertainty and imprecision. A statement should be less uncertain if the attribute of the object of concern is expressed in an imprecise way, but it may be less meaningful. The more an attribute of the object is expressed in precise numbers, the less the certainty about these numbers should be. That is because any quantification encompasses inaccuracies. That may be one of the reasons for confusion between imprecision and uncertainty. However, contrary to common

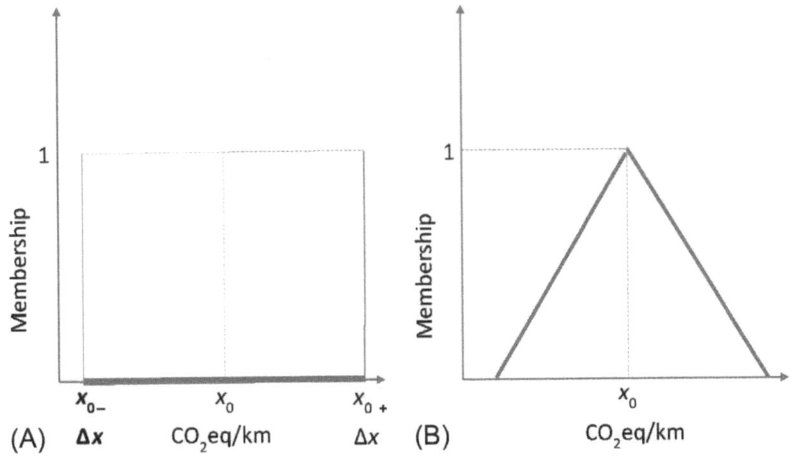

FIG. 2.6 Interval (A) and fuzzy number (B) modeling of imprecision.

sense, as was explained before, the higher the imprecision of the attribute, the lower should be the uncertainty. At the extreme, giving to an attribute a value of a universal set makes the statement certain but the imprecision is the highest and the meaningfulness the lowest. Another reason of confusion is that imprecision can also be caused by random errors and in that case it should be modeled by probability functions. For a set of available observations or information related to an object of study, the most relevant strategy is to maximize the meaningfulness, which means trade-offs between the embedded imprecision and the uncertainty.

The way to handle imprecision should be the following: sensitivity analyses, assessment of several variants based on consistent combinations of point estimations of the parameters and determination of the range of the results, assessments based on the arithmetic of fuzzy numbers and expressing the results as fuzzy numbers. Whereas, uncertainty can be modeled using scenario techniques, Monte Carlo simulation of precise or imprecise probabilistic functions, or estimation using possibility theory.

2.6.2 Different Sources of Imprecision and Uncertainty in LCA

Several sources of imprecision and uncertainty are present in LCA. Following the different stages presented in Fig. 2.1, it is worth noting that the goal and scope of the LCA can be stated with imprecision. The goal can be defined in an ambiguous way. The boundaries of the system can also be imprecise due to the ambiguity of the goal and that may produce inconsistencies. Imprecision also affects the LCI stage with respect to the inventory data that cannot be estimated with a perfect precision. Some of these data result from measurements and are affected by measurement errors. Others are the results of models and are affected by the lack of completeness of the models as well as by the imprecision of the inputs to the models. The inputs are often from multiple sources and the possible inconsistency between the ways they have been generated adds to the imprecision of each data source. As an example, such inconsistency can stem from different modes of allocation from one data source to the other and within the same data source. Another example is related to the assumptions and choices regarding the aspects that are considered or not with respect to the inputs (cut-offs); nuclear waste disposal is an illustration of this example, as it is considered in some LCI databases and not in others.

The results of the LCIA stage are also affected by the same sources of imprecision as for LCI. The results of both stages are impacted by the imperfect knowledge of the background physical processes, imperfectability of the available information, value choices and by the granularity of the modeling, such as the definition of the unit processes and elementary flows.

The interpretation stage generally proceeds with a deductive approach and may be affected by imprecision due to the reasoning process and imprecision in the three stages that are sources of inconsistencies and intractable results. Uncertainty can also be present due to lack of knowledge of the models used in the LCI and LCIA or lack of additional knowledge that is necessary for deducing the meaning of the results.

2.6.3 Handling of Imprecision and Uncertainty in LCA

Despite numerous existing works on uncertainty in LCA, this field of research is still open for more investigation. First, imprecision and uncertainty are treated under the umbrella of "uncertainty" and imprecision is often called "parameter uncertainty." Second, fuzzy numbers when used are termed "uncertainty" and are often considered as an alternative to probability distribution functions. Uncertainty with respect to imperfections of knowledge has not been considered up to now, due to the lack of elaboration or implementation of theories of uncertainty or to a choice to not bother the LCA practitioners by adding more complexity to a tool that is already considered as complex. However, much effort has been made at different levels, such as monitoring data quality from providers of datasets where data sources are assessed using a pedigree matrix [51] according to criteria such as temporal, geographic and technological correlations as well as reliability. The pedigree matrix is converted into uncertainty that is added to the "parameter uncertainty." Concerning the latter uncertainty, in Ecoinvent database for instance, default values of the variance of the probability distribution functions of the "parameter uncertainty" can be used. Several probability functions are available, such as normal, lognormal, binomial, gamma, and uniform. Range and triangular representations are used as well. Based on the chosen probability distribution function, Monte Carlo simulations can be conducted. These simulations allow estimation of the significance level when comparing the performance of a system under study with a reference system.

2.7 EXTENSION OF ENVIRONMENTAL LIFE CYCLE ASSESSMENT

2.7.1 General Considerations on Extension of Environmental Life Cycle Assessment

The environmental life cycle assessment (E-LCA) is evolving towards the following directions: developing background knowledge in order to enforce scientific foundations, collecting consistent data for feeding LCI databases, harmonizing existing characterization models and providing guidelines to practitioners, and extending assessments to economic and social dimensions within an integrated and consistent life cycle framework. The standardized LCA is evolving from an environmental LCA towards an integrated LCSA that would combine life cycle costing (LCC) and social life cycle assessment (S-LCA) to the E-LCA.

2.7.2 Life Cycle Costing

LCC is a method consisting of estimating the total cost of a product, taking into account the whole life cycle of the product as well as the direct and external costs. Actually estimating the cost of a product along with its life cycle or the life cycle environmental resources consumption and releases to the environment can refer to the same conceptual framework. The goals and scope have to be properly defined and data inventory including allocation issues must be considered. There is no need to consider impact assessment through characterization models in the case of LCC, as all estimates are in the same currency unit. However, it is possible to define concepts of risk cost that would be aligned to midpoint and endpoint impacts. Finally, the interpretation stage is also relevant in the case of LCC. There are several challenges in combining LCA with LCC, such as harmonizing the goals and the scope; where the goals of the interested public may be different, LCC is often undertaken for a particular actor. Therefore, the context in which the combination of both assessments is relevant must be stated. Such context may be concerned with public decisions in which the environmental impact of a new product must be complemented by its costs to the whole community, including owner cost, user cost and external cost. Fair information to consumers is another relevant context.

Another challenge is avoiding double counting or inconsistent assessment. For instance, economic allocation in the case of E-LCA requires the use of cost allocation factors, whereas external costs in the case of LCC requires internalizing the environmental impacts. Fully adapting both requirements along with the life cycle is challenging.

2.7.3 Social Life Cycle Assessment

Along the life cycle of a product there are social and socioeconomic impacts related to the human resources considered as inputs and impacting the area of any economic activity. These potential impacts have been for a long time addressed within other initiatives, such as corporate social responsibility standardized through ISO 26000 and implemented in the global reporting initiative that promotes sustainability reporting by organizations. The specificity of the S-LCA stems from the intention to align the methodology of S-LCA on that of the environmental LCA consisting of the four interrelated stages: goals and scope, inventory, impact assessment, and interpretation. One of the issues discussed is the protection areas that must be considered in S-LCA. Dreyer et al. [52] proposed to add "human well-being" and "human dignity" as new areas of protection. The midpoint impact categories to be considered depend on the case under study and the stakeholders. UNEP/SETAC [53] identified five categories of stakeholders, namely workers, consumers, local communities, society, and the value chain actors, excluding consumers. For each of these stakeholder group, they listed selected impact categories. Incorporating S-LCA into an integrated LCSA involves much the same challenges as LCC, including context-dependent issues, multiplicity of actors and related specific issues, and double counting avoidance.

2.7.4 Organizational Life Cycle Assessment

Environmental LCA can be undertaken from an organizational perspective, which supposes that the concerned organization adopts a life cycle thinking approach for its activities. Then the organizational LCA (O-LCA) consists of compiling, analyzing and evaluating from an environmental point of view the inputs and outputs of the organization along the life cycle, following the four-stage approach defined by ISO 14044. Contrary to the E-LCA, LCC and S-LCA, the O-LCA as defined in the initiative of UNEP/SETAC is not a component of the life cycle sustainability assessment. It is rather the restriction of the E-LCA to the scope of an organization. Another concept of organizational LCA can be defined following a general theory of system and focusing on the organizational issues of activities related to a product system. In that sense, the organizational LCA would evaluate the potential economic, environmental and social LCA of the managerial and procedural dimensions of the product system complementarily to the processes that focus on technology chains. However, these two dimensions of the product system are inextricably linked.

2.8 CONCLUSION AND PERSPECTIVES

Despite the efforts toward standardization, much work is still required in order to build-up consensus on the methods used in LCA within the scientific communities. While the importance of assessing the impact of a product system along with its life cycle is clear, a number of topics remain to be critically discussed, such as: homogeneousness of the various data used, relevance of temporal and spatial scopes, consistency of the mass and materials balances, depth of the scientific knowledge encoded in the impact models, imprecision and uncertainty. While avoiding allocation is advised by ISO, the trend towards policy requirements on sustainability of products would impose allocation in the case of multifunctional systems. The desperate expectation of getting the same results from different methods of LCA should be turned toward more effort in contextualization, relevance and consistency analyses as well as in progress of scientific knowledge supporting the assessment of the impacts. In that sense, the perspective of including environmental, social and economic LCAs in the same framework, while attractive, would increase inextricability and data requirements; homogenizing the frameworks of these different LCAs may be achieved but be detrimental to the richness of the analyses. The choice of the authors of this book is to undertake economic, environmental and social LCA in interrelated but separate frameworks linked throughout a multiple criteria assessment.

References

[1] Meadows HD, Meadows LD, Randers J, Behrens III WW. Limits to growth. New York, NY: New American Library; 1972.

[2] Hunt GR, Sellers DJ, Franklin EW. Resource and environmental profile analysis: a life cycle environmental assessment for products and procedures. New York, NY: Elsevier Science Publishing Co, Inc; 1992.

[3] Midwest Research Institute. Resource and environmental profile analysis of nine beverage container alternatives. Washington, DC: U.S. Environmental Protection Agency; 1974. EPA Report 530/sw-90c.

[4] Midwest Research Institute and Franklin Associates, Ltd. Resource and environmental profile analysis of five milk container systems. Washington, DC: U.S. Environmental Protection Agency; 1977.
[5] Argonne National Laboratory. Energy and materials use in production and recycling of consumer-goods packaging. Washington, DC: U.S. Department of Energy; 1981.
[6] Society of Environmental Toxicology and Chemistry (SETAC). Guidelines for life cycle assessment: a "code of practice". In: From the SETAG Workshop held at Sesimbra, Portugal, 31 March 3. Edition 1, Brussels and Pensacola, FL; 1993.
[7] International Standard (ISO). Environmental management—life cycle assessment—principles and framework. ISO 14040:2006 (E), Switzerland.
[8] International Standard (ISO). Environmental management—life cycle assessment—requirements and guidelines. ISO 14044:2006 (E), Switzerland.
[9] International Standard (ISO). Environmental management—life cycle impact assessment—examples of application of ISO 14042 ISO/TR 14047:2003 (E), Switzerland.
[10] International Standard (ISO). Environmental management—life cycle assessment—data documentation format. Technical specification ISO/TS 14048:2002 (E), Switzerland.
[11] International Standard (ISO). Environmental management—life cycle assessment—examples of application of ISO 14041 to goal and scope definition and inventory analysis. Technical report ISO/TR 14049:2000 (E), Switzerland.
[12] International Standard (ISO). Greenhouse gases—carbon foot print of products—requirements and guidelines for quantification and communication. Technical specification ISO/TS 14067:2013, Switzerland.
[13] Klöpffer W, Grahl B. Life cycle assessment (LCA). A guide to best practice. Weinheim: Wiley-VCH; 2014.
[14] International Standard (ISO). Environmental management—life cycle assessment—critical review processes and reviewer competencies: additional requirements and guidelines to ISO 14044:2006. Technical specification ISO/TS 14071:2014, Switzerland.
[15] International Standard (ISO). Environmental management—life cycle assessment—requirements and guidelines for organizational life cycle assessment. Technical specification ISO/TS 14072:2014, Switzerland.
[16] International Standard (ISO). Environmental management—water footprint—illustrative examples on how to apply ISO 14046. Technical report ISO/AWI TR 14073, May 2016, Switzerland.
[17] von Bertalanffy L. General system theory. New York, NY: George Braziller; 1968.
[18] Ciroth A, Finkbeiner M, Hildenbrand J, Klöpffer W, Mazijn D, Prakash S. Towards a life cycle sustainability assessment. UNEP/SETAC Life Cycle Initiative, 2011.
[19] Bernesson S, Nilsson D, Hansson PA. A limited LCA comparing large—and small—scale production of rape methyl ester (RME) under Swedish conditions. Biomass Bioenergy 2004;26:545–59.
[20] Guinée JB, Heijungs R. Calculating the influence of alternative allocation scenarios in fossil fuel chains. Int J Life Cycle Assess, 2007;12:173–80.
[21] Kamidelivand M, Gnansounou E. Life cycle environmental impacts of a prospective palm-based biorefinery in Parà State, Brazil. Bioresour Technol 2013;150:438–46.
[22] Panichelli L, Dauriat A, Gnansounou E. Life cycle assessment of soybean-based biodiesel in Argentina for export. Int J Life Cycle Assess 2009;14:144–59.
[23] Cherubini F, Jungmeier G. LCA of a biorefinery concept producing bioethanol, bioenergy and chemicals from switch grass. Int J Life Cycle Assess 2010;15:53–6.
[24] Madival S, Auras R, Singh SP, Navayan R. Assessment of the environmental profile of PLA, PET and PS clamshell containers using LCA methodology. J Clean Prod 2009;17(13):1183–94.
[25] Coltro L, Mourad AL, Oliveira APLV, Baddini JPOA, Kletecke MR. Environmental profile of Brazilian green coffee. Int J Life Cycle Assess 2006;11(1):16–21.
[26] Frischknecht R. Allocation in life cycle inventory analysis for joint production. Int J Life Cycle Assess 2000; 5(2):85–95.
[27] Weidema BO. Avoiding co-product allocation in life-cycle assessment. J Ind Ecol 2001;4(3):11–33.
[28] European Union. Directive 2009/EC of the European Parliament and the Council on 23 April 2009 on the promotion of the use of energy from renewable sources. Off J Eur Union 05.06.2009.
[29] Ekvall T, Azapagic A, Finnveden G, Rydberg T, Weidema Bo P, Zamagni A. Attributional and consequential LCA in I LCD handbook. Int J Life Cycle Assess 2016;21:293–6.
[30] Heijungs R, Suh S. The computational structure of life cycle assessment. Dordrecht: Kluwer Academic Publishers; 2002.

[31] Frischknecht R, Rebitzer G. The ecoinvent database system: a comprehensive web-based LCA database. J Clean Prod 2005;13:1337–43.
[32] http://www.umberto.de/en/ecoinvent. Consulted on 31 March 2016.
[33] Weidema BP, Bauer C, Hischier R, Mutel C, Nemecek T, Reinhard J, et al. Overview and methodology, data quality guideline for the ecoinvent database version 3. St-Gallen: Ecoinvent Centre—Swiss Centre for Life Cycle Inventories; 2013.
[34] Baitz M, Colodel MC, Kupfer T, Florin J, Schuller O, Kokborg M. Gabi database & modelling principles. Germany: Thinkstep AG; 2016.
[35] Jolliet O, Frischknecht R, Bare J, Boulay AM, Bulle C, Fantke P. Global guidance on environmental life cycle impact assessment indicators: finding of the scoping phase. Int J Life Cycle Assess 2014;19:962–7.
[36] Frischknecht R, Frantke P, Tschümperlin L, Niero M, Antón A, Bare J. Global guidance on environmental life cycle impact assessment indicators: progress and case study. Int J Life Cycle Assess 2016;21:429–42.
[37] European Commission—Joint Research Centre—Institute for Environment and Sustainability. ILCD handbook—analysis of existing environmental impact assessment methodologies for use in life cycle assessment. European Union; 2010.
[38] Forster PV, Ramaswamy P, Artaxo T, Berntsen T, Betts R, Fahey DW. Changes in atmospheric constituents and in radiative forcing. Climate change 2007: the physical science basis. Cambridge/New York, NY: Cambridge University Press; 2007.
[39] Guinée JB, Gorrée M, Heijungs R, Huppes G, Kleijn R, de Koning A. Handbook on life cycle assessment. Operational guide to the ISO standards. Dordrecht: Kluwer Academic Publishers; 2002.
[40] Goedkoop MJ, Spriensma R. The Eco-indicator 99—a damage oriented method for Life cycle impact assessment—methodology report. Pré Consultants B.V.; 2000. www.pre.nl.
[41] Goedkoop MJ, Heijungs R, Huijbregts M, Schryver AD, Struijs J, van Zelm R, ReCiPe 2008. Report I: characterization. Ruimte en Milieu; May 2013.
[42] De Schryver AM, Brakkee KW, Goedkoop MJ, Huijbregts AJ. Characterization factors for global warming in life cycle assessment based on damage to humans and ecosystems. Environ Sci Technol 2009;43(6):1689–95.
[43] IPCC. Climate change 2001: impacts, adaptation, and vulnerability. Cambridge: Cambridge University Press; 2001.
[44] Meinshausen M. Emission & concentration implications of long-term climate targets. Dissertation 15946, Zurich: Swiss Federal Institute of Technology; 2005.
[45] Murray CJL. Quantifying the burden of disease: the technical basis for disability-adjusted life years. Bull World Health Organ 1994;72:429–45.
[46] Anand S, Hanson K. Disability-adjusted life years: a critical review. J Health Econ 1997;16:685–702.
[47] Murray CJL, Vos T, Naghavi M, Flaxman AD, Michaud C, Ezzati M. Disability-adjusted life years (DALYs) for 291 diseases and injuries in 21 regions, 1990–2010: a systematic analysis for the Global Burden of Disease Study 2010. Lancet 2012;380:2197–223. www.thelancet.com.
[48] Thomas CD, Cameron A, Green RE, Bakkenes M, Beaumont LJ, Collingham YC. Extinction risk from climate change. Nature 2004;427(6970):145–7.
[49] French S. Uncertainty and imprecision: modelling and analysis. J Oper Res Soc 1995;46:70–9.
[50] Smets Ph. Imperfect information: imprecision-uncertainty. UMIS Var Unc.; 1999.
[51] Ciroth A, Muller S, Weidema B. Empirically based uncertainty factors for the pedigree matrix in ecoinvent. Int J Life Cycle Assess 2013;1–11.
[52] Dreyer LC, Hauschild MZ, Schierbeck JA. Framework for social life cycle assessment. Int J Life Cycle Assess 2006;11(2):88–97.
[53] Andrews ES, Barthel L-P, Beck T, Benoît C, Sylvatica C, Cucuzzella C, et al. Guidelines for social life cycle assessment of products. UNEP/SETAC life cycle initiative; 2009.

CHAPTER

3

Life-Cycle Assessment of Agricultural Feedstock for Biorefineries

K. Dufossé, W. Ben Aoun*,†, B. Gabrielle**

*Université Paris-Saclay, Thiverval-Grignon, France †Terres Innovia, Thiverval-Grignon, France

3.1 INTRODUCTION

3.1.1 Biomass Feedstock Supply as a Key Component in the LCA of Biorefineries

The supply of biomass feedstock to the biorefinery plant gate plays a key role in the overall environmental performance of biobased products. Feedstock production, harvesting, and delivery may account for more than half of the life-cycle greenhouse gas (GHG) emissions of first-generation biofuels [1], a share reaching 80% or more for combustion pathways providing heat and/or electricity [2,3]. The production of biomass from lignocellulosic crops interacts with a host of environmental, ecological, economic, and social issues, together with human health [4]. Environmental impacts encompass water availability and quality, soil and air quality, biodiversity and climate through the emissions of GHGs and C sequestration in soils [5,6]. Following the controversy on the GHG benefits of first-generation biofuels [7], concerns have also been raised for lignocellulosic crops [8], mostly pointing at our limited knowledge of their environmental and economic performances.

This chapter starts by categorizing agricultural feedstock sources and reviewing the main trends regarding their life-cycle impacts, before highlighting the main methodological challenges in the assessment of biomass supply chains and presenting state-of-the-art examples on how to address them and provide guidance in the selection of optimal supply mixes in accordance with local potentialities.

3.1.2 A Typology of Biomass Feedstocks According to Their Environmental Performance

Agricultural biomass feedstocks may be broadly categorized into dedicated (or purpose-grown) energy crops, byproducts such as cereal straw or secondary waste streams arising

http://dx.doi.org/10.1016/B978-0-444-63585-3.00003-6

from biomass processing (e.g., rice husk, molasses, or sugar-cane bagasse). Dedicated energy crops may be further broken down into food crops which are currently used for first-generation biofuels, and lignocellulosic plants (ranging from sorghum to short rotation coppice (SRC)) producing a generic biomass (lignocellulose) utilizable for various biorefinery or bioenergy pathways.

Table 3.1 summarizes the performance criteria of major biomass production systems across the world regions, whether using dedicated crops or residues [4]. The management of energy plants includes the provision of seeds or seedlings, stand establishment and harvest, soil tillage, and various rates of irrigation, fertilizer and pesticide inputs. The latter depend on crop requirements, target yields, and local pedoclimatic conditions and may vary across world regions for a similar species. Purpose-grown crops may be harvested several times a year (for forage-type feedstocks such as hay or alfalfa), once a year (for annual species such as wheat, or perennial grasses), or every 2–8 years (for SRC). In terms of logistics, biomass is typically transported to a collection point on the farm or at the edge of the road before transport to the biorefinery unit or an intermediate storage. It may be preconditioned and densified to make storage, transport, and handling easier. Transportation distances for raw biomass may vary from a few kilometers to over 100 km depending on the size of the biorefinery and its position with respect to the availability of energy crops or residues.

Overall, primary residues are generally considered the best option given their low cost and environmental burdens compared to their coproducts of primary interest. However, their collection raises a trade-off with soil quality in the long run [9], and their actual availability, bearing in mind sustainability constraints and other competing uses, is difficult to assess [10]. These aspects will be addressed and discussed through several examples in Section 3.2.

TABLE 3.1 Typical Characteristics of the Production Technologies for Agricultural Biomass Crops and Their Primary Residues

			Management				Costs
Feedstock Type	**Region**	**Yield (GJ/ha/year)**	**Fertilizers**	**Water needs**	**Pesticides**	**Coproducts**	**USD_{2005}/GJ**
Oil crops		As oil					
Oilseed rape	Europe	40–70	+++	+	+++	Rape cake, straw	7.2–16.0
Soybean	N America	16–19	++	+	+++	Soy cake, straw	11.7
	Brazil	18–21	++	+	+++		NA
Palm oil	Asia	135–200	++	+	+++	Fruit bunches, press fibers	NA
	Brazil	169	++	+	+++		12.6[a]
Jatropha	World	17–88	++	+	+	Seed cake, shells	3.2
Starch crops		As ethanol					
Wheat	Europe	54–58	+++	++	+++	Straw, stillage[b]	5.2
Maize	N America	72–79	+++	+++	+++	Straw, stillage	10.9

TABLE 3.1 Typical Characteristics of the Production Technologies for Agricultural Biomass Crops and Their Primary Residues—cont'd

Feedstock Type	Region	Yield (GJ/ha/year)	Management			Coproducts	Costs
			Fertilizers	Water needs	Pesticides		USD_{2005}/GJ
Cassava	Thailand	40–70	++	+	++	Stillage	3.3–4.0
Sugar crops		As ethanol					
Sugar cane	Brazil	116–149	++	+	+++	Bagasse, straw	1.0–2.0[a]
	India	95–112					NA
Sugar beet	Europe	116–158	++	++	+++	Molasses, pulp	5.2–9.6
Sweet Sorghum	China	105–160	+++	+	++	Bagasse	4.4
Lignocellulosics							
Miscanthus	Europe	190–280	+/++	++	+		4.8–16
Switchgrass	Europe	120–225	++	+	+		2.4–3.2
	N America	103–150	++	+	+		4.4
Short rotation	S Europe	90–225	+	++	+	Tree bark	2.9–4
Eucalyptus	S America	150–415	+/++	+	+		2.7
Short rotation Willow	Europe	140					4.4
Primary residues							
Wheat straw	Europe	60	+			Not applicable	1.9
	USA	7–75					NA
Sugar cane straw	Brazil	90–126	+				NA
Corn stover	N America	15–155	+				NA
	India	22–30	+				0.9
Sorghum stover	World	85	+				NA

[a] *Market price.*
[b] *Stillage: protein-rich byproduct of ethanol production.*
Yields are expressed as GJ of energy content in biomass prior to conversion to energy, or of ethanol end-product for sugar and starch crops. Costs refer to private production costs, or market price when costs were unavailable. Key to management inputs: +, Low; ++, Moderate; +++, High requirements.
Adapted from Chum H, Faaij APC, Moreira J, et al. Bioenergy. In: Edenhofer O, Pichs-Madruga R, Sokona Y, et al., editors. IPCC special report on renewable energy sources and climate change mitigation. United Kingdom, New York: Cambridge University Press; 2011.

Crops are needed with low input requirements and/or that can grow easily on marginal lands in order to decrease the pressure on food crops. Among lignocellulosics, perennial crops such as miscanthus and switchgrass present interesting features, combining high yields with low inputs (Table 3.1).

3.1.3 A Generic Framework for the LCA of Agricultural Feedstocks

This chapter focuses on the environmental balance of feedstock production and procurement. Biomass logistics (including transportation and storage) introduce a large number of possible options to supply biomass to the biorefinery plant gate. It is not central to this chapter but is considered in some of the examples given that it interacts with biomass production and harvesting systems and cannot therefore be disconnected from the latter. Thus, system boundaries will extend from cradle to farm- or plant gate, depending on the examples. The functional unit will be expressed either as tons of biomass for feedstocks with similar characteristics (e.g., lignocellulose) and/or GJ of biomass energy content to account for variations in heating values across feedstocks.

Bioenergy feedstock production is usually viewed as competing with other usages of natural resources, chief of which is land, with possible negative effects on biodiversity, water availability, soil quality, and climate. However, synergistic effects may also emerge through the design of integrated production systems, which also provide additional environmental services. Intercropping and mixed cropping are interesting options to maximize the output of biomass per unit area farmed. Mixed cropping systems result in increased yields compared to single crops and may provide both food/feed and energy feedstock from the same field. Double-cropping systems have the potential to generate additional feedstocks for bioenergy and livestock utilization and potentially higher yields of biofuel from two crops in the same area in a year. Lastly, agroforestry systems make it possible to use land for both food, timber and energy purposes with mutual benefits for the associated species. The associated land equivalent ratios may reach up to 1.5, meaning a 50% savings in land area when combining trees with arable crops compared to monocultures [4]. Unfortunately, these combinations and options have been little assessed or implemented so far and will therefore not be addressed in this chapter that follows the traditional monospecific approach for feedstocks.

3.2 AGRICULTURAL RESIDUES

The main advantage of residues lies in their nature of byproduct, alleviating their environmental burden and mitigating their costs. However, this comes with a set of potential trade-offs:

- their low yields per hectare (see Table 3.1), implying higher impacts for the collecting stage per ton of biomass, compared to purpose-grown crops,
- their removal rate has to be carefully regulated to maintain (or improve) soil carbon and organic matter stocks,
- there are other competing end uses which may be of higher priority locally (in particular for livestock).

3.2.1 Assessing Feedstock Availability Taking Into Account Soil Carbon Stocks and Competing Usages

Among agricultural residues, cereal straw (from wheat, barley, or maize) is preferred, and oil seed rape straw generally disregarded because of its brittle nature, making it hard to collect. Indeed, due to the high nitrogen content, the latter would preferably return to the soil in order to improve or maintain soil nitrogen (N) stock [9]. In order to maintain soil carbon (C) and organic matter stocks, only a portion of cereal straw should be removed for biorefinery purposes. This fraction or ratio varies widely according to soil type, the climatic conditions (rainfall and temperature) as well as the type of crop sequence and expected yields [11]. Long-term measurement or computer simulations (which are more accessible) can help identify the amount of potentially available straw that sustains soil organic C stocks [12]. Some long-term field experiments have focused on the relationship between soil C variations and the amounts of straw removed from different cropping systems, for various soil and climate conditions. From these data, mathematical models of C dynamics (such as RothC, Century, or AMG) [13] may be calibrated to simulate more combinations of soil-climate-crop and thus providing advice to farmers depending on the local context [12]. In an example in the Picardy region in Northern France [11], optimized straw removal ratios were derived from a simple soil C model for all types of pedoclimatic conditions and main crop successions occurring within the region. These ratios correspond to the fraction of crops within the succession for which the straw is exported, while straw is returned to the soil for the remaining crops. At regional scale, these ratios varied within a wide range, from 66% (i.e., straws removed for two crops out of three) down to less than 25% (removal for one in four crops).

To derive the tonnage of straw available from such ratios, regional agricultural statistics may be used to infer the main crop successions within the region, together with average crop yields that determine the amount of straw produced per hectare. Overlaying this information with soil maps, gridded climate data and administrative boundaries (for the level at which agricultural statistics are available) results in a set of geographical units to which straw removal ratios can be applied. Among all cereal straws, wheat and barley straws are generally preferred because of their high straw yields. From these potentially available straw stocks, it is essential to subtract the quantities of straw required to meet local or regional livestock needs. There are several methods to estimate these needs, based on cattle statistics and described in the following examples.

The emergence of a market for agricultural byproducts, driven in particular by an increasing demand for energy and bioproducts, may lead to straw-producing crops being favored over other crops (oil and protein crops in particular, and grasslands) to generate an additional income. However, this trend is not desirable since it would further reduce crop diversity, which is a key to ensuring the provision of ecosystem services.

3.2.2 LCA Methodological Issues for Agricultural Residues

In most life-cycle assessments (LCAs) applied to agricultural residues, only the collecting stage is considered within the inventory, and residues come with a zero-environmental burden. This assumption is also used from a policy perspective, since for instance the renewable energy directive in Europe recommends using it when assessing the GHG emissions of biofuel chains [14]. Impacts of residue removal with respect to a baseline scenario where they are returned to soils include a forgone sink effect in soils (as addressed in the preceding section), along with the exportation of nutrients, which should be accounted or compensated for [15]. Alternatives include an allocation of the crops' environmental burdens to the different coproducts, whether on a mass, energy, or economic basis [16]. The latter is for instance recommended when dealing with main products having a disproportionately high value compared to their coproducts (e.g., flax fibers and flax shives) [16]. Mass or energy-based allocation would attribute more burden to the residues and may impact the ranking of residues with respect to dedicated crops.

3.2.3 A Case Study for Two Regions of France (Burgundy and Picardy)

The previously described methodology was applied to two different regions in France, with different contexts. Picardy is a cereal and sugar beet producer, located in Northern France, whereas Burgundy, located in North-Eastern France, includes a larger part of livestock and pastures. The main crop sequence farmed in these regions is described in Table 3.2, while straw removal ratios are detailed in Table 3.3. The agroecosystem model CERES-EGC [15] was run for 20 years with climate data predicted for the 2010–2030 time period with a general circulation model to estimate crop and straw yields, along with environmental emissions for C and reactive N compounds. The simulation units (SUs) obtained by overlaying the

TABLE 3.2 Main Crop Sequences in Burgundy and Picardy, as Derived From Regional Statistics and Expert Knowledge [17]

Region	Main Crop Sequences
Burgundy	Winter barley—winter oilseed rape—winter wheat
Picardy	Winter oilseed rape—winter wheat—sugar beet—winter wheat—winter barley

TABLE 3.3 Breakdown of SU According to the Frequency of Straw Removal and Region, in Percentage of the Total Cropland Area

	Share of SU Per Management Type (%)	
Frequency of Straw Removal (%)	Picardy	Burgundy
25		12.3
33	47.8	38.5
50	52.2	25.8
67		23.3

regional soil map, grid points for simulated weather data and administrative boundaries for land-use data are depicted in Fig. 3.1. In terms of size, SUs range from 2.5 to 20,000 ha, with a median of 1210 ha. Potential amounts of straw that could be removed from arable fields without jeopardizing soil C stocks are presented in Table 3.4. Regional livestock supply requirements for this table were extracted from regional statistics [17] a database compiled in a previous research project [18].

Whatever the calculation methods (based on empirical straw removal ratios or biophysical modeling), regions differ strongly in terms of potential to produce and use agricultural residues for biorefinery purposes, taking into account soil quality constraints and the needs of local farms for livestock. Small conversion units may easily be supplied, but large-scale units (e.g., second-generation ethanol plants) would require more detailed studies, probably leading to the conclusion that complementary sources of lignocellulosic feedstocks should be brought into the supply mix. From a sustainability perspective, it should also be noted that a low potential for straw at the regional level implies a low density of available feedstock

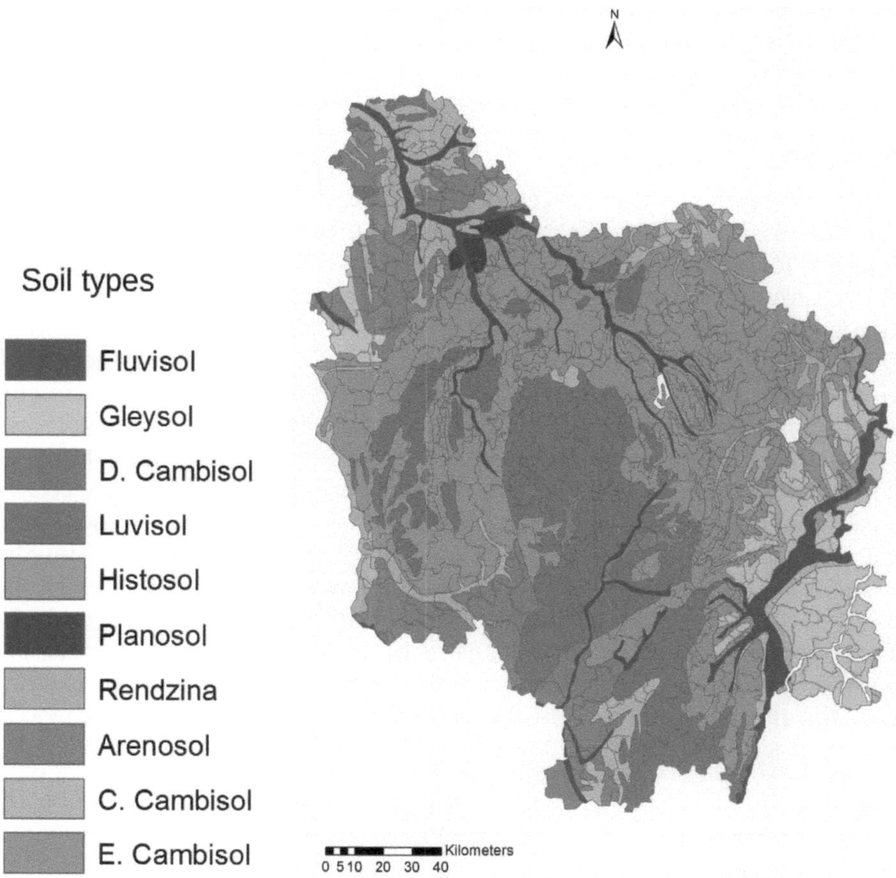

FIG. 3.1 Map of soil units used in the simulation of straw removal in the Burgundy example.

TABLE 3.4 Estimates of Straw Availability Obtained From the ECOBIOM Project (Also Including Livestock Requirements; Ref. [18]) and the Alternative Method Proposed Here Involving the Agroecosystem Model CERES-EGC, in Picardy and Burgundy

Source		Picardy	Burgundy
Share of the reference succession in the region	(% of UAL)	61	33
	(% of cropland)	72	65
ECOBIOM project	Livestock requirements (t DM)	−292,637	−748,131
	Availabity for energy purposes (t DM)	739,173	384,831
CERES-EGC	Straw produced (t DM)	1,084,827	1,120,022
		±3914	±2988
	Availabity for energy purposes (t DM)	792,190	371,891

around the conversion plant and costlier logistics. This would be worsened if straw was to be imported from a neighboring region, as may happen between livestock-dominated and cropland-dominated regions.

3.2.4 Synthesis

Agricultural byproducts such as cereal straws clearly present interesting features for substituting for fossil energy, with environmental burdens comparable to forest residues and significantly lower than agricultural crops, on a mass or energy basis [16]. However, their use should be carefully planned to avoid detrimental effects on soil organic C or exacerbate competition with other end uses, in particular for livestock needs. This results in relatively lower average yields of harvested straw (0.79 and 1.24 tDM ha^{-1} for Burgundy and Picardy, respectively), compared to any other dedicated crop yields in the case studies presented here. This makes logistics costlier since a large area has to be collected to supply biorefineries, increasing the impacts of biomass transportation. Competition is also likely to occur within the bioeconomy sector, since a growing number of bioenergy/biorefinery projects are based on straw as a main feedstock, at local to regional scales. As the pressure on such resources increases, the value of residues is likely to increase [19]. At this point, it is interesting to rethink straw no longer as a waste but as a coproduct of grain, with a monetary value. From an LCA perspective, this could mean that the impacts of growing food crops should be shared between the main product (grains) and coproducts, using for instance economic, or mass- or energy-based allocation ratios. This inevitably increases the environmental burden of straw, while the choice of the allocation method drastically affects the performance of residues [16].

3.3 AGRICULTURAL CROPS

3.3.1 First-Generation Biofuel Feedstocks

Current biofuels use a human-edible fraction of agricultural crops as feedstocks, whether in the form of simple or complex sugars or vegetable oil. The major crop types involved are listed in Table 3.1, along with a selection of characteristics from the point of view of crop management, yields and costs. Given their large input requirements compared to lignocellulosic feedstocks, these crops generally have several-fold higher emissions and resource consumptions at the plant gate, whether on a mass or energy basis [3]. Sugarcane and sugar beet provide an exception due to their comparatively larger yields per hectare compared to the other feedstocks [20]. At a whole-chain level, the higher farm-gate impacts are somewhat mitigated for biofuel generation because of lower energy requirements to convert oil or sugars to liquid fuels compared to second-generation pathways [3].

The main trends and ranges in terms of well-to-wheels (WtWs) GHG savings incurred by the substitution of fossil-based fuels by biomass are depicted in Fig. 3.2. The climate mitigation potential of bioenergy pathways appears very variable and highly related to the biomass yields of crops per unit area (Table 3.1). For purpose-grown crops, there is a major contribution from fertilizer inputs and N_2O emissions occurring during crop growth [21], implying that crops with low requirements and/or nutrient concentrations should be favored. Given the current uncertainties on N_2O emissions and their impact on the GHG balance of biofuels [1,22], their estimation is discussed in detail in the next section. The other main methodological issue for LCA application to biobased products derived from food crops is the handling of coproducts during the conversion process, which is also emphasized in the following section.

3.3.2 Main Methodological Issues With LCA: N_2O Emissions and Byproduct Handling

Emissions of N_2O (a potent GHG) result from biogeochemical processes occurring in the soils where crops are grown or further down the N "cascade" of transport and transformation characterizing the fate of fertilizer N [1]. These emissions are termed direct and indirect,

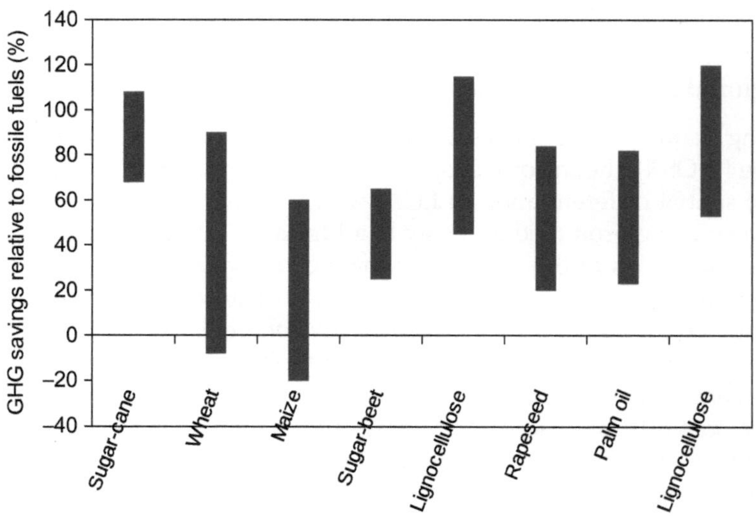

FIG. 3.2 WtW GHG emissions of a range of 1G and 2G biofuels compared to fossil equivalents (excluding indirect LUC effects). *Source: Reference Chum H, Faaij APC, Moreira J, et al. Bioenergy. In: Edenhofer O, Pichs-Madruga R, Sokona Y, et al., editors. IPCC special report on renewable energy sources and climate change mitigation. United Kingdom, New York: Cambridge University Press; 2011.*

respectively. In most LCAs of 1G biofuels, direct and indirect emissions are calculated from generic emission factors assuming a fixed relationship between fertilizer input rates and N_2O efflux. This corresponds to the Tier 1 method proposed by IPCC for national GHG inventories [23]. Since biogeochemical processes are strongly influenced by local pedoclimatic conditions, the use of Tier 1 emission factors generates large uncertainties, with a relative range that can reach 100% [22]. Agroecosystem modeling provides a promising alternative to better capture the direct emissions of N_2O [24], following the Tier 3 approach of GHG reporting [23]. It is exemplified as follows for biodiesel from oilseed rape in France. A similar approach may also be used for indirect emissions of N_2O, but it requires modeling at a larger scale (landscape rather than arable field). Preliminary simulations with such a model showed that indirect emission factors could vary substantially due to the interactions between the processes governing the fate of reactive N in its cascade [25].

Since the conversion of 1G feedstocks to biofuels generates a large fraction of byproducts (making up as much as 50% in mass), the handling of byproducts is a critical issue in the outcome of the LCA. The various alternatives are detailed in Chapter 2 of this book. Briefly, ISO norms on LCA stipulate that allocation should be avoided as much as possible by system subdivision. However, this is not feasible in practice for 1G biofuels, since the feedstock is fractionated to produce biofuels (from sugar, starch, or oil), on the one hand, and animal feed, on the other hand, as the result of a unit process. When comparing biofuels with fossil fuels, which is one of the main purposes of such LCAs, system expansion is usually considered best practice. It consists of keeping the coproduct in the biofuel production system, and introducing an alternative to these coproducts in the fossil-based value chain (for instance, soybean cake as an equivalent of rapeseed cake for biodiesel from oilseed rape). However, allocations usually dominate in LCAs because they are easier to implement. This choice has a major impact on life-cycle GHG emissions for 1G biofuels (see Fig. 3.3) and remains a sensitive issue [21].

3.3.3 A Regional Approach to the LCA of Biodiesel From Oilseed Rape in France

The following example presents a regionalized approach to the LCA of biodiesel from winter oilseed rape (WOSR), the major feedstock for biofuel production in France. It follows the recent trend of spatial differentiation in LCA, which mainly pertains to impact assessment models but is also being promoted as a way to address the large variability in crop management practices and yields at the scale of a whole country [26]. This is particularly true of WOSR, a crop with a long growing cycle and large N requirements.

A recent study commissioned by the French Environment Agency (ADEME) provided a set of references for the LCA of biodiesel and other first-generation biofuels [27]. In this study, the life-cycle inventory data concerning feedstock production for WOSR was provided by the French technical institute for oilseeds and hemp (Terres Inovia), in the form of representative country-wide means for agricultural practices and crop yields. The latter was also used to derive national averages for the field emissions of reactive N (Nr), in particular for nitrous oxide (N_2O) on the basis of fixed, generic emission factors following the Tier 1 methodology of IPCC guidelines [23]. Drawing national averages for the LCA of biofuels may be questioned by the fact that the environmental performance of bioenergy chains varies largely

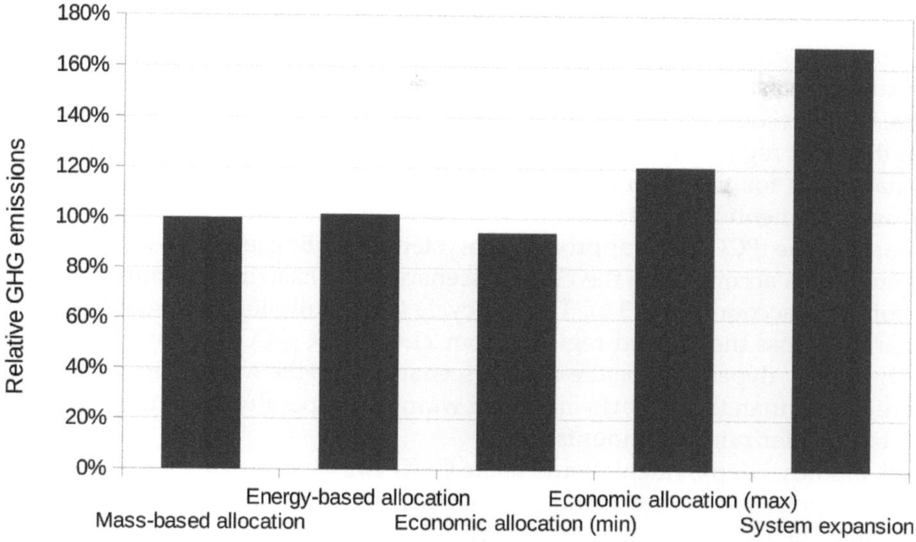

FIG. 3.3 Comparison of life-cycle GHG emissions of bioethanol from winter wheat in France using either allocation methods (mass-based, energy-based, or economical), and system expansion. The emissions are relative to those calculated with the mass-based allocation. *Source: Reference BioIS, ADEME, MEDAD, MAP, ONIGC, IFPEN. Elaboration d'un référentiel méthodologique pour la réalisation d'Analyses de Cycle de Vie appliquées aux biocarburants de première génération en France. Biointelligence service, Paris, 2008.*

from one region to another [28]. Pending adequate data being available, it is thus preferable to narrow down the geographical scope of the LCA in order to ensure a greater level of accuracy. LCA results on a national scale may still be obtained by aggregating regional results. As discussed in the previous sections, the use of agroecosystem modeling offers the opportunity to achieve such accuracy by taking into account local factors [23]. Since LCA is very sensitive to management practices [29], it appears essential to combine agroecosystem modeling (for Nr emissions) with a detailed description and accounting of the actual management practices for current biofuels.

Thus, the first objective of this example was to improve the reliability and representativeness for the LCA data of WOSR production in the context of France by combining LCA with agroecosystem modeling and a typology of N management practices at regional scale. Because of ongoing and future changes in climate conditions, the ecosystem model was run for two series of near-term future weather data, pertaining to different GHG emission trajectories, and the effect of the latter on LCA results was also examined. The second objective of this study was to test the effect of WOSR fertilization practice improvements on LCA outputs and their potential contribution to the production of a more environmentally friendly biodiesel.

Data from a 2012 survey from the eight major WOSR production regions in France were analyzed to derive a typology of current fertilization practices. A recommended rate of inorganic fertilizer N inputs was calculated for each type of fertilization practices, based on a currently used decision-support system, involving a simple N budget approach. Second, the agroecosystem model CERES-EGC was parameterized and run to simulate crop yields and gaseous nitrogen emissions for all current and recommended fertilization practices, thus taking into account the heterogeneity of soil and climate conditions. The reliability of the decision-support system was finally assessed by comparing the simulated yields obtained with current and recommended fertilization practices. The crop yields and Nr emissions simulated in the eight regions investigated were fed into an LCA of biodiesel production, using various databases for the inventory phase. Impacts were calculated using the CML 2000 method, as implemented in the SimaPro v7.2 package (Pré Consultants, NL).

Compared to the IPCC Tier 1 approach, ecosystem modeling led to 29–41% lower estimates for N_2O emissions according to the climate scenario, emphasizing the importance of taking local factors into account (Fig. 3.4). The life-cycle GHG emissions of biodiesel from WOSR varied starkly across the studied regions, from 21.6 to 31.6 g CO_2 eq MJ^{-1} and from 25.5 to 40.3 g CO_2 eq MJ^{-1} depending on the climate scenario: with the wetter climate scenario, emissions were higher than the scenario involving warmer temperatures compared to the current situation but similar rainfall amounts.

Overall, biodiesel deployment would abate GHG emissions by 67–74% compared to fossil diesel, not accounting for indirect land-use change (LUC) effects. Improved fertilization practices mitigated the LCA impact indicators by 3–10% (Fig. 3.5).

3.3.4 Lignocellulosic Biomass From Purpose-Grown Crops

Second-generation biofuels made from lignocellulose are often portrayed as a promising alternative to using food crops as feedstock for biofuel production, alleviating competition with food production given the nonedible nature of their raw material. While this holds

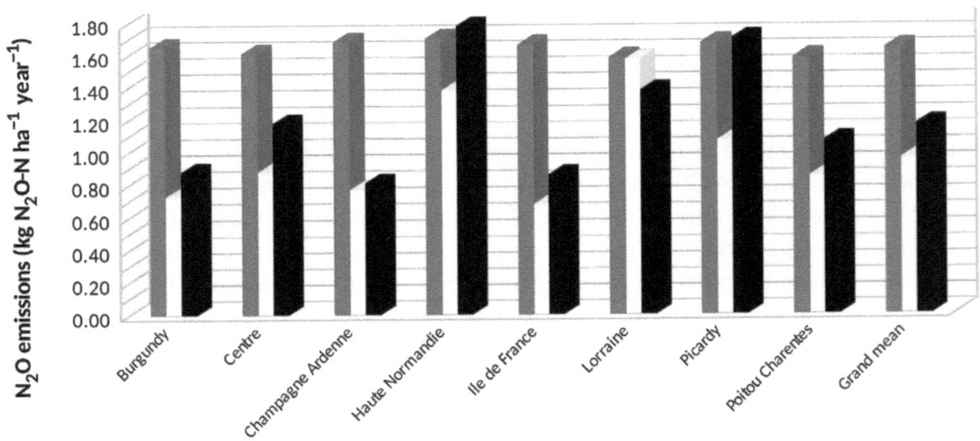

FIG. 3.4 Direct N_2O emissions simulated with an agroecosystem model in the eight oilseed rape production regions in France, with two scenarios for future climate change: a slightly warmer climate (+0.6°C temperature increase) and similar rainfall compared to current conditions (SC1), and a twice warmer climate with drier winter and spring seasons but wetter autumn (SC2). The gray bars indicate estimates derived from the Tier 1 methodology of IPCC.

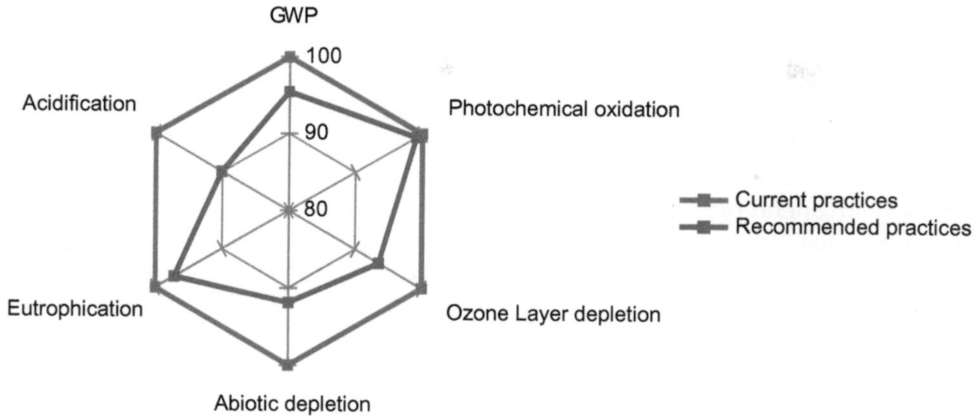

FIG. 3.5 Effect of using a decision-support system to optimize *N* inputs to WOSR in France on the life-cycle impacts of biodiesel in France [30]. *GWP*, global warming potential.

for agricultural or forest residues, it is less obvious for dedicated crops or trees for which supplementary land is still required to meet a particular demand for biorefinery purposes. Marginal lands may be a prime target for the establishment of these crops [31], but raise significant challenges in terms of implementation and have not been deployed on a large scale so far. Thus, LUC scenarios and effects are paramount to the overall performance of lignocellulosics [32]. Aside from this issue, which is covered in Chapters 8 and 9 of this book, the main methodological issues when using LCA to assess the environmental performance of lignocellulosic feedstocks are the estimation of N_2O emissions and soil C variations under perennial crops [33].

This section reviews the main candidate feedstocks for producing lignocellulose, and the previously discussed methodological challenges. It continues with a detailed example on the development of miscanthus in the Burgundy region (similar to the example of Section 3.2 with cereal straw), highlighting the benefits of using detailed land-use and climate data to design and assess miscanthus supply scenarios.

Three different types of lignocellulosic crops can be distinguished: annual crops, perennial herbaceous crops, and SRC. Since they can easily be integrated into a sequence of arable food crops, annual crops are of interest for fast and large-scale deployment. They can also be selected in accordance with the pedoclimatic characteristics of the region to optimize their yields and minimize the ratio of agricultural inputs to biomass production. Triticale (a cross between wheat and rye), fodder sorghum, fiber linen, kenaf, and hemp are the best candidates in temperate or Mediterranean areas. Green harvest (before grain filling) allows higher yields per hectare than agricultural byproducts, producing between 8 and 16 t DM ha^{-1} for triticale. However, as they are grown on arable land, these crops may directly displace food crops and thus present a high potential of indirect LUC or intensification of food crop management [34].

Perennial herbaceous crops have a life-cycle of 20 years or longer. Among all of these, miscanthus and switchgrass draw particular attention. The first (*Miscanthus x Giganteus*) is a sterile hybrid from Asia, whereas the latter (*Panicum virgatum*) is a species widely spread in the United States, that can be found under many latitudes. These two crops are from the Poacea family and present particularly high yields, around 20 t DM ha^{-1} for miscanthus [35]. Meanwhile, they require very few inputs [36], thanks to a large development of below-ground biomass as rhizomes and a dense root system as well as an efficient nutrient cycling [37,38]. Low GHG emissions were measured compared to annual crops [39] or pastures [40], whereas an important nitrate leaching occurs during the first few years of cropping [41]. These crops are established or sown once, then annually harvested, which minimizes the number of field operations compared to annual crops. They also can be established on marginal, degraded, or low-productive lands so that they minimize their competition with food crops, minimizing LUC impacts [42]. They also have the potential to store C in soils [43] and can potentially offset the life-cycle GHG emissions of the biofuels they are converted to [33]. This further enhances their environmental benefits over fossil fuels, although depending on the type of land they are grown on, there may be indirect LUC effects to take into account [32].

SRC is a fast-growing tree species (poplar, willow, eucalyptus, or black locust) cropped similar to perennial agricultural crops. They are established at higher stem densities than regular plantations, and harvested every 3–5 years over a total life cycle of 20–30 years. Thus, potential sources of lignocellulose are numerous and can be adapted to a wide range of pedoclimatic conditions.

3.3.5 Main Methodological Issues With LCA for Lignocellulosic Feedstocks

Similar to food crops, the estimation of N_2O emissions is a key issue in the life-cycle inventory. Although they are generally lower with these species because of lower fertilizer inputs and emission factors [43], there are only a few data sets available, especially for perennial crops and SRC. Still, they cannot be ignored altogether and should be carefully handled [24]. The following example details a possible approach to addressing them.

Another critical item in the GHG balance of energy crops lies in the LUC hypothesis. Growing such crops on marginal land (land not used for agricultural production) is the favored option to reduce the competition with food crops, but locating these lands is unfortunately not trivial [31]. Since ultimately farmers will select where to establish energy crops, it is important to gain more insight into their criteria and decision-making process. A recent study focused on this issue in Burgundy and extracted decision rules based on measurable land features (such as parcel size, slope, and distance to farm headquarters [44]).

The resulting model makes it possible to map fields that would be potentially suitable for perennial crops, and the associated yields and GHG emissions for feedstock production, for instance with a crop model. However, the latter are only available for few species, and therefore can only address a limited portfolio of candidate feedstocks. This is the reason why the following case study focuses on miscanthus, as a representative for perennial grasses in general, to illustrate a novel LCA methodology.

3.3.6 An Example on Miscanthus Supply in Burgundy (France)

This study follows up on the example on cereal straw developed in Section 3.2. First, potential miscanthus yields were mapped using the agroecosystem model CERES-EGC [24],

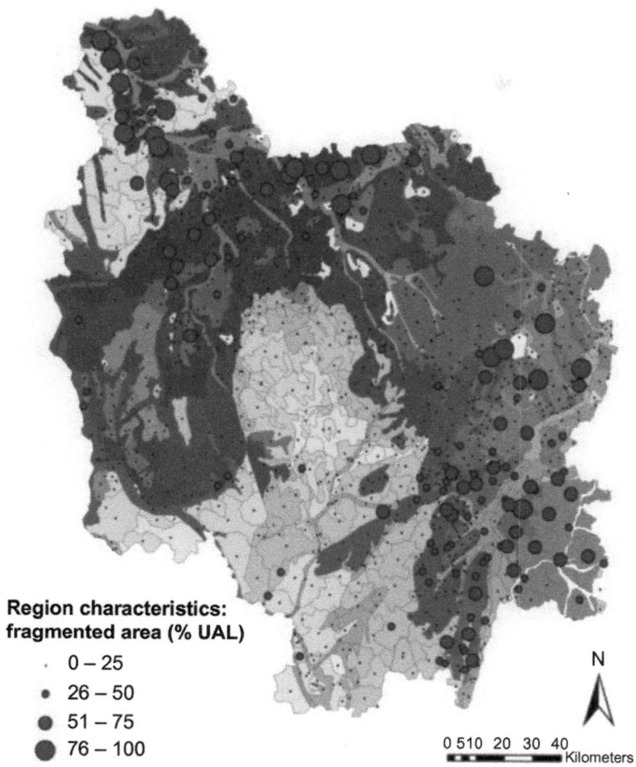

FIG. 3.6 Proportion of the fragmented and set-aside land-use classes in Burgundy (background shows the soil types).

based on soil maps, gridded weather data, and land-use data from the agricultural census (Fig. 3.6). The rates of N_2O emissions, soil C sequestration and the net GHG balance of miscanthus were also assessed. Yields for a reference succession of arable crops were also mapped, to identify areas with lower productivity which could be termed "marginal" from the point of view of agricultural production (Fig. 3.7).

These layers of information may be used to predict the location of potential new fields of miscanthus, by eliciting farmers' decision rules. The work by Martin et al. [44] evidenced the role of biophysical characteristics in decisions such as:

- the productivity of land parcels (low-productivity parcels being more suitable for energy crops),
- parcels that are fragmented or have an irregular shape, and
- land that is under fallow.

A scenario for potential miscanthus production and GHG emission in Burgundy was derived based on these criteria, by further excluding all areas under ecological protection (i.e., Natura 2000 sites), and including only arable land from fragmented and set-aside areas, according to the Corine Land Cover 2006 land-use mapping (Fig. 3.6). It resulted in a total production potential of about one million tonnes of biomass from miscanthus per year,

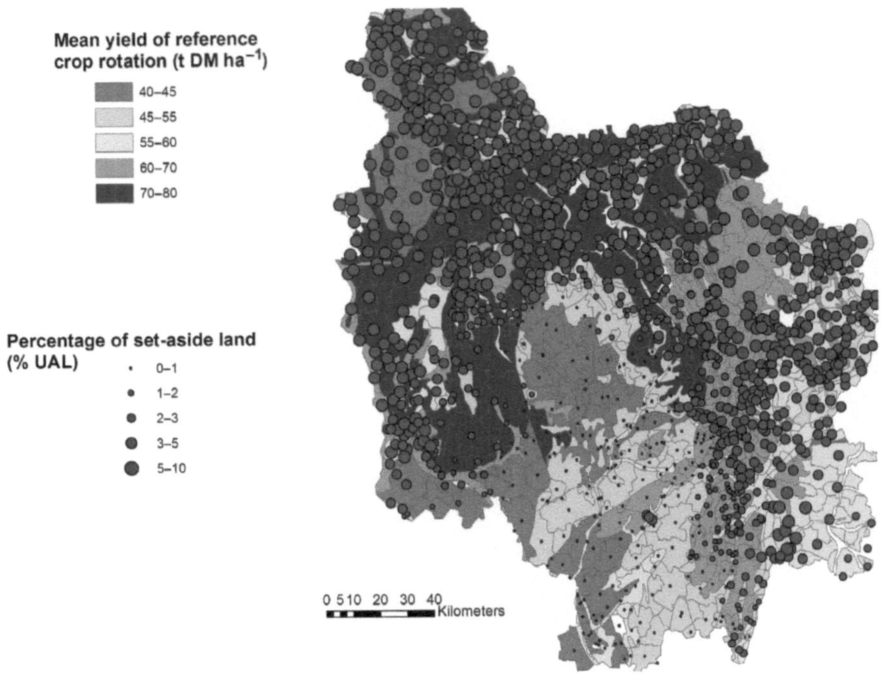

FIG. 3.7 Map of yields for the reference succession of arable crops in Burgundy (yields are cumulated over 5 years).

through the conversion of 61,249 ha, making up 3.2% of the utilizable agricultural area of the region (Table 3.5). Overall, the C sequestered under the miscanthus more than offset the emissions of N_2O related to fertilizer inputs to the crops.

3.4 OVERALL COMPARISON OF FEEDSTOCKS AND LAND-USE CHANGE EFFECTS

Among the range of candidate suitable raw material types for a given biorefinery process, LCA provides a unique means of comparing the environmental performance of feedstocks and thus minimizing the environmental impacts associated with biomass procurement [16]. In the examples detailed in this chapter, based on several regions in France, there appeared a clear ranking between the three categories of agricultural biomass investigated: agricultural residues such as cereal straw had generally lower impacts per tonne of biomass procured than perennial grasses, and arable crops (such as WOSR) had the highest impacts. This is illustrated in Fig. 3.8, which compares different sources of lignocellulose in the Burgundy region of France.

This type of ranking is somewhat sensitive to the biomass yields per unit area, and also crop management and prior land use [45], making it hard to generalize. At the value-chain level, there may be further trade-offs between LCA impacts per tonne of biomass and the

TABLE 3.5 Overall Miscanthus Biomass Production in Burgundy, GHG Emissions Per and Arable Area Converted to Miscanthus

Biomass produced (tDM $year^{-1}$)	1,004,262	±3859
N_2O emissions (direct and indirect) (kg CO_{2eq} tDM^{-1})	55	±7
Net GHG emissions (kg CO_{2eq} tDM^{-1})	−27	±29
Area converted (ha)	61,249	
(% UAL)	3.3%	
Mean yield (tDM ha^{-1})	16.4	±0.1

UAL, utilizable agricultural area.

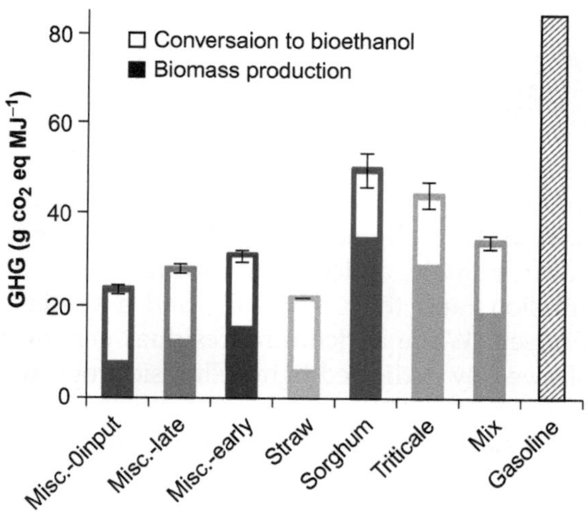

FIG. 3.8 WtW emissions of GHG for ethanol produced from different sources of lignocellulose in Burgundy, an optimal supply mixture (Mix) and the reference gasoline fuel.

impacts of biomass processing, possibly mitigating the differences across feedstock types. The approaches exemplified in this chapter provide an avenue to address these key issues by: (1) designing land-use scenarios based on the modeling of farmers' decision-making processes as to the establishment of biomass crops, (2) taking into account local pedoclimatic factors via ecosystem modeling, and (3) implementing a spatially explicit approach to LCA at the feedstock supply-area scale. Even though modeling is an attractive route, the collection of field data (in particular regarding the yields and location of commercial plots) is also indispensable to calibrate the models and ensure realistic projections of the biomass supply chains in practice.

Since LUCs are paramount to the LCA of purpose-grown crops, in particular for the GHG budgets via the impacts on soil C dynamics [43,45], their effects warrant special care in the LCA. Direct LUC occur upon the conversion of cropland, grassland, or fallow to energy crops, while indirect effects are associated with market mechanisms compensating for the forgone production of the former land use [30]. While the former is relatively easy to assess based on default values of soil C stocks associated with various land uses in the major world biomes [7], indirect effects are much more controversial and give rise to a wide range of estimates depending on exogenous assumptions on, for example, diet changes or future yield pattern for food crops (see Chapters 8 and 9 of this book). Fig. 3.9 illustrates how the inclusion of both direct and indirect LUC affects the life-cycle GHG savings incurred when displacing fossil fuels with biofuels produced from the same feedstock types as depicted in Fig. 3.8 in the Burgundy case study. Direct LUC effects are negative to neutral for the sorghum and straw, and positive for triticale and miscanthus, thus offsetting some of the GHG emissions of feedstock cultivation. Indirect effects result in a reduction of GHG savings for all feedstocks except straw, especially for the sorghum crop. Indirect emissions were taken from the meta-analysis by De Cara et al. [42]. Given the uncertainties around these estimates, the numbers in Fig. 3.9 should be considered as indicative and would not be suitable for the certification of GHG abatements permitted by these biofuels.

3.5 CONCLUSIONS AND PERSPECTIVES

Feedstocks are responsible for a significant part of the life-cycle environmental impacts of biofuels. These impacts arise from field operations to grow or collect the raw material as well as logistics to handle and transport the biomass to the biorefinery gate. This chapter compared different sources of agricultural feedstocks, namely, agricultural residues and dedicated crops. Specific issues such as N_2O emissions and their modeling, and a spatially explicit regional approach of LCA were addressed. While agricultural residues generally come out as the least-impacting feedstock, followed by dedicated lignocellulosic crops and

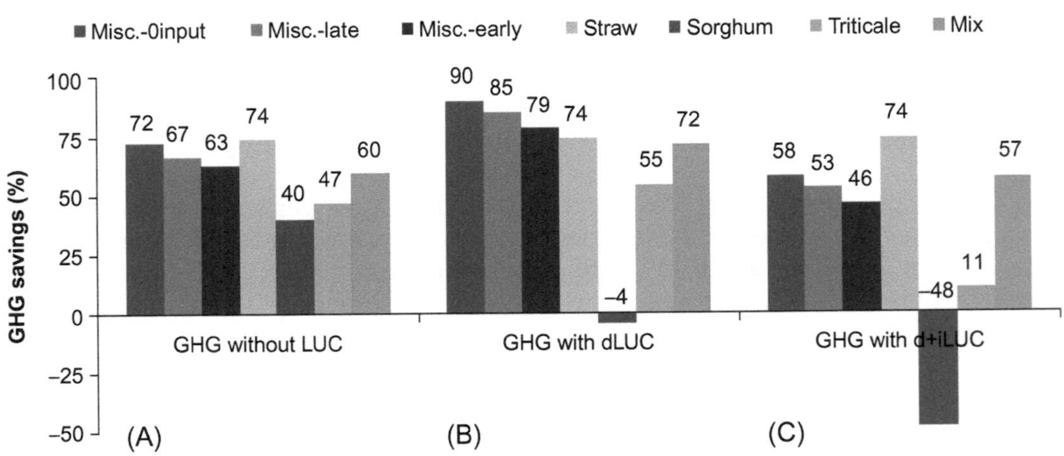

FIG. 3.9 GHG savings of second-generation ethanol from various feedstocks in Burgundy, including or not the effects of direct and indirect LUCs on soil C stocks. Key to labels: misc-0input: unfertilized miscanthus, winter harvest; misc-late: fertilized miscanthus, winter harvest; misc-early: fertilized miscanthus, autumn harvest; mix: optimal supply mix (taking into account production constraints) including equal shares of miscanthus, straw, and triticale.

finally by first-generation arable crops, the results are sensitive to factors such as crop management practices, harvesting systems, and crop yields. LUC patterns are also a key driver of LCA results. Improving the LCA of agricultural feedstocks requires enhancing knowledge on those issues, as presented in this chapter following ongoing lines of work in research.

References

[1] Bessou C, Ferchaud F, Gabrielle B, Mary B. Biofuels, greenhouse gases and climate change. A review. Agron Sustain Dev 2011;31:1–79.

[2] Sebastian F, Royo J, Gomez M. Cofiring versus biomass-fired power plants: GHG (greenhouse gases) emissions savings comparison by means of LCA (life cycle assessment) methodology. Energy 2011;36:2029–37.

[3] Gabrielle B, Gagnaire N, Massad R, Dufossé K, Bessou C. Environmental assessment of biofuel pathways in Ile de France based on ecosystem modelling. Bioresour Technol 2014;152:511–8.

[4] Chum H, Faaij APC, Moreira J, et al. Bioenergy. In: Edenhofer O, Pichs-Madruga R, Sokona Y, et al., editors. IPCC special report on renewable energy sources and climate change mitigation. United Kingdom, New York: Cambridge University Press; 2011.

[5] Von Blottnitz H, Curran MA. A review of assessments conducted on bio-ethanol as a transportation fuel from a net energy, greenhouse gas, and environmental life cycle perspective. J Clean Prod 2007;15:607–19.

[6] Boehmel C, Lewandowski I, Claupein W. Comparing annual and perennial energy cropping systems with different management intensities. Agric Syst 2008;96:224–36.

[7] Searchinger T, Heimlich R, Houghton RA, Dong F, Elobeid A, Fabiosa J, et al. Use of U.S. croplands for biofuels increases greenhouse gases through emissions from land use change. Science 2008;319:1238–40.

[8] Robertson GP, Dale VH, Doering OC, Hamburg SP, Melillo JM, Wander MM, et al. Sustainable biofuels redux. Science 2008;322:49–50.

[9] Wilhelm WW, Johnson DME, Karlen DL, Lightle DT. Corn stover to sustain soil organic carbon further constrains biomass supply. Agron J 2007;99:1665–7.

[10] Scarlat N, Martinov M, Dallemand JF. Assessment of the availability of agricultural crop residues in the European Union: potential and limitations for bioenergy use. Waste Manag 2010;30:1889–97.

[11] Picardie F, Picardie, C. INRA Laon-Mons, ARVALIS Institut du végétal, LDAR. Exporting straw without jeopardizing soil organic matter—guidelines for field management. Technical report, FRCA Picardie, Laon, 2009 (in French).

[12] Saffih-Hdadi K, Mary B. Modeling consequences of straw residues export on soil organic carbon. Soil Biol Biochem 2008;40:594–607.

[13] Smith P, Smith JU, Powlson DS, McGill WB, Arah JRM, Chertov OG. Comparison of the performances of nine soil organic matter models using datasets from seven long-term experiments. Geoderma 1997;81:153–225.

[14] European Parliament. Directive 2009/28/EC of the European Parliament and of the Council of 23 April 2009 on the promotion of the use of energy from renewable sources and amending and subsequently repealing Directives 2001/77/EC and 2003/30/EC; 2009.

[15] Gabrielle B, Gagnaire N. Life-cycle assessment of straw use in bio-ethanol production: a case-study based on deterministic modeling. Biomass Bioenergy 2008;32:431–41.

[16] Godard CH, Boissy J, Gabrielle B. Life-cycle assessment of local feedstock supply scenarios to compare candidate biomass sources. GCB Bioenergy 2013;1:16–29.

[17] Agreste. Enquête sur les pratiques culturales en 2006 (No. 200), Chiffres et Données—Série Agriculture. Ministère de l'agriculture, de l'agroalimentaire et de la forêt, Paris, 2008.

[18] Le Net, E. A socio-economic and environmental approach to the supply of lignocellulosic biomass. Final report, ECOBIOM project ANR PNRB 05, 2009, FCBA, Paris (in French).

[19] Sourie JC, Gabrielle B, Bamière L, Gagnaire N, Silveira JD. An economic and environmental approach of advanced biofuel pathways based on the use of cereal straw through enzymatic hydrolysis and combined heat and power. Technical report, INRA, UMR Economie Publique, Grignon; 2006. (in French).

[20] Quirin M, Gartner SO, Pehnt M, Reinhardt GA. CO_2 Mitigation through biofuels in the transport sector: status and perspectives, main report. Heidelberg, Germany: Institute for Energy and Environmental Research (IFEU); 2004.

[21] Cherubini F. GHG balances of bioenergy systems—overview of key steps in the production chain and methodological concerns. Renew Energy 2010;35:1565–73.
[22] Smeets EMW, Bouwman LF, Stehfest E, Van Vuuren DP, Posthuma A. Contribution of N_2O to the greenhouse gas balance of first-generation biofuels. Glob Chang Biol 2009;15:1–23.
[23] Tanabe K, Ngara T, Miwa K, Buendia L, Eggleston H, van Amstel A, et al. IPCC guidelines for national greenhouse gas inventories. Hayama, Kanagawa: Prepared by the National Greenhouse Gas Inventories Programme IGES; 2006.
[24] Dufossé K, Gabrielle B, Drouet JL, Bessou C. Using agroecosystem modeling to improve the estimates of N_2O emissions in the life-cycle assessment of biofuels. Waste Biomass Valoriz 2013;4:593–606.
[25] Duretz S, Drouet JL, Durand P, Hutchings NJ, Theobald MR, Salmon-Monviola J, et al. NitroScape: a model to integrate nitrogen transfers and transformations in rural landscapes. Environ Pollut 2011;159:3162–70.
[26] Loiseau E, Junqua G, Roux P, Bellon-Maurel V. Environmental assessment of a territory: an overview of existing tools and methods. J Environ Manage 2013;112:213–25.
[27] ADEME. Analyses de cycle de vie appliquées aux biocarburants de première génération consommés en France—Final report ADEME, Angers; 2010.
[28] Van der Hilst F, Lesschen JP, van Dam JMC, Riksen M, Verweij PA, Sanders JPM, et al. Spatial variation of environmental impacts of regional biomass chains. Renew Sustain Energy Rev 2012;16:2053–69.
[29] Grillo Reno ML, Silva Lora EE, Escobar Palacio JC, Venturini OJ, Buchgeister J, Almazan O. A LCA (life cycle assessment) of the methanol production from sugarcane bagasse. Energy 2011;36:3716–26.
[30] Ben Aoun, W. The development of bio-diesel from oilseed rape in France: environmental impacts and consequences on land-use. Ph.D. Dissertation, AgroParisTech, Paris-Saclay University, Paris, 2015.
[31] Biofuels: environmental consequences and interactions with changing land use, Howarth R, Bringezu S, editors. Proceedings of the scientific committee on problems of the environment (SCOPE) international biofuels project rapid assessment 22–25 September 2008, Gummersbach, Germany; 2009.
[32] Havlık P, Schneider UA, Schmid E, Bottcher H, Fritz S, Skalsky R. Global land-use implications of first and second generation biofuel targets. Energy Policy 2010;39:5690–702.
[33] Agostini F, Gregory A, Richter G. Carbon sequestration by perennial energy crops: is the jury still out? Bioenergy Res 2015;8:1057–80.
[34] Brunelle T, Dumas P, Souty F. The impact of globalization on food and agriculture: the case of the diet convergence. J Environ Dev 2014;23:41–65.
[35] Lesur C, Jeuffroy MH, Makowski D, Riche AB, Shield I, Yates N. Modeling long-term yield trends of *Miscanthus x giganteus* using experimental data from across Europe. Field Crop Res 2013;149:252–60.
[36] Cadoux S, Riche AB, Yates NE, Machet JM. Nutrient requirement of *Miscanthus x giganteus*: conclusions from a review of published studies. Biomass Bioenergy 2012;38:14–22.
[37] Amougou N, Bertrand I, Cadoux S, Recous S. *Miscanthus x giganteus* leaf senescence, decomposition and C and N inputs to soil. GCB Bioenergy 2012;4:698–707.
[38] Strullu L, Cadoux S, Preudhomme M, Jeuffroy MH, Beaudoin N. Biomass production and nitrogen accumulation and remobilisation by *Miscanthus x giganteus* as influenced by nitrogen stocks in belowground organs. Field Crop Res 2011;121:381–91.
[39] Drewer J, Finch JW, Lloyd CR, Baggs ME, Skiba U. How do soil emissions of N_2O, CH_4 and CO_2 from perennial bioenergy crops differ from arable annual crops. GCB Bioenergy 2011;4:408–19.
[40] Roth B, Jones M, Burke J, Williams M. The effects of land-use change from grassland to *Miscanthus x giganteus* on soil N_2O emissions. Land 2013;2:437–51.
[41] Lesur C, Bazot M, Bio-Beri F, Mary B, Jeuffroy MH, Loyce C. Assessing nitrate leaching during the three-first years of *Miscanthus x giganteus* from on-farm measurements and modeling. GCB Bioenergy 2013;6:439–49.
[42] De Cara S, Goussebaïle A, Grateau R, Levert F, Quinemer J, Vermont B. Critical review of studies evaluating the effect of land-use changes on the environmental balances of biofuels. Environment and Energy Agency (ADEME), Angers, France; 2012. [in French].
[43] Don A, Osborne B, Hastings A, Skiba U, Carter MS, Drewer J, et al. Land-use change to bioenergy production in Europe: implications for the greenhouse gas balance and soil carbon. GCB Bioenergy 2011;4:372–91.
[44] Martin L, Wohlfahrt J, Le Ber F, Benoît M. Perennial biomass crop cultivation and its territorial patterns [L'Espace géographique (English Edition)]. 41:133–47.
[45] Davis SC, Boddey RM, Alves BJR, Cowie AL, George BH, Ogle SM, et al. Management swing potential for bioenergy crops. GCB Bioenergy 2013;5:623–38.

CHAPTER

4

Life Cycle Assessment of Sugar Crops and Starch-Based Integrated Biorefineries

P. Vaskan, E. Ruiz Pachón*, E. Gnansounou†*

*Ecole Polytechnique Fédérale de Lausanne (EPFL), Lausanne, Switzerland

†Bioenergy and Energy Planning Research Group (BPE), Ecole Polytechnique Fédérale de Lausanne (EPFL), Lausanne, Switzerland

4.1 INTRODUCTION

World consumption of petroleum and other liquid fuels is forecasted to rise from 87 million barrels per day (MMbbl/d) in 2010 to 98 MMbbl/d in 2020 and 119 MMbbl/d in 2040 [1]. This growing fossil fuel consumption leads to a rapid increase in greenhouse gas (GHG) emissions, and according to the International Energy Outlook [2], CO_2 emissions are projected to rise from 29 billion tons in 2006 to 40.4 billion in 2030. The problem of mitigating GHG emissions as well as the perspective of oil depletion and energy security motivate the development of renewable and environmentally friendly alternative sources of energy.

Nowadays, one of the most promising renewable energy alternatives is biomass. The term *biomass* refers to organic matter that has stored energy through the process of photosynthesis. This energy can be converted to satisfy everyday human needs such as food, transport, power, heat, etc. In addition, biomass can be a primary source of hydrocarbons that can be used in the petrochemical industry to replace fossil hydrocarbons. The clear advantages of the biorefinery, compared to the petroleum process, are a reduction of the potential risks associated with climate change and a shift away from the reliance on nonrenewable energy sources.

The type of biorefinery depends on the local context and availability of the resource. The sources of biomass can vary significantly from algae to wood. Nevertheless, sugar and starch crops are the most often used biomass feedstock. The main product of sugar/starch-based biorefineries is bioethanol, which is an important material for the food industry and transport

http://dx.doi.org/10.1016/B978-0-444-63585-3.00004-8

sector. Bioethanol is an alcohol made by fermentation, mostly from 6-carbon sugars such as glucose; thus, biomass sources such as sugar and starch, containing a large amount of glucose or precursors to glucose, are favorable to be converted into ethanol.

In general biomass materials containing fermentable sugars can be divided into two groups. The first group includes the sugar crops and includes sugarcane, sugar beet, or sweet sorghum, highly concentrated in glucose sugar. The sugar crops are the most desired feedstock to produce ethanol as they provide a high yield for ethanol production. However, the food industry also employs these materials for food production, which causes competition for the resource. Another group of biomass sources for ethanol production is starch-based feedstock. It includes such crops as cereal grains (corn, rice, wheat, barley, oats, ...), potato, sweet potato, and cassava. The starch is composed of long chains of glucose molecules that can be broken into simple glucose molecules to be converted into ethanol. Therefore, an additional hydrolysis reaction is required in starch-based ethanol production. The bioethanol obtained from these two groups of feedstock is known as first-generation (1G) bioethanol. The main drawback of this bioethanol is competition with food production, which increases prices for sugar-based feedstock and limits availability of the resources.

A novel way of producing bioethanol involves the use of lignocellulosic material as a feedstock. This product is known as second-generation (2G) bioethanol and is considered a biofuel with a large potential to replace fossil fuels [3,4]. One of the main advantages of the lignocellulosic material is that it is abundant and it has limited competition with food production, and thus it does not compromise food security [5,6]. However, despite the fact that 2G biofuel production has shown techno-economic potential [7], the cost of bioethanol production from lignocellulosic materials remains too high. This challenge has led to a search for further process improvements, delaying its implementation [8].

One of the most promising ways to reduce production cost is to implement the 2G process into 1G ethanol production. An integrated 1G-2G biorefinery has many potential improvements since 2G bioethanol production may share part of the infrastructure where 1G ethanol production already exists. Recent studies on integration are related to the sugarcane agroindustry, because in this sector the lignocellulosic feedstock has been considered refinery feedstock for a long time. Dias et al. [9] performed a simulation of a traditional distillery where bioethanol is produced from sugarcane juice and sugarcane bagasse. In addition the potential improvements in 1G-2G integration are presented in several works [10,11]. Another sugarcane-based integrated biorefinery study [12] concludes that the use of bagasse as 2G feedstock implies a considerable opportunity revenue. In general, the 1G-2G integration leads to energy savings through improvements in waste handling, heat integration, or flexibility in feedstock allocation and substitution.

The economic and environmental viability of integrated 1G-2G biorefineries for both ethanol production and electricity generation has already been demonstrated in several studies [11,13]. However, an integrated biorefinery may allow the primary production of sugar from sugar crops along with ethanol/electricity generation, and the use of byproducts as supplements or replacements, e.g., molasses, C5 syrup and dried distillers grains with solubles (DDGS) for animal feed [14]. While there are many studies that analyze biorefineries producing 1G or 2G ethanol and electricity generation, there is a lack in the investigation of an integrated 1G-2G process incorporating food, feed (and other sugar-starch crop–derived products) production alongside biofuel.

The evaluation of the environmental sustainability of a technological option requires the consideration of the whole product life cycle, which can be achieved through a life cycle assessment (LCA). Numerous studies have been recently presented concerning the evaluation of the life cycle impacts in bioenergy systems devoted to the production of bioethanol and coproducts. Cherubini and Strømman [15] showed that about 90% of the reviewed biorefinery studies accounted for GHG emissions. It proves that mitigation of climate change is one of the main driving forces for investigation and incorporation of bioenergy systems into the industrial process.

Comparing different bioethanol types produced from various feedstocks, the one from sugarcane is the most efficient fuel regarding mitigating climate change impact [15]. Bioethanol from other feedstocks, in particular based on sugar and starch crops in temperate climates, shows 40–65% of the GHG emissions of conventional fossil fuels [16]. However, most of the reviewed studies on bioethanol from sugar or starch crops do not take into account the land use changes (LUCs) effect when sugar crop plantations replace tropical or temperate forests and therefore deplete initial carbon storages. Including LUC in the assessment may even counterbalance the final results. Gnansounou et al. [16] showed that, depending on preferred assumptions and LUC parameters, the range of life cycle GHG emissions reduction in comparison with gasoline may change from −112% to +120% for a given bioethanol production process.

The comparison between the first and second generations of transportation bioethanol shows that usually 2G ethanol provides higher GHG savings than 1G biofuels [17,18]. However, the situation can change drastically depending on preferred assumptions. For example, if the energy required for bioethanol production comes from C-intensive fossil sources (e.g., coal), the bioenergy system can issue more GHG emissions than its fossil alternative [19,20].

Despite the abundance of LCA studies dedicated to 1G and 2G ethanol production, there are only few applied to integration of 1G-2G biorefineries. Dias et al. [21] investigated and compared economic and environmental performance of integrated and stand-alone 2G ethanol production from sugarcane bagasse and trash. Results showed that integration of 1G and 2G processes provides the best environmental performance among all the evaluated alternatives. However, the LCA in this work neglected several important factors such as LUC emissions and well-to-wheel (WtW) LCA that is more appropriate for an ethanol fuel evaluation than cradle-to-gate methodology. In a recent work Gnansounou et al. [22] went beyond the state of the art and presented a novel concept of LCA of an integrated 1G-2G sugarcane-based biorefinery. The proposed methodology comprises WtW LCA, value-based economic allocation of final products and intermediate process streams, LUC emissions and finally comparison of LCA results from a biorefinery with those of a relevant reference system (RS). In general, results showed that a configuration producing the largest amount of ethanol leads to the best economic performance, while the scenario providing the widest range of products presents the lowest GHG emissions among the alternatives.

In general, there are wide variations in LCA methodology used to evaluate the environmental performance of a sugar/starch-based biorefinery, mainly due to the selection of RS, system boundaries, allocation procedures, functional unit, inclusion of LUC effects and others. This chapter addresses the discussion and analysis of recent achievements and key issues in LCA of sugar/starch-based integrated biorefineries. The process design and technological concepts of such biorefineries are explained in detail as well.

4.2 OBJECTIVES AND SCOPE

This chapter performs a comprehensive analysis of the existing achievements in the LCA methodology applied to evaluate environmental performance of sugar- or starch-based integrated biorefineries. The general information about LCA can be found in Chapters 1 and 2. The purpose of the present chapter is to focus on the LCA of sugar- or starch-based integrated biorefineries. The overview of process design and the technological achievements in integrated biorefineries are presented as well in order to provide a comprehensive guide to LCA case studies.

4.3 PROCESS DESIGN

The simulation of different configurations of prospective integrated biorefineries producing 1G and 2G bioethanol and other coproducts is performed using Aspen Plus—aspenONE. The simulations are run in order to generate specific outputs that are used later to examine each of the configurations from an economic and environmental perspective (via LCA). Sugar cane and wheat are selected as feedstocks for the sugar- and starch-based biorefineries, respectively.

4.3.1 The Processing of Sugarcane

Sugarcane has been processed for sugar production for a long time, and for ethanol production in the last few decades. However, data for a sugarcane integrated biorefinery, where bioethanol is produced from sugarcane juice and sugarcane bagasse, are unavailable. Because of this, the simulation of a prospective sugarcane integrated biorefinery is performed using Aspen Plus.

Four scenarios have been considered based on the coproducts obtained along with 1G-2G bioethanol. The feed-to-boiler in the different configurations also includes the green harvesting residues (GHRs) of the sugarcane plantation in order to satisfy the energy requirements of the process [22].

The selected configurations evaluated in this case study are:

- *Scenario ED FF:* An ethanol distillery producing sugar juice that is used for bioethanol production (1G). Along with the distillery, a 2G plant is considered to process the lignocellulosic material (the sugarcane bagasse in this case). The 2G plant produces C5-based syrup to feed animals and 1G-2G bioethanol.
- *Scenario ED OF:* An ethanol distillery producing sugar juice that is used for bioethanol production (1G). The 2G plant produces only bioethanol.
- *Scenario SM FF:* A sugar mill producing B molasses that is used for bioethanol production (1G), raw sugar, and C molasses as direct 1G products. Along with the sugar mill, the 2G plant processes the sugarcane bagasse to produce C5-based syrup for animal feed and bioethanol.
- *Scenario SM OF:* A sugar mill producing B molasses that is used for bioethanol production (1G), raw sugar, and C molasses as direct 1G products. The 2G plant produces only bioethanol.

A cogeneration system is annexed to all the configurations, providing the electricity required to run the process as well as some electricity in excess (if any) being sold to the grid.

The four scenarios considered for the Aspen Plus process simulations are listed in Table 4.1.

The first step for designing the process is to establish the capacity of the plant and the feed to the cogeneration system. In the case study, the capacity of the plant is 13,000 tons (wet basis—w.b.) of sugarcane per day. It is assumed that 60% of the sugarcane bagasse is used for 2G bioethanol production, while the remainder is sent to feed the boiler. Regarding the GHR, 50% is sent to cogeneration, leaving the rest on the soil as fertilizer. The rate of sugarcane entering the plant, along with the availability of bagasse and GHR, are listed in Table 4.2.

The assumed composition of sugarcane, bagasse and GHR in the Aspen model is shown in Table 4.3.

Some of the components of the lignocellulosic material are not available in the Aspen Plus databank, so their properties and structures are modified as used in the NREL model developed by Aden et al. [23]. Depending on the physical state of the components in the material streams, the stream class is defined in the Setup Specifications Global sheet of the Aspen Plus simulation. In this case, the streams are defined as MIXCISLD representing streams with conventional solids with no particle distribution. When the MIXCISLD stream class is selected, there are two valid substreams: MIXED (vapor-liquid) and CISOLID (solids). The liquid or vapor components are presented in the mixed substream, while the fiber components (cellulose, hemicellulose, and lignin) are inserted as solids. After running the simulation, Aspen Plus automatically generates results for both substreams.

All the unit operations required to represent the process were inserted into the correspondent hierarchy block. As an example, the simulation flowsheet developed to represent the ED FF scenario is illustrated in Fig. 4.1.

TABLE 4.1 Technological Integrated Scenarios

Ethanol distillery (ED)—2G feed and fuel (FF)	ED FF
Ethanol distillery (ED)—2G only fuel (OF)	ED OF
Sugar Mill (SM)—2G feed and fuel (FF)	SM FF
Sugar Mill (SM)—2G only fuel (OF)	SM OF

TABLE 4.2 Capacity of the Plant and Availability of Bagasse and GHR Produced From Sugarcane

Material	Value
Sugarcane processed	13,000 ton/day (w.b.)
Sugarcane bagasse available	1820 ton/day (d.b.)
Sugarcane GHR available	2000 ton/day (d.b.)

TABLE 4.3 Composition of Sugarcane, Sugarcane Bagasse, and Green Harvesting Residues in the Plant

	Weight Percent (%)		
	Sugarcane	**Sugarcane Bagasse**	**Green Harvesting Residues (in Plant)**
H_2O	71.9	50.0	30.0
Sucrose	13.9	3.00	0.00
Glucose	0.80	0.10	0.00
Extract	0.41	2.00	6.00
Cellulose	5.90	20.0	25.0
Xylan	3.00	10.0	15.0
Galactan	0.10	0.30	0.50
Mannan	0.04	0.10	0.00
Arabinan	0.25	1.00	2.00
Lignin	3.00	11.5	15.0
Acetate	0.00	0.00	1.00
Ash	0.70	2.00	6.00

4.3.1.1 *Simulation of an Ethanol Distillery*

The ethanol distillery is named 1G plant in ED- simulations. In this area (see Fig. 4.2), 541.7 ton/h (w.b.) of sugarcane enters the 1G plant, where it is first washed with 606.7 ton/h of water at 40°C to remove dirt and impurities that may be present [24].

The cleaned material is then chopped and shredded to a fiber size ready for sugar extraction. The technology involves roller mills running in series with counter current juice flow to leach soluble solids [25]. Imbibition water (103.9 ton/h) is added to the last unit at 65°C for improving the sugar extraction in the final juice [26]. This process extracts the sugar and the plant material left out is known as bagasse. The sugarcane bagasse is conveyed to the 2G plant for ethanol production, having a specific composition as presented in Table 4.3. The mechanical power needed for the mill is supplied by the cogeneration system annexed to the plant. Bagacillo or cush-cush particles are set aside for later use as mud-cake compactor. On this basis, the sugar extraction yield (i.e., the proportion of initial sugar present in the juice after extraction) reaches 94% (see Table 4.4). The water and soluble solids represent about 85% of the total fresh stem weight [27].

After sugar extraction, the juice is treated by defecation to settle out some impurities from the raw juice and to prevent the sugar from changing form. The juice is heated near to the boiling point (103.5°C) and limed (1.2 kg CaO/ton of cane, 6.5 kg of water per kg of lime) at 15° Baumé to raise the juice pH from 4.5 to 8 [28,29]. This stream is then flashed and sent to the clarifier where the precipitate is separated. The settled solids from the clarifier are filtered and washed. The final residue is known as cachaza (mud cake).

After filtration, the purified juice or thin juice contains 14% of sucrose with a purity of 94.2% [26]. This juice directly enters the fermentation area of the 2G plant.

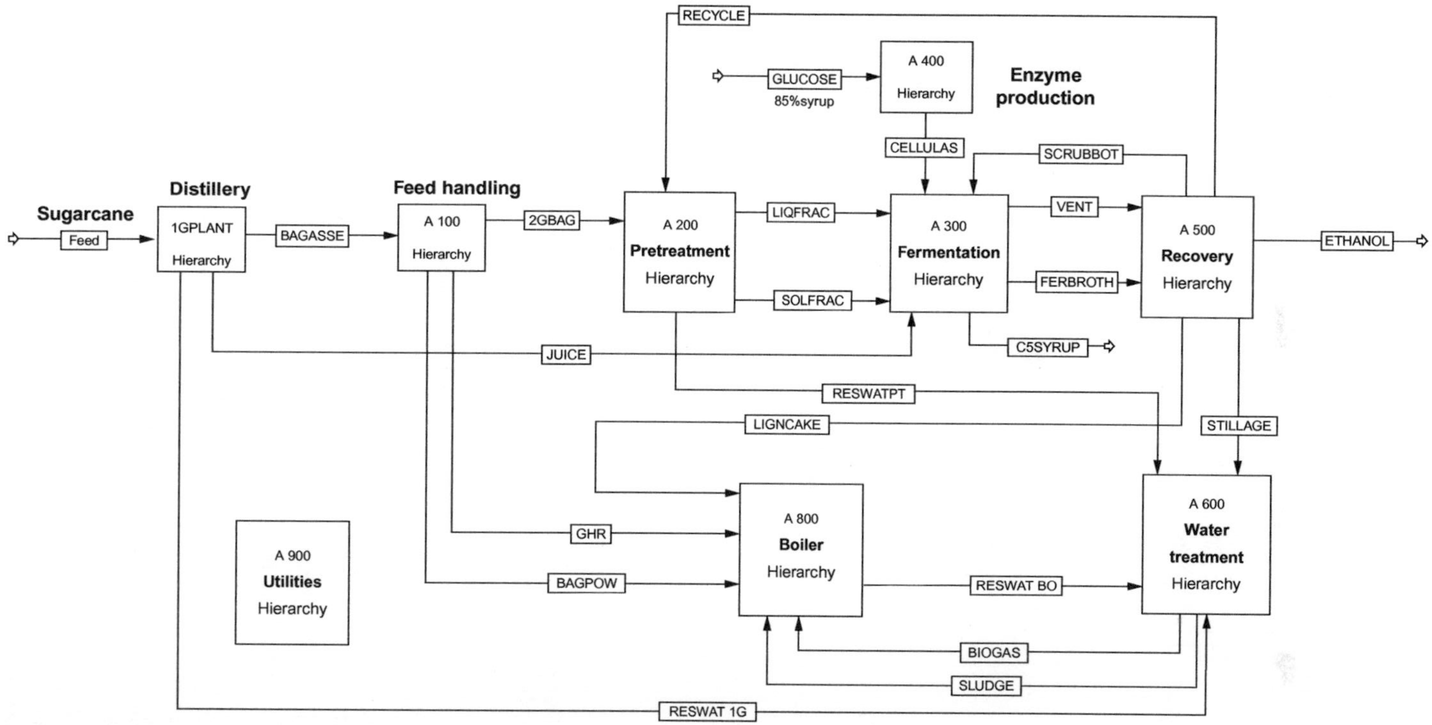

FIG. 4.1 Simplified simulation mass flowsheet performed in Aspen Plus for the ED FF scenario: integrated first and second generation ethanol and C5 syrup production from sugarcane.

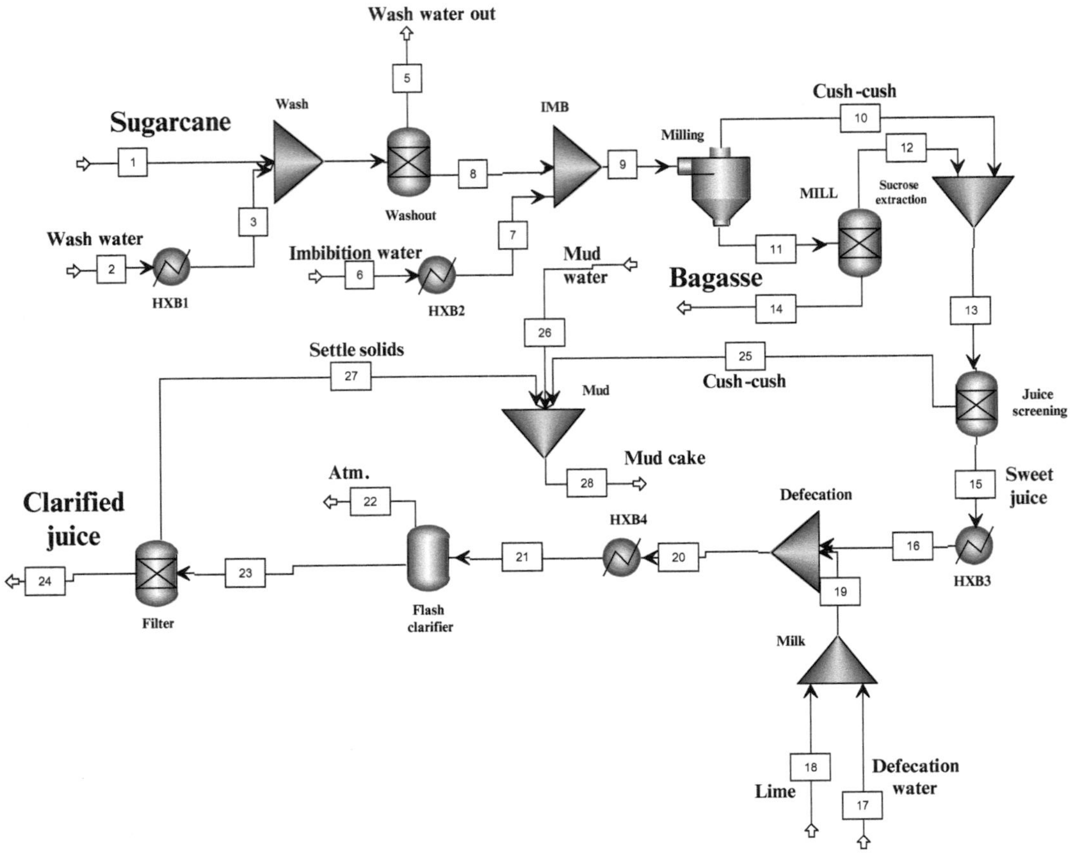

FIG. 4.2 Aspen Plus process flow diagram (material streams only) used in the current simulations for evaluating the distillery part of the biorefineries, i.e., 1G plant in ED- scenarios.

TABLE 4.4 Specifications in the Process Simulation of the 1G Plant (ED- or SM- Scenarios)

Parameter	Value
Sucrose in juice	14%
Efficiency of juice extraction	94%
Moisture in mud cake	75%
1G plant electricity consumption	17,951 kWh

The mass balances calculated by Aspen Plus in the 1G plant area of the ethanol distillery–based scenarios are listed in Table 4.5.

Most power consumption of the 1G plant is due to the mechanical power needed for the mill (17,951 kW).

Regarding heat utilities in the 1G plant (42 Gcal/h), the heat exchangers use the low pressure (LP) steam produced in the boiler.

TABLE 4.5 Mass Balance of Sugarcane Juice Extraction (Calculated by Aspen Plus in the 1G plant Area of the Ethanol Distillery–Based Scenarios)

Mass Flow (ton/h)	Sugarcane	Sweet Juice	Clarified Juice	Bagasse
Water	391	416.8	420.7	75.8
Sugars	75.8	71.4	71.4	4.3
Cellulose	30.3	0.1	0	29.7
Hemicellulose	18.1	0.1	0	17.7
Lignin	18.1	0.1	0	17.8
Others	8.3	1.8	1.8	6.4
Total	541.7	490.2	493.8	151.7

4.3.1.2 Simulation of a Sugar Mill

The sugar mill is also referred to as 1G plant in SM- simulations. In a sugar mill, the sugar production from sugarcane is basically the continuation of the distillery process explained before (see Fig. 4.3). In this case, the clarified juice obtained after milling, extraction, juice treatment, and filter is thickened in a multieffect evaporator [25].

The thin juice (or clarified juice) that has been diluted with water during extraction and purification enters the evaporating section with 14% sugar while the thick juice leaving the evaporators contains approximately 60% sugars [30]. The simulated multieffect evaporator includes five flash separators operating between 1.41 and 0.13 atm. The condensed steam used for evaporation is returned as residual water.

Two percent of the thick juice is separated and sent to the fermentation area of the 2G plant in order to increase the ethanol yield. The remaining 98% of the thick juice enters the crystallization area for obtaining sugar. Crystallization is performed by vacuum pans at reduced conditions of pressure and temperature. After that, the obtained mixture of crystals (mainly sucrose) and mother liquor (green syrup) is separated by centrifugation. The sugar produced is first washed with hot water, dried, screened and finally stored after cooling. The sugar industrial yield achieved is 722 kg per ton of clarified juice processed, and the efficiency (expressed as the ratio of white sugar produced to the initial sugar content) is over 72% [25].

The green syrup left over by centrifuges is known as molasses. This stream passes through an additional boiling stage to extract some of the remaining sugars. Two types of molasses, known as B and C molasses, are produced. The B molasses enters the fermentation area of the 2G plant along with the 2% of thick juice. Due to the high concentration of nonsugars in the C molasses, no further crystallization is economically feasible, so it is stored to be used by other industries.

The mass balances calculated by Aspen Plus in the 1G plant area of sugar mill–based scenarios are listed in Table 4.6.

As in the ED- scenario, the mechanical power needed for the mill (17,951 kW) is provided by the power cogeneration part. LP steam produced in the boiler provides the heat required for the juice treatment, evaporation, and crystallization (71 Gcal/h).

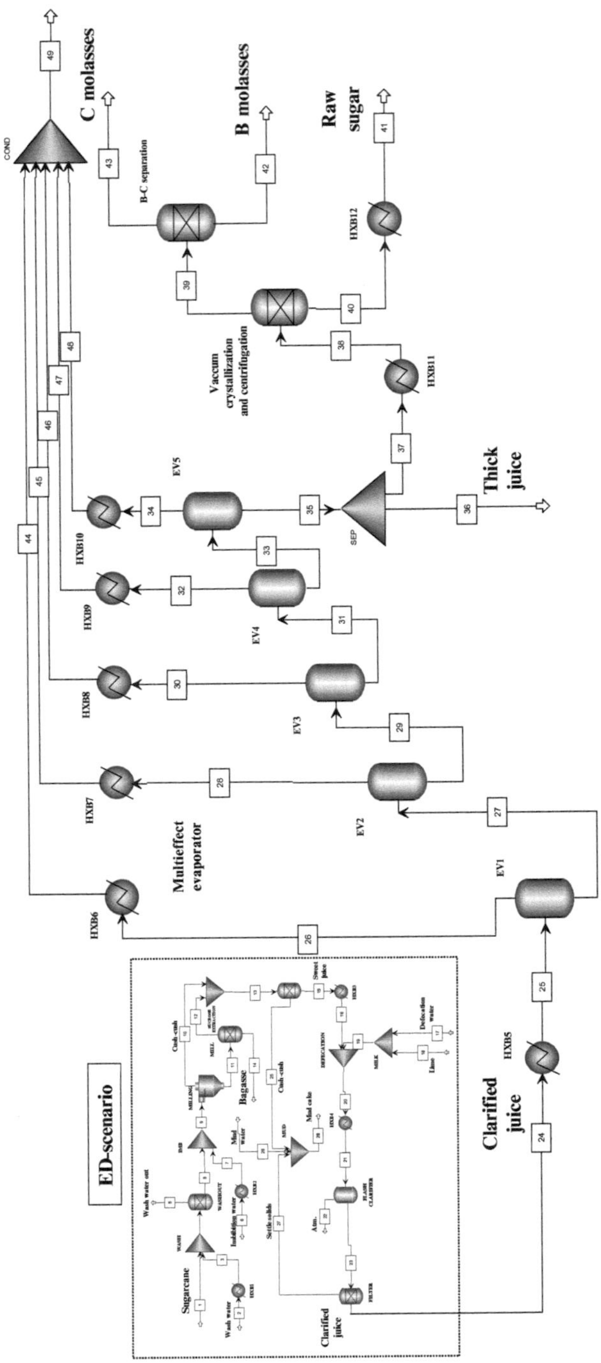

FIG. 4.3 Aspen Plus process flow diagram (material streams only) used in the current simulations for evaluating the sugar mill part of the biorefineries, i.e., 1G plant in SM- scenarios.

TABLE 4.6 Mass Balance of Sugarcane Juice Extraction (Calculated by Aspen Plus in the 1G plant Area of Sugar Mill–Based Scenarios)

Mass Flow (ton/h)	Sugarcane	Thick Juice	B Molasses	C Molasses	Raw Sugar
Water	390.97	0.96	2.04	45.21	0.03
Sugars	75.83	1.43	8.51	9.87	51.55
Others	74.87	0.04	0.78	0.78	0.15
Total (ton/h)	541.67	2.4	11.3	55.9	51.7

4.3.2 The Processing of Sugarcane Bagasse

In this chapter, the integration of a 2G plant in a normal distillery or sugar mill plant is modeled basing on the model developed for a corn stover by NREL [31].

The 2G plant model can be divided into six main areas: (1) feed handling, (2) pretreatment (and molasses feed production in -FF scenarios), (3) saccharification and fermentation, (4) distillation and dehydration, (5) wastewater treatment (WWT), and (6) combined power and heat generation. Even if the enzymes could be purchased by the company, the enzyme production section is also included in this model as it must be taken into account for LCA. Therefore, seven areas in total are modeled for the 2G plant (see Fig. 4.1).

The two configurations considered to process the conversion of sugarcane bagasse into different products are:

- the 2G plant producing feed and fuel (-FF)
- the 2G plant producing only fuel (-OF)

The IBUS base-case system [32,33] implemented in Inbicon A/S has been chosen to model the FF scenarios. In these scenarios, yeast is used for fermenting saccharified cellulose into fuel (ethanol). Moreover, the liquid fraction resulting after a liquid hot water (LHW) pretreatment is concentrated to produce syrup rich in C5 sugars used to feed animals.

On the other hand, the target of the OF scenario is the simultaneous fermentation of both pentoses and hexoses to obtain ethanol. In this configuration, a harsher pretreatment is used to completely hydrolyze the hemicellulose. This is a dilute acid pretreatment accompanied with an ammonia detoxification unit. The simulation of this process is very close to the 2011 NREL design developed for corn stover [31].

Both OF and FF processes are very similar; however, some details of each will be discussed in the following paragraphs.

(1) *Feed handling*

The sugarcane bagasse (50% moisture) is convoyed to the feed handling area from the 1G plant. The amount of bagasse entering the 2G part determines the ethanol production as well as the steam consumption/generation of the process. The 60% of the bagasse available is assumed to be processed in the 2G plant.

Power consumption in this area is limited to the transport and manipulation of the biomass before pretreatment.

(2) *Pretreatment*

During pretreatment, the hemicellulose carbohydrates presented in bagasse are converted to soluble sugars (xylose, mannose, arabinose, and glucose) by hydrolysis reactions. Fresh water is added to liquefy the lignocellulosic material until a specific percentage of solids is achieved in the hydrolysate slurry (see Fig. 4.4). To get this, a solid loading of 30% for pretreatment has been specified as a design spec in the general flow-sheeting options of the models. The breakdown of the biomass after pretreatment facilitates the following enzymatic hydrolysis of the cellulose (saccharification).

The pretreatment technique depends on the scenario selected:

- LHW pretreatment (-FF scenarios)

In the -FF configurations, the hydrolysis of biomass is catalyzed by LHW. The bagasse is also soaked with process water recycled from distillation. The stream is heated with HP steam (13 atm) to destroy the protecting lignin structure and make the cellulose more digestible. The pretreatment reactions are carried out in two stages. The bagasse is first fed into a stoichiometric reactor operating at 180°C boiling point to produce soluble sugars. The second stage is a lower temperature "oligomer-conversion" reactor operating at 130°C and 5.7 atm, converting the oligomers leaving the first stage to monomers.

The hydrolysate slurry is flash-cooled to 100°C and 1 atm, releasing a large amount of residual water. The flash vapor is condensed and sent to the WWT area.

The hydrolysate slurry containing 30% of solids is conveyed to a Pneumapress pressure filter to separate liquid and solids as described in [23]. After pretreatment both fiber and liquid fractions are produced.

- The fiber fraction contains more than 90% of cellulose in bagasse, lignin, and most of the nonreacted hemicellulose. The fiber fraction is diluted in water to 20% of solids to ensure miscibility throughout the next enzymatic hydrolysis. A solid loading of 20% for saccharification is specified as a design spec in the general flow-sheeting options of the model.
- The liquid fraction contains C5 solubles from hemicelluloses, some cellulose and hemicellulose not recovered in the fiber fraction and the majority of the fermentation inhibitors arising from pretreatment. Due to the considered mild conditions, the production of inhibitors is low [32].

The pretreated liquid fraction is directed to a three-effect evaporator operating at 0.6, 0.3, and 0.2 atm (Fig. 4.5). The final C5 syrup leaves the system with 65% of dry matter (DM) content. Vent from the last effect is condensed with cooling water.

- Dilute acid pretreatment (-OF scenarios)

Pretreatment in -OF scenarios should be strong enough to hydrolysate most of the hemicellulose carbohydrates in the sugarcane bagasse. In this case, hydrolysis reactions are catalyzed using dilute sulfuric acid and heat from steam (Fig. 4.6). The first step of the reaction is carried out at 158°C boiling point; the second stage works at the same conditions as in -FF. After the pretreatment reactors, the hydrolysate slurry is flash-cooled, vaporizing a large amount of water along with some sulfuric acid and furfural. The flash vapor is condensed and sent to the WWT. The hydrolysate slurry is sent to a conditioning reactor, where a mixture of gas ammonia and water is added to raise the pH from 1 to 5. The liquid is then adjusted to the optimal fermentation pH of 4.5 by adding a slight amount of sulfuric acid.

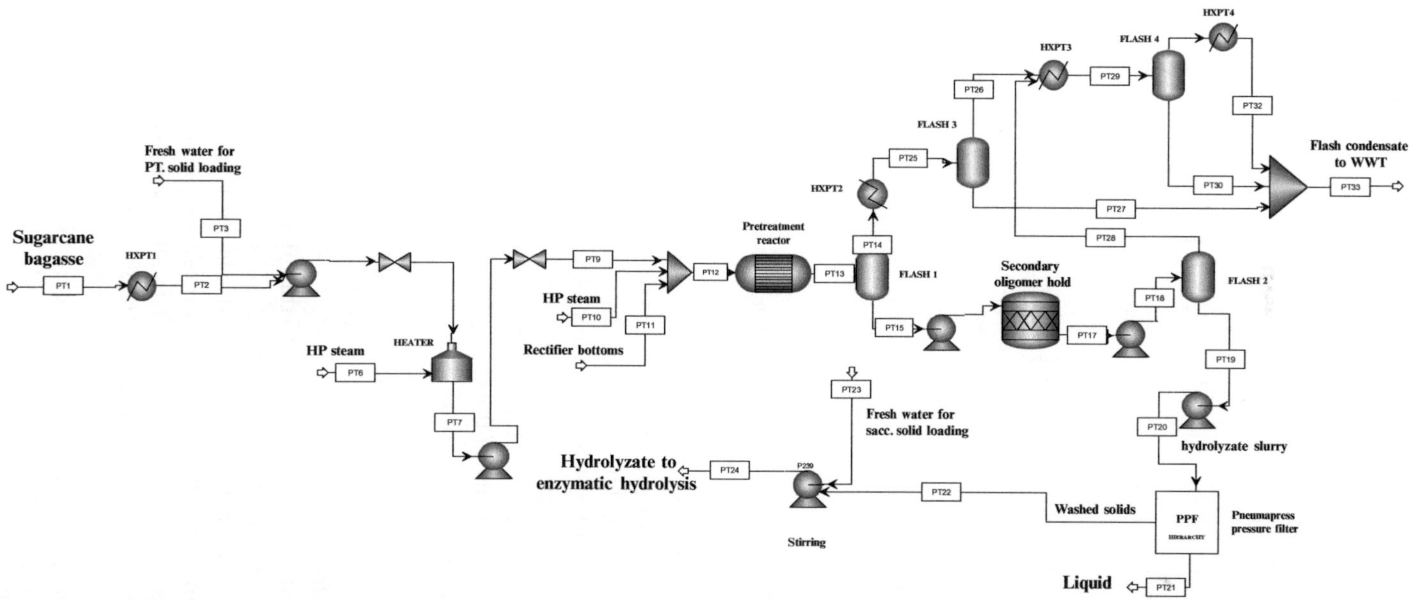

FIG. 4.4 Aspen Plus process flow diagram (material streams only) used in the current simulations for evaluating the pretreatment of the 2G plant in -FF scenarios.

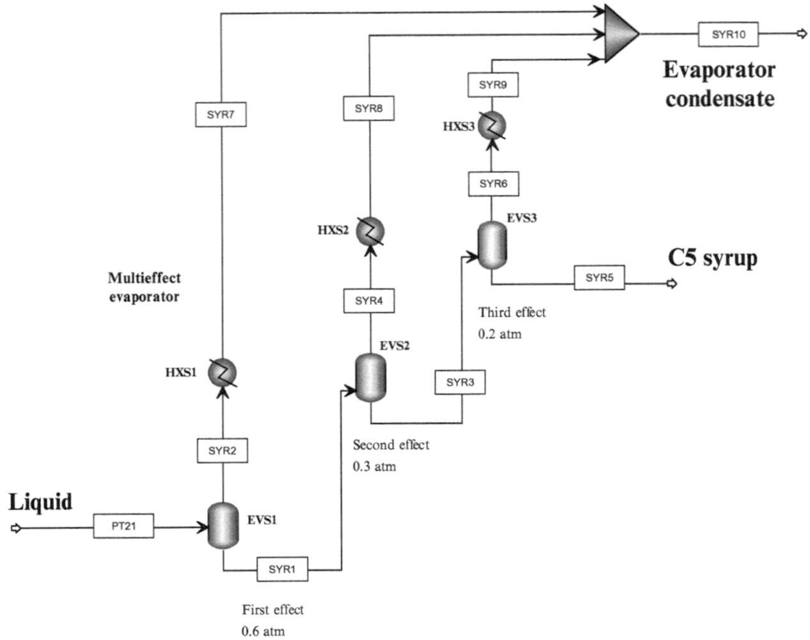

FIG. 4.5 Aspen Plus process flow diagram (material streams only) used in the current simulations for evaluating the syrup production area of the 2G plant in -FF scenarios.

(3) *Saccharification and fermentation*

Two different operations are performed in this process area: saccharification of the cellulose to glucose using cellulase enzymes, and fermentation of the resulting glucose (and other sugars in the case of -OF scenarios) to ethanol.

The high viscosity of the pretreated solid fraction (-FF) or pretreated hydrolysate (-OF) hinders the pumping. A liquefaction before saccharification partially hydrolyses the cellulose, decreasing the viscosity of the stream. This is performed using *Trichoderma reesei*, a cellulase enzyme produced in situ [31].

The liquefied material is saccharified in a continuous reactor at 20% water insoluble DM content (Fig. 4.7) where a pumpable viscous liquid is produced. After that, the saccharification continues in one of several parallel batch bioreactors where 90% of the cellulose is converted to glucose.

The saccharified slurry is last mixed with the juice coming from the 1G plant and cooled to optimal temperature for fermentation (i.e., around 32 °C); 10% of this stream is sent to the seed fermenter reactor where *Zymomonas mobilis* (ZYMO) is produced adding some nutrients (corn steep liquor (CSL) and diammonium phosphate (DAP) in Fig. 4.7).

Once the slurry is inoculated with ZYMO, the fermentation is carried out in the following batch bioreactor. In the Aspen model, 3% of the inoculated slurry is considered to be lost (stream SF15 in Fig. 4.7), leading to the formation of lactic acid. Yeast converts the 95% glucose (-FF) or 95% glucose and 85% C5 sugars (-OF) into ethanol in the last bioreactor. The obtained fermentation broth includes an ethanol concentration close to 7%, ready to enter the beer distillation.

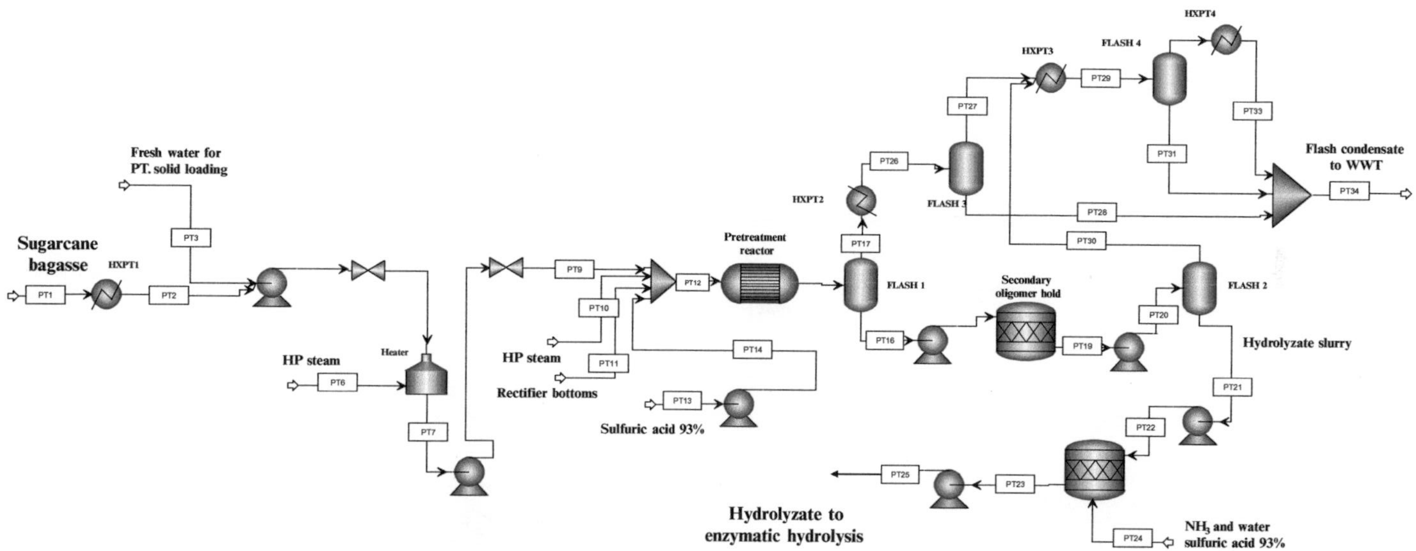

FIG. 4.6 Aspen Plus process flow diagram (material streams only) used in the current simulations for evaluating the pretreatment of the 2G plant in -OF scenarios.

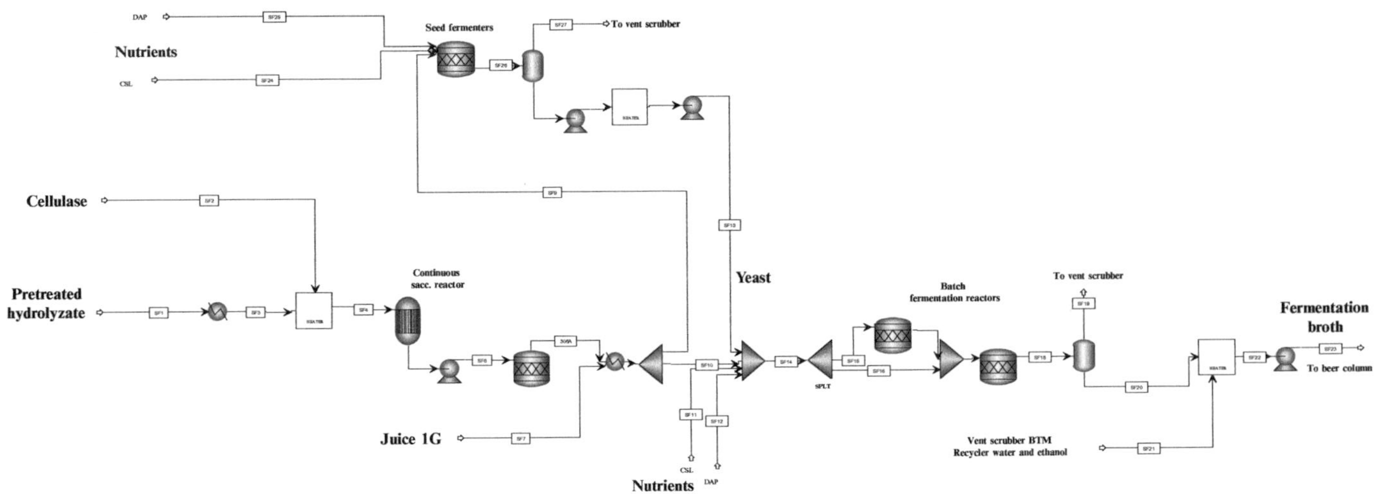

FIG. 4.7 Aspen Plus process flow diagram (material streams only) used in the current simulations for evaluating the saccharification and fermentation part of the 2G plant.

(4) *Distillation and dehydration—ethanol production*

Fermentation broth containing lignin and other soluble and insoluble solids is fed to a steam distillation column, where a nearly azeotropic mixture of ethanol and water evaporates from the fiber fraction. A beer distillation column is modeled in Aspen Plus (Fig. 4.8) using a RadFrac equilibrium column accomplished in 21 stages, a partial-vapor condenser and a kettle reboiler. The feed enters above the seventh stage from the top. Ninety-nine percent of the ethanol is removed as a vapor-side drawn on the eighth tray. The pressure and temperature in stage 1 is specified at 2 atm and 60°C, with a 0.2 atm pressure drop throughout the entire column. The insoluble solids and solubilized sugars remain in the column bottoms along with 0.05% of the ethanol. Overhead from the beer column is fed to a water scrubber in order to remove ethanol from CO_2.

After distillation, the ethanol-water mixture is separated in a rectification column. This is simulated with a RadFrac equilibrium column using 35 stages, a partial-vapor condenser and a kettle reboiler operating at 1.6 atm. The vapor-side of the beer column is fed above the 16th stage (from the top). Ethanol product leaves rectification at 92.5% mass purity (Stream D6). The rectifier bottom (saturated water stream) is recirculated to the pretreatment step.

The ethanol product is further purified to 99.5% through vapor-phase molecular sieve adsorption. The 99.5% of the water remains in the rectified stream (Stream D6) and the rest of the ethanol is recirculated above the eighth stage of the rectification column.

Beer distillation bottoms are subject to a decantation process in order to separate lignin from the nonfermentable solubles of the stillage. This separation process is modeled as a Pneumapress pressure filter, as in the NREL 2011s model. The distillation bottom is separated in both solid and liquid streams. Fifty percent of the nonfermentable solubles end up in the thin stillage directed to a WWT area. Ninety-eight percent of the solids and the other 50% of solubles fall in the lignin cake (35% moisture) to combustion.

(5) *Wastewater treatment*

The WWT section treats process water to reduce the plant makeup water requirement. It is modeled as the WWT area in NREL's model [31]. This step involves anaerobic and aerobic digestion processes. Anaerobic digestion produces biogas to feed the combustor. Aerobic digestion produces clean process water and a sludge also fed to the combustor.

(6) *Combined power and heat generation*

The lignin cake coming from distillation, the excess bagasse of the sugarcane and the GHR removed from the soil are fed to a combustor (Fig. 4.9). The biogas and the dirty air produced in the distillation area combined with air are also added to the combustion reactor operating at 870°C and 0.97 atm. The flue gas from the combustor exits to the baghouse releasing particulate ash. The scrubbed gas is exhausted through a stack.

In -OF scenarios, flue gas from the combustor enters first to a spray dryer for flue gas desulfurization (because of the dilute acid pretreatment). All of the sulfur (in -OF scenarios) entering the combustor is converted to sulfur dioxide. Lime is sprayed into the flue gas as 20 wt% slurry at 20% stoichiometric excess. This reactor converts 92% of SO_2 into sulfate, which falls out the bottom of the spray dryer. The water in the slurry is vaporized and exits with the flue gas to the baghouse.

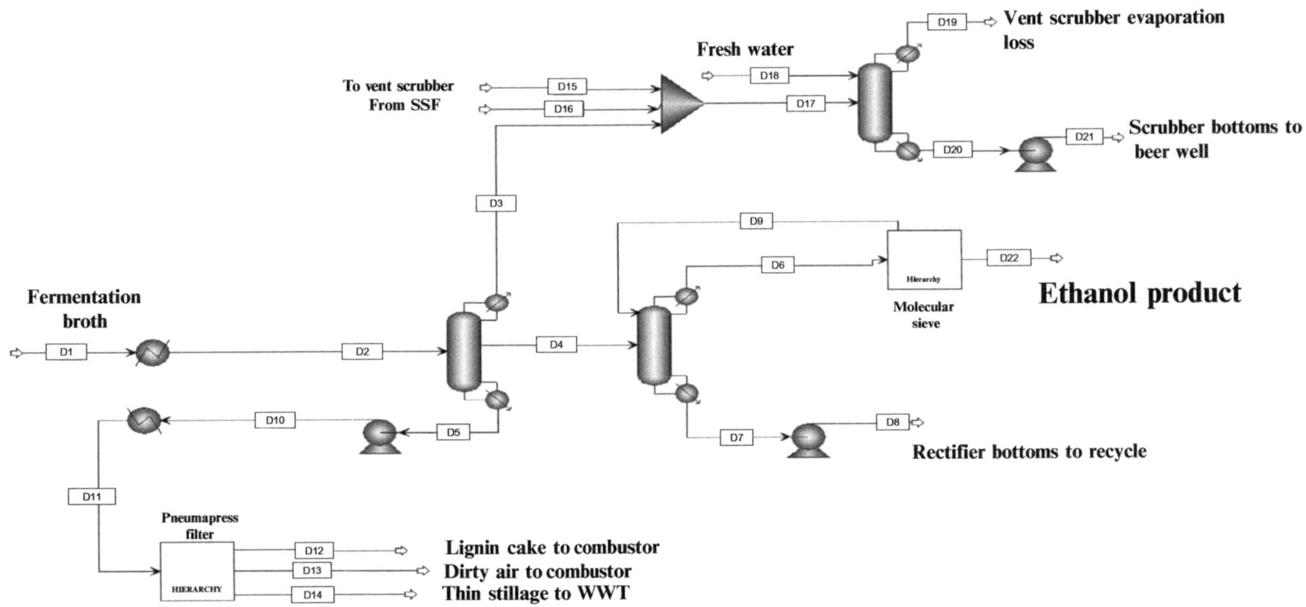

FIG. 4.8 Aspen Plus process flow diagram (material streams only) used in the current simulations for evaluating the ethanol production area of the 2G plant.

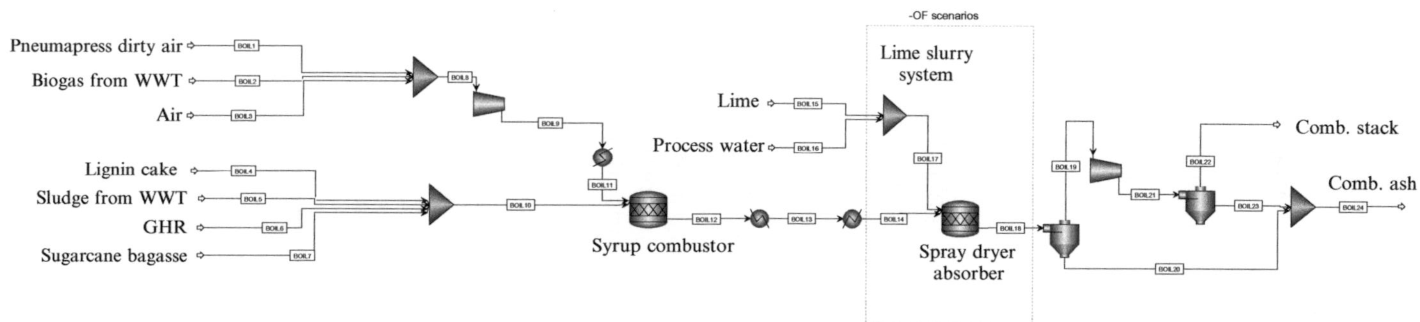

FIG. 4.9 Aspen Plus process flow diagram (material streams only) used in the current simulations for evaluating the cogeneration system of the 2G plant. The dotted line represents an area available in -OF scenarios, but is deactivated in -FF scenarios.

The heat released during combustion is used to produce steam at 454°C and 60 atm. The steam is passed through an isentropic turbine operating at 13 atm (here, HP steam is extracted for pretreatment), 9.5 atm (LP steam is extracted for the rest of the steam requirements), and 0.1 atm. The specifications for mechanical and isentropic efficiencies of the turbines are 0.96 and 0.85, respectively. The steam not used is condensed and recycled. More information for simulating the steam production can be found in the NREL report [31].

Burning the fuel streams to generate steam and electricity allows the plant to be self-sufficient in energy. This reduces solid waste disposal costs, and generates additional revenue through sale of excess electricity. The excess energy is served to grid after satisfying the plant's demand.

4.3.3 Process Simulation Results

The sugarcane models simulated in Aspen Plus were used to evaluate the performance of the considered sugarcane-based biorefinery scenarios. The required inputs to assess the life cycle were obtained from the heat, work and materials output streams of the Aspen Plus models. These values are the mass rate of the stream (mass balance) as well as the heat duty and power of every stream (energy balance).

The energy required in the different areas of a sugarcane plant (1G and 2G) is shown in Table 4.7 in terms of electricity and steam wastes [22].

It can be shown that in the 1G plant area the electricity required is lower compared to the 2G plant: i.e., processing sugarcane is less of a power consumer than sugarcane bagasse. This is mainly because of the higher solid content presented in the 2G plant feedstock (bagasse).

Electricity demand among the four scenarios does not vary significantly because most of the electricity consumption in the 1G pathway is due to the milling and preparation process, being common to both ED- and SM- scenarios. Juice conditioning is also common to both scenarios. However, fermenting and distilling the juice involves a greater demand of steam than extracting the sugar from the juice and fermenting and distilling a more concentrated substrate such as B molasses, as can be seen in Fig. 4.10 [22].

TABLE 4.7 Electricity and Steam Consumption in the Biorefinery Configurations Obtained by Process Simulation

Configuration		Electricity (kWh)	Steam (Gcal/h)
1G plant	Sugar mill	17,951	71
	Distillery	17,951	42
2G plant	SM FF	31,251	120
	SM OF	33,480	72
	ED FF	46,812	167
	ED OF	49,242	122

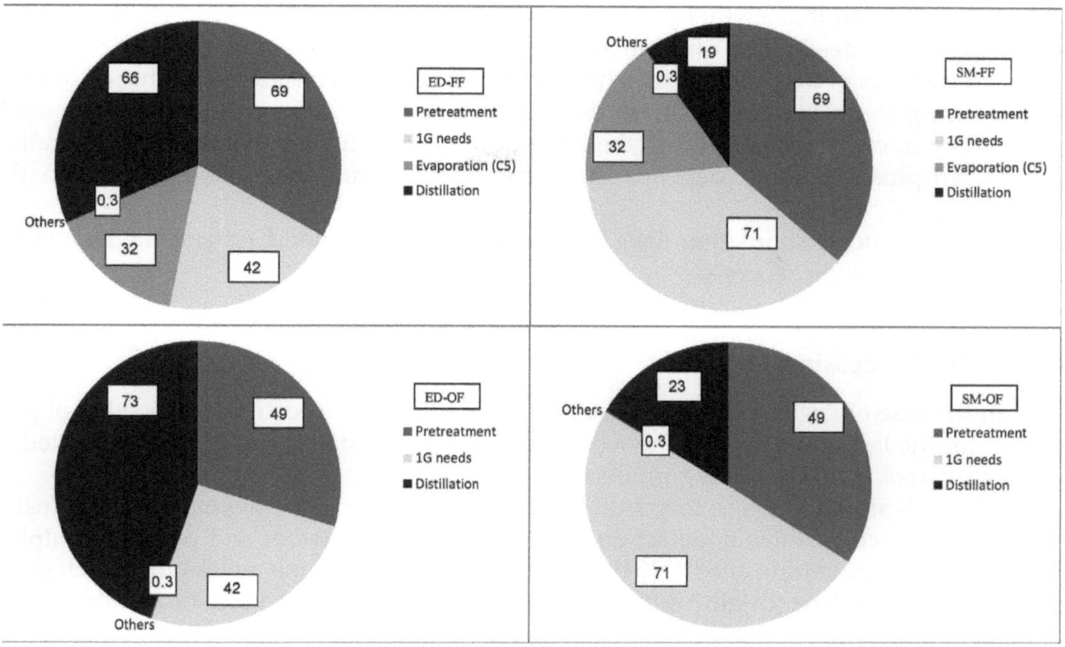

FIG. 4.10 Heat demand distribution in percentage values for the options evaluated. Ethanol distillery (ED) and sugar mill (SM) producing only fuel (OF) or feed and fuel (FF).

Regarding the integrated scenarios, it can be said that distillery plants require more energy than sugar mills, so the excess electricity produced in ED- scenarios will be lower than that of SM- scenarios (Table 4.8).

Besides disaggregate energy consumption of the plant, the main variables regarded to evaluate the biorefinery scenarios are the ethanol and C5 syrup yields, the sugar yield achieved in sugar mills, the total electricity required and produced by the process, and the steam demand in the whole plant. The main results obtained by process simulation for the four different scenarios are shown in Table 4.8 [22].

TABLE 4.8 Process Simulation Results for the Biorefinery Scenarios Considered

Configuration	Ethanol (kg/ton SC)	Syrup (kg/ton SC)	Sugar (kg/ton SC)	C Molasses (kg/ton SC)	Electricity Demand (kWh/ton SC)	Steam Demand (Gcal/ton SC)	Surplus Electricity (kWh/ton SC)
SM FF	24	27	96	103	91	191	48
SM OF	34	–	96	103	95	143	60
ED FF	82	27	–	–	120	209	13
ED OF	92	–	–	–	124	164	25

The yields and duties are calculated per ton of sugarcane (w.b.) entering the factory.

It can be observed that ethanol yield increases as the number of byproducts decreases, i.e., depending on the destination of C5 molasses.

Self-consistency of the processes is achieved by the scenarios considered, so that all of them are producing electricity being sold to the grid. A difference between them is the amount of electricity in excess produced. As the amount of GHR and bagasse sent to cogeneration is fixed, surplus electricity depends on both electricity and heat requirements in the process [22].

Outputs of process simulation have been used to perform the LCAs and the results are explained in following sections.

4.3.4 The Processing of Wheat

As in the case of sugarcane, wheat has been employed as a feedstock for bioethanol production for the last few decades. The wheat kernels include starch that can be fermented to produce ethanol by a dry milling process.

The wheat is stored in the factory until is milled. After milling the raw material, the dust is introduced in a continuous liquefaction tank mixed with hot water and purchased alpha-amylase enzyme operating at 88°C [34]. In liquefaction, this enzyme attacks the starch polymer, producing maltose (di-glucose) and higher oligomers. Lime and caustic are also added to the tank for controlling pH (6) and calcium for the enzyme.

After liquefaction, the mash is heated to 110°C and then cooled to 60°C for the continuous saccharification process [34]. In this stage, the glucoamylase enzyme is added to produce the glucose monomers. Sulfuric acid is also added to control pH to 4.4. The saccharified material is cooled to fermentation temperature (32°C) and sent to the fermentation area of 2G plant.

The use of wheat straw as feedstock in a lignocellulosic biorefinery for 2G ethanol production is considered. The process design for wheat straw is similar to the 2G plant and already explained for sugarcane bagasse.

4.4 COMPARATIVE LCA

The final decision on the type of biorefinery configuration depends on the results of the assessment of the configurations according to certain criteria, such as economic, environmental, social, etc. Nowadays the LCA is the most common use methodology for evaluation of environmental performance of industrial processes and activities. The application of LCA methodology to an integrated sugar/starch-based biorefinery is presented and illustrated in the particular examples in the next sections. The following methodology and examples are based on the work presented in a previous paper [22]. The general information about LCA can be found in Chapters 1 and 2.

4.4.1 Goal and Scope Definition, System Boundaries

The goal and scope phase includes definition of basic technical details of LCA work such as goals, system frames, type of allocation, functional units, etc. The goal of the LCA of an integrated sugar- or starch-based biorefinery could be process design, operation-oriented goals, or policy-oriented goals. The general recommendation is to consider more detailed systems if the goal is a design or an operation improvement and use more simplified system definitions for policy-oriented LCA. For example, if the goal is to compare different configurations of bioethanol production, a well-to-tank (WtT) LCA is suitable since the performance of the fuel combustion in the vehicle's engine is not affected by bioethanol production. However, if the aim is to compare bioethanol (e.g., bioethanol blends) with their fossil substitutes, the situation changes drastically. The problem is that the combustion performance mainly depends on the type of blend (E5, E10, E85, etc.) and neglect of utilization stages of the fuel in transport can lead to significant change in the final LCA results. In this case WtW LCA is more suitable than WtT LCA since WtW incorporates both utilization and fuel production stages. Therefore it is recommended to employ the WtW LCA for evaluating environmental performance of an integrated biorefinery aiming to produce bioethanol.

The important issue addressed for a sugar/starch-based multiproducts biorefinery is inclusion or not of coproducts of bioethanol into system boundaries. In such a case the main rule is to include all coproducts; however, the question of which outputs can be recognized as coproducts and which as wastes is relevant. The answer, which obviously depends on decision-makers' preferences, affects the choice of allocation type. For example, if economic allocation is chosen in the analysis, all outputs that have economic values should be considered as coproducts and therefore be included in the system boundaries. Another important challenge is the evaluation of the LUC factor for the LCA results. In general, the LUC is highly recommended to be considered in the analysis of integrated sugar/starch biorefineries. The issue is that sugar/starch crops mainly grow on land recently converted from tropical forests and pastures (in the case of sugarcane) or temperate grasslands (in the case of wheat) that causes a change in carbon pools above and below the ground. This shift in land use could have a strong influence on the GHG balance of the whole biorefinery system.

The systems boundaries of selected sugar-based integrated biorefineries can be seen in Figs. 4.11 and 4.12. Fig. 4.11 represents systems boundaries of two "fuel" scenarios ED FF (Fig. 4.11A) and ED OF (Fig. 4.11B), while Fig. 4.12 shows system definitions of integrated ethanol distillery-sugar mill configurations SM FF (Fig. 4.12A) and SM OF (Fig. 4.12B). The details of the scenarios can be found in Section 4.3.1. As can be noticed from Figs. 4.11 and 4.12, system boundaries include plantation and harvesting parts with land use, transportation of feedstock, simplified pathway of bioethanol production (see details in Sections 4.3.1 and 4.3.2) with intermediated process streams, chemical inputs, and final use of products in appropriate services [22]. The passenger car maintenance and fabrication are also considered in the system boundaries.

4.4.2 Functional Unit

The functional unit in the LCA of multiproduct biorefinery systems is an important basis that enables alternative goods, or services, to be compared and analyzed. In general, there are four groups of functional units that can be used in the LCA of a sugar/starch-based integrated biorefinery [15]: input of biomass (in mass or energy units); output of relative service (kilometers of transport, heat, power); agricultural land (hectares of land needed to produce biomass); year or other time units. For comparative LCAs of various biorefineries, it is desirable to use inputs of biomass units, in particular 1 ton of feedstock for WtT LCA and related output units, in particular kilometers of transportation service for WtW LCA.

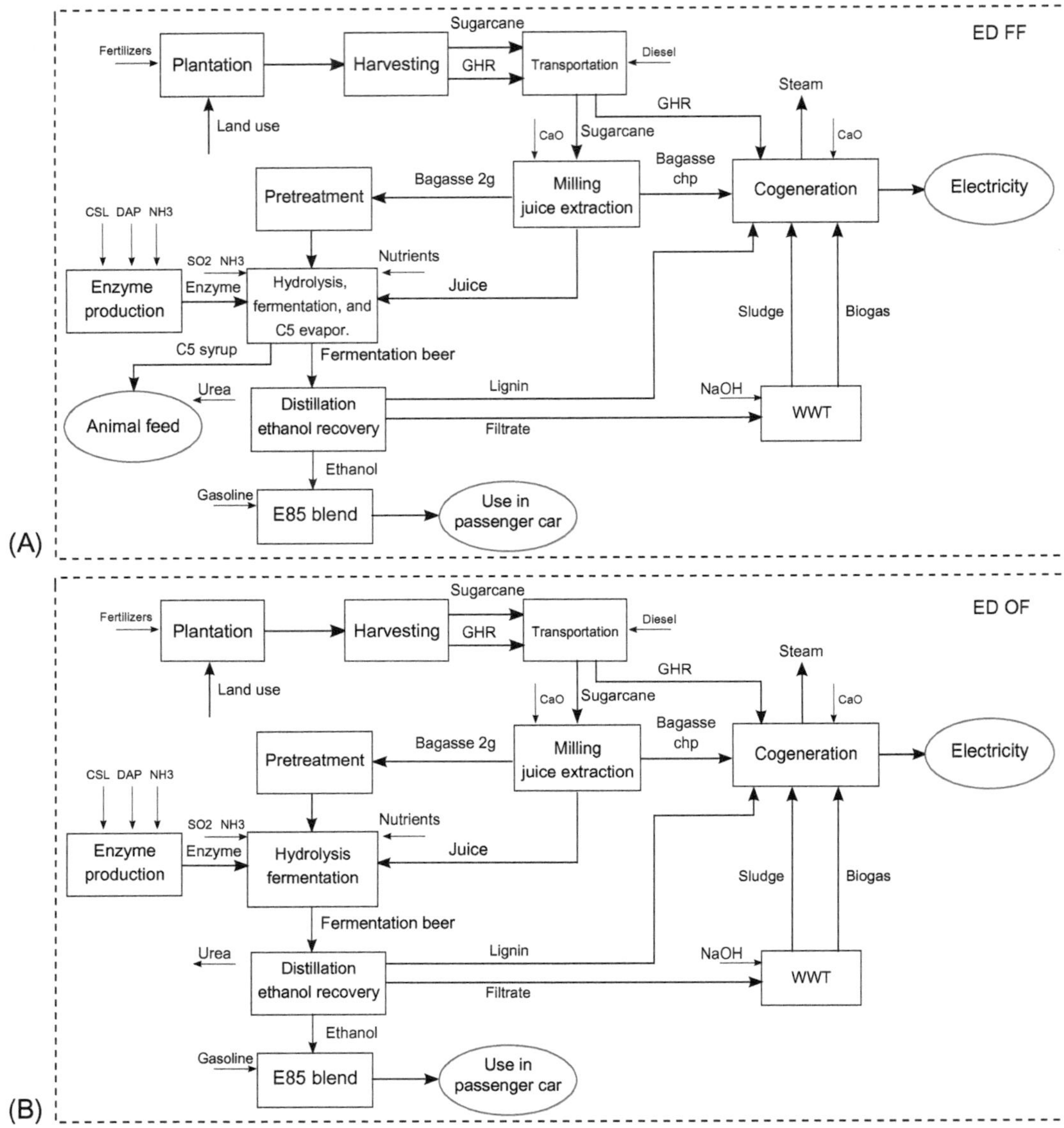

FIG. 4.11 System boundaries of partly integrated sugarcane-based biorefinery scenarios ED FF (A) and ED OF (B).

4.4.3 Reference System

The choice of an adequate RS is a complicated task in the context of a multiproduct biorefinery. The main challenge is that various coproducts aimed at different purposes should be evaluated and compared within one system. Traditionally sugar/starch crop-based bioethanol production systems compare their environmental impacts with those from a fossil fuel pathway (e.g., gasoline or diesel). However, this approach alone is not suitable for

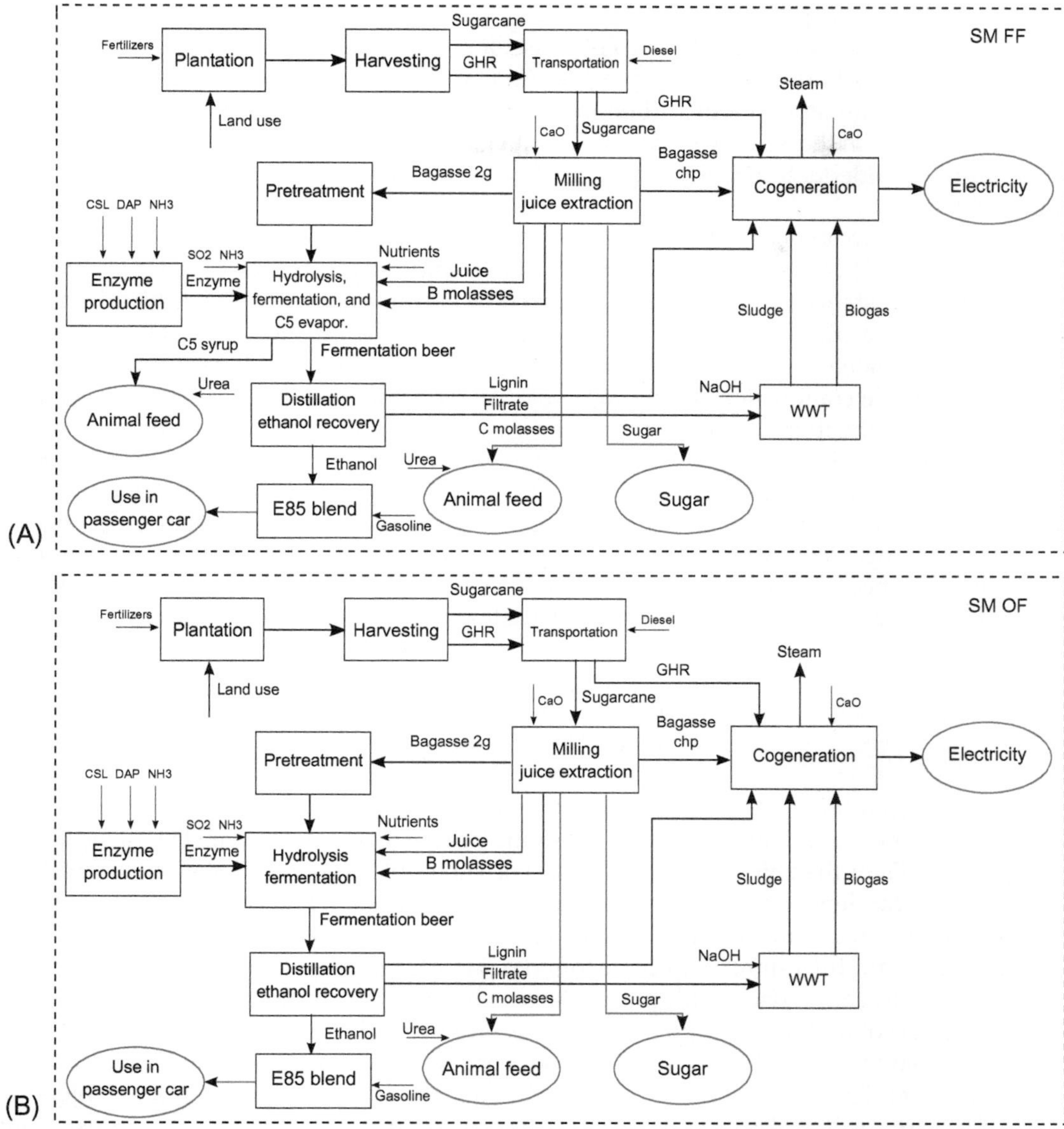

FIG. 4.12 System boundaries of fully integrated sugarcane-based biorefinery scenarios SM FF (A) and SM OF (B).

analysis of multiproduct biorefinery systems whose products could replace an existing counterpart, not necessarily fossil based, whose environmental impacts are significantly different. In addition, it is highly recommended to compare impacts associated with services where the product could be used. Thus in the case of the multiproduct biorefinery, the relevant substituted products and provided reference services should be defined. Moreover, as was previously mentioned, the production of feedstock can use land that was previously dedicated to other purposes (e.g., storing carbon, such as forest or grassland); therefore, the correct RS should include such previous land use. It is especially important for the sugarcane-based biorefinery whose feedstock is produced on land recently converted from tropical forest that stores huge amounts of carbon either above or under the ground. Neglecting the previous land use in the RS may lead to incorrect LCA results. Next, an approach is presented for building an adequate RS, illustrated in the particular example presented early in the paper [22].

Integrated sugarcane-based biorefineries selected for analysis in this chapter (Section 4.3.1, Figs. 4.11 and 4.12) can issue up to five products: ethanol, sugar, C molasses, C5 syrup, and electricity. The distance traveled service provided by gasoline and ethanol fuel blends is used for comparing the ethanol product with substituted fossil fuel. It should be noticed that the performance of bioethanol as a vehicle fuel strongly depends on the part of ethanol in the fuel blend. Indeed, the combustion performance of different ethanol blends varies since the ethanol has less combustion efficiency than pure gasoline. Therefore, the fuel efficiency of the ethanol component in fuel blends should be defined in order to evaluate the WtW LCA of bioethanol fuel. Gnansounou et al. [16] proposed a methodology that estimates the fuel efficiency of ethanol in the fuel blend. This method is based on the idea that fuel properties of gasoline in the fuel blends are equal to the properties of standard gasoline fuels, and therefore the main difference between a blend and pure fuels is associated with the presence of biofuel in the fuel blends. Eq. (4.1) presents the calculations of fuel efficiency of biofuel for 1 km traveled:

$$FE_{\text{biofuel}} = \frac{BF_{\text{biofuel}} FE_{\text{fuelblend}}}{1 - (1 - BF_{\text{biofuel}}) \dfrac{FE_{\text{fuelblend}}}{FE_{\text{fossilfuel}}}} \tag{4.1}$$

where $FE_{\text{fuelblend}}$, FE_{biofuel}, $FE_{\text{fossilfuel}}$ represent fuel efficiency of fuel blend, biofuel (bioethanol in our case), and fossil fuel (gasoline), respectively, in L/km. The blend factor (BF_{biofuel}) is 0.85 for the E85 blend considered. The choice of blend type, substituted gasoline, and particular car depends on local context and the most popular patterns in the region. There are no particular recommendations about such choices; however, in the work proposed by Gnansounou et al. [16] a summary can be found of increasing and decreasing of several blend consumptions in comparison with pure gasoline. In the Brazilian context, the average fuel efficiency of gasoline for a new passenger car is around 7.48 L/100 km [35]. In comparison with pure gasoline, according to [16], the fuel consumption for E85 ethanol blend has increased to about 35%. According to Eq. (4.1) the fuel efficiency of ethanol in a fuel blend is 10.76 L/100 km for E85. As fuel efficiency is known, the travel distance of a car consuming a fuel blend can be calculated. Thus, after applying a WtW LCA, environmental impacts associated with car operation consuming a fuel blend and gasoline can be compared for the same traveled distance.

As was mentioned before, the substituted products and services for coproducts also must be defined. The sugar product produced at a sugarcane biorefinery is compared with the same amount of sugar from a sugarcane mill (in kilograms) in Brazil. To better validate the comparison, it is recommended to make the production conditions of reference sugar as close as possible to the conditions of the biorefinery studied.

The C molasses and C5 syrup can be compared with an animal feed service. Cattle feed based on C molasses and C5 syrup is a valuable source of energy, vitamins, and several minerals for livestock. C molasses and C5 syrup can be used as binding agents to stick feed granules together or as a liquid feed supplement enriched by urea (a protein source). The C molasses and C5 syrup differ in their total DM content. In the present example, a cattle feed supplement produced at a sugarcane mill in Brazil contains 8% urea and 92% molasses with DM content of 73%. To equalize the amount of services provided by biorefinery products and the referenced cattle feed, the feed supplement equivalent was calculated based on DM content of C5 syrup and C molasses.

Finally, to evaluate the environmental impact of surplus electricity generated at the biorefinery, the Brazil country mix of electricity (in kilowatt-hours) is included in the RSs. As in previous services, the substituted product should be defined in the local context. Indeed, in the case of electricity, results can be widely different depending on the context. For example, reference electricity can be assumed to be produced from hydropower, natural gas, coal, or other sources, all of which have different GHG emission factors. Clearly, savings could be much larger if coal electricity is a reference rather than hydroelectricity.

Figs. 4.14–4.17 represent the concept of a RS and the link between the RS and the evaluated system for different scenarios. In particular, Figs. 4.13 and 4.14 depict a 1G-2G ethanol integrated biorefinery without sugar and molasses production (ED OF and ED FF scenarios, respectively). Figs. 4.15 and 4.16 show a sugar mill integrated with ethanol production (SM FF

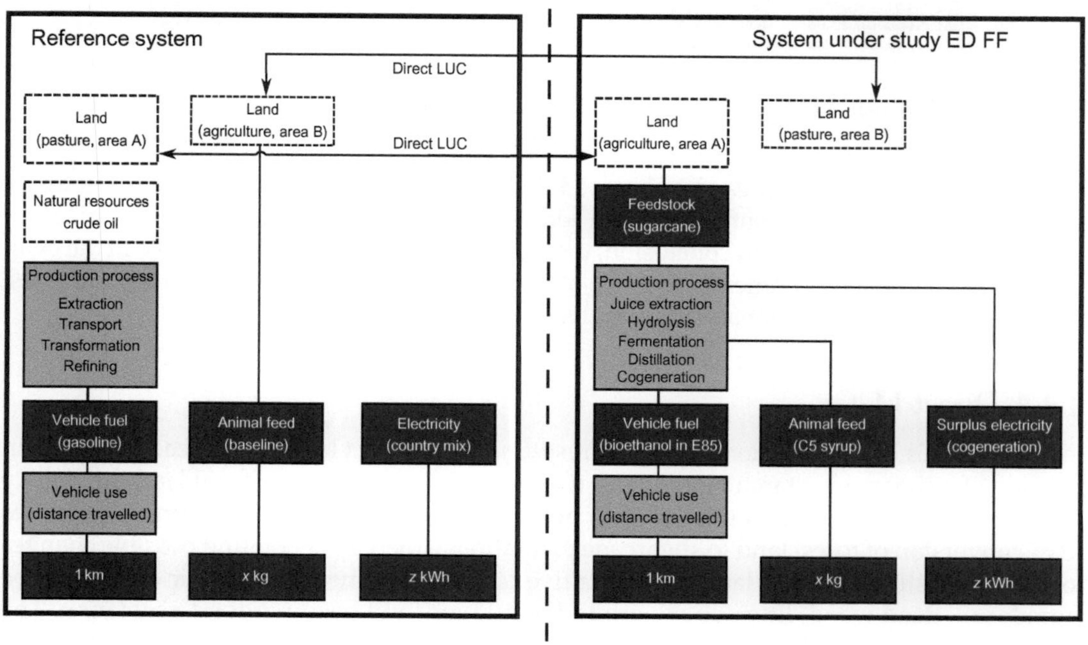

FIG. 4.13 System definition and boundaries (from reference system to system studied) for ED FF scenario.

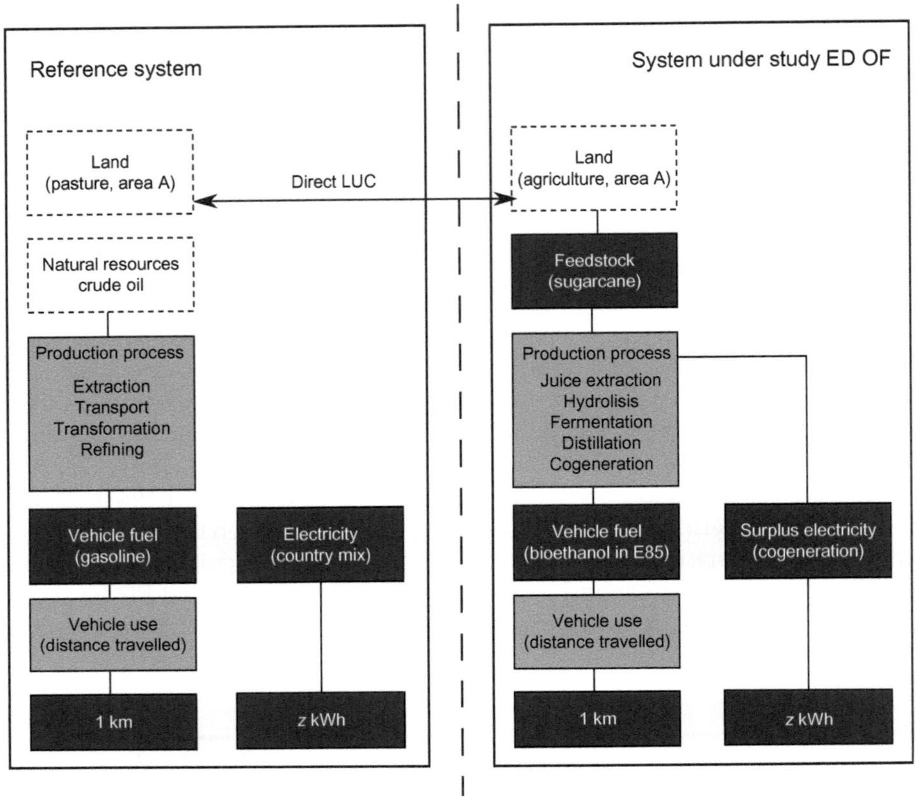

FIG. 4.14 System definition and boundaries (from reference system to system studied) for ED OF scenario.

and SM OF, respectively). It can be observed that land use, traveled distance, and electricity service are included in all configurations. Feed services are presented in ED FF scenarios and in both SM- configurations. In conclusion, the definition of the RS plays a key role in the evaluation of potential environmental impacts of integrated sugar- or starch-based biorefineries. According to the assumptions made, results can be very different.

4.4.4 Direct LUCs

Direct LUCs cover emissions of GHGs resulting from direct human-induced land use. In the case of a sugarcane-starch biorefinery, it appears when new agricultural land previously used for other purposes is converted into production of feedstock for biorefinery purposes (e.g., conversion of forest land to sugarcane plantations), thereby generating possible changes to the carbon storage of that land. There are five carbon pools in which organic carbon can be stored: soil, dead wood, litter, above-ground vegetation, and below-ground vegetation. The common approach used to evaluate the LUC emissions is to calculate the increase or decrease of a carbon pool as a consequence of LUC and then convert this difference of carbon storage into CO_2 emission.

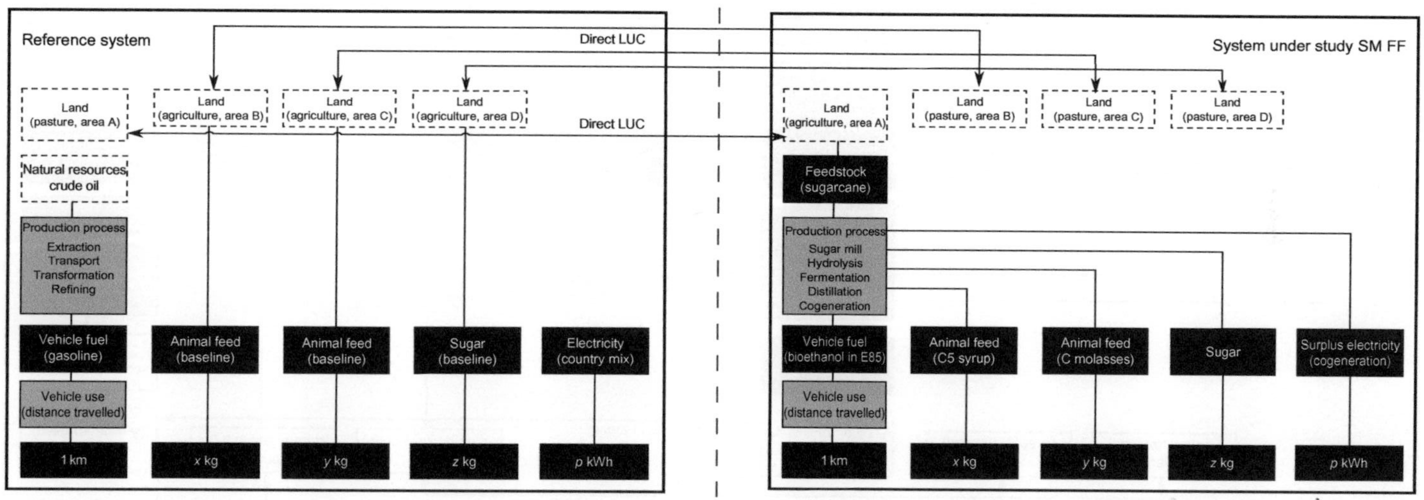

FIG. 4.15 System definition and boundaries (from reference system to system studied) for SM FF scenario.

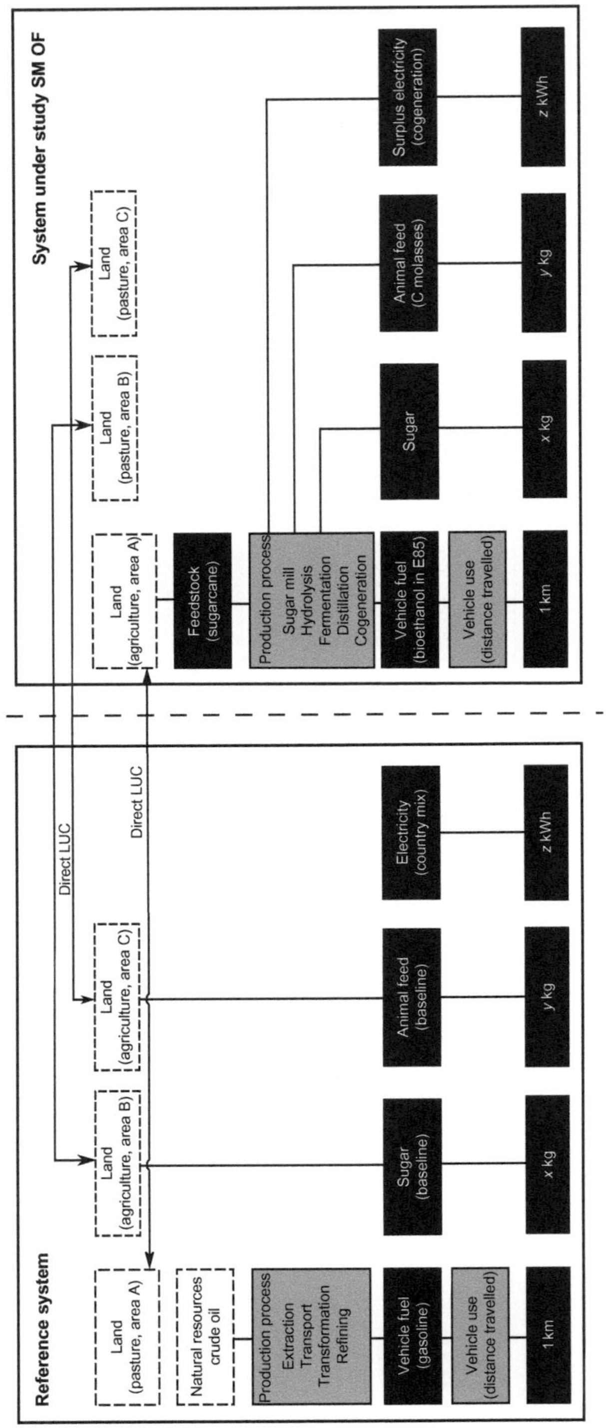

FIG. 4.16 System definition and boundaries (from reference system to system studied) for SM OF scenario.

In this work it is proposed to use the gain-loss method for land recently converted to a new land-use category proposed by IPCC [36]. The following equations allow the calculation of the annual change in carbon stock:

$$\Delta C_{\text{change}} = \Delta C_{\text{soil}} + \Delta C_{\text{biomass}} = \Delta C_{\text{soil}} + \Delta C_{\text{biomass}}^{\text{gain}} - \Delta C_{\text{biomass}}^{\text{conversion}} - \Delta C_{\text{biomass}}^{\text{looses}} \quad (4.2)$$

where:

ΔC_{soil} = annual change in carbon stocks in soil, in tons of C per year
$\Delta C_{\text{biomass}}^{\text{gain}}$ = annual growth in carbon stocks in biomass, in tons of C per year
$\Delta C_{\text{biomass}}^{\text{conversion}}$ = initial change in carbon stocks in biomass on land due to conversion of land category, in tons of C per year
$\Delta C_{\text{biomass}}^{\text{losses}}$ = annual deduction in biomass carbon stocks due to various losses of biomass, in tons of C per year.

$$\Delta C_{\text{biomass}}^{\text{conversion}} = \sum_i \{(B_{\text{before}_i} - B_{\text{after}_i})\Delta A_{\text{to\_others}_i}\} CF \quad (4.3)$$

B_{before_i} = biomass stocks on land category i before the conversion, in tons of d.m. per ha
B_{after_i} = biomass stocks on land category i immediately after the conversion, in tons of d.m. per ha
$\Delta A_{\text{to\_others}_i}$ = total area of converted land i, in ha/year
CF = carbon fraction of DM
i = land category.

The two most popular types of natural land converted to sugar-starch crop plantations are grasslands and forests. Tables 4.9 and 4.10 provide general data on carbon stores in grassland and forests in different climate types and ecological zones [36]. It can be noticed that the carbon pool of the biomass for forest is much bigger than for grassland.

The annual soil carbon stock change for different climates, agronomic practices and soil characteristics can be retrieved from an IPCC tool produced by Colorado State University [37]. In the example considered based on sugarcane feedstock, a pasture is considered to

TABLE 4.9 Biomass Stocks Present on Grassland

Climate Zone	Above-Ground Biomass (tons of d.m. per ha)	Total (Above and Below Ground) Biomass (tons of d.m. per ha)	Error (%)
Boreal—dry and wet	1.7	8.5	±75
Cold temperate—dry	1.7	6.5	±75
Cold temperate—wet	2.4	13.6	±75
Warm temperate—dry	1.6	6.1	±75
Warm temperate—wet	2.7	13.5	±75
Tropical—dry	2.3	8.7	±75
Tropical—wet	6.2	16.1	±75

TABLE 4.10 Biomass Stocks Present on Forest

Ecological Zone	Above-Ground Biomass (tons of d.m. per ha)	Error (%)
Tropical rain forest, Asia	280	±75
Tropical rain forest, North and South America	300	±75
Tropical dry forest, Asia	130	±75
Tropical dry forest, North and South America	210	±75
Temperate continental forest, Asia, Europe	120	±75
Temperate continental forest, North and South America	130	±75

TABLE 4.11 Parameters of Land Use Changes

Native land	Brazil, tropical wet climate, medium improved pasture
Converted land	Long-term cultivation, full tillage, high with manure
ΔC_{soil}, carbon stock changes in soil, tons of C/ha/year	−0.82
$C_{biomass}^{sugarcane}$, carbon stock changes in biomass of sugarcane, tons of C/ha/year	11.88
$C_{biomass}^{pasture}$, carbon stock changes in biomass of pasture, tons of C/ha/year	10.14
ΔC, total annual carbon stock changes of converted land, tons of C/ha/year	0.92

be an initial category of land use. After making assumptions that the carbon pool of the biomass immediately after the conversion is equal to zero ($B_{after_i} = 0$) and the carbon fraction of the biomass is near 50%, Eq. (4.2) is simplified to the following equation:

$$\Delta C_{change} = \Delta C_{soil} + \Delta C_{biomass} = \Delta C_{soil} + C_{biomass}^{sugarcane} - C_{biomass}^{pasture} \quad (4.4)$$

The summary of land use and carbon stock parameters used in Eq. (4.4) is presented in Table 4.11. The biomass carbon stock for pasture and sugarcane was taken from the literature [38,39]. Eq. (4.4) provides the total annual carbon stock change of converted land (ΔC_{change}) equal to 0.92 ton C/ha/year. To calculate GHG emission the value of ΔC_{change} is multiplied by the parameter 44/12. The obtained result for GHG emission is 3.37 ton CO_2/ha/year. It has a positive sign, which means that 3.37 ton of CO_2 can be taken from the atmosphere from 1 ha of cultivated land per year.

4.4.5 Allocation Method and Value-Based Approach

The sugar- or starch-based integrated biorefinery consists of different stages that provide various intermediate products interconnected with each other during the process chains (see Figs. 4.11 and 4.12). For a correct LCA application, these intermediate products at each stage demand an allocation. Despite the fact that there are several methods investigated in the literature [16], there is no unique approach that can be recognized as the best allocation practice. In general, the choice of the allocation method for a particular case depends on the context and the decision maker's preferences.

The general recommendation for a multiproduct biorefinery is to use an economic-based allocation, because an economic profit is the main driving force of such a production system. One of the most effective ways for correctly calculating economic allocation factors is the value-based approach proposed by Gnansounou and Dauriat [7]. This method of analysis involves different techo-economic characteristics such as product prices, capital investments, stream rates, engineering approaches, etc. The value-based methodology calculates the values for all process streams step by step at every process stage and then estimates the allocation factors of output products. The method consists of several steps:

1. Split the production chain into process stages (index i) (see Figs. 4.11 and 4.12).
2. Determine the competitive prices for biorefinery products and operational features.
3. Calculate for every stage the annuity for capital recovery A_i (\$/year) (Eq. 4.5) and operation cost O_i (\$/year).
4. Estimate the value of every intermediate product stream at different stages by using Eq. (4.6).

$$A_i = \frac{i \times (1+i)^n}{(1+i)^n - 1} \times \text{TIC} \tag{4.5}$$

$$V_{i,k'} = \frac{\sum_{i=1}\left(Q_i P_i + \sum_{k \in M} V_{i,k} m_{i,k} - O_i - A_i\right)}{m_{i,k'}} \quad \forall k' \in N \tag{4.6}$$

where index k is an index for the intermediate product streams; Q_i represents the amount of end product at a process unit, ton/year (kWh for electricity); P_i is the current market price for end products issued at a process stage i, \$/ton (\$/kWh); $m_{i,k}$ is the flow rate of intermediate product streams, ton/year; $V_{i,k}$ represents the value of the intermediate product streams, \$/ton; $M(i,k)$ and $N(i, k')$ are the sets of intermediate streams k leaving unit I and streams k' entering at unit I, respectively; TIC is the total investment cost, \$.

As can be noticed, calculation of stream values is based on the principle that every process stage is economically independent and for each of them the sum of values of entering intermediate streams equals the sum of the values of leaving streams minus operation and capital recovery costs (Eq. 4.6). The market prices of biorefinery products and feedstocks in the context of the Brazilian region are presented in Table 4.12.

4.4.6 Inventory Analysis, Impact Assessment and Interpretation

Life cycle inventory (LCI) is the methodology step that involves creating an inventory of input and output flows for a product system. Such flows include inputs of water, energy, and raw materials, and releases to air, land, and water. The inventory can be based on literature analysis or on process simulation. In the present work, the inventory analysis is based on the Ecoinvent v. 3 database for agricultural and logistic parts and Aspen Plus v. 8.6 software for process design (Sections 2.1–2.3). Table 4.13 shows the data inventory of process design for

TABLE 4.12 Prices and Project Parameters Used in the Economic Analysis

Project lifetime, years	25
Reference year	2010
Start operation year	2020
Discount rate, %	10
Sugarcane price (MSP), $/ton [21]	23.25
Sugarcane GHR price, $/ton [21]	17.05
Ethanol price, $/ton [21]	760
Sugar price, $/ton [40]	320
Molasses C price, $/ton [41]	36.30
Syrup C5 price, $/ton [41]	123.26
Electricity price, $/kWh [21]	0.08488

TABLE 4.13 Data Inventory per 1 ton of Sugarcane for SM FF Scenario

Stage	Inputs	kg (kWh, Gcal)	Outputs	kg (kWh, Gcal)	Emissions	kg
Harvesting	Feedstock	1109.888	Sugarcane	1000.00		
			GHR	109.889		
Milling, juice extraction	Sugarcane	1000.000	Bagasse to 2 g	168.000		
	Electricity (kWh)	33.140	Bagasse to CHP	112.000		
	Lime	1.200	Juice + b molasses	65.237		
	Steam (Gcal)	0.131	Sugar	95.505		
			C molasses	103.138		
Hydrolysis	Bagasse	168.000	Hydrolyzate	640.004		
	Electricity (kWh)	9.203				
	Steam (Gcal)	0.128				
Fermentation	Hydrolyzate	640.004	Beer from fermentation	368.138		
	Enzyme	16.651	Syrup, C5	27.238		
	Corn steep liquor	0.894				
	DPA	0.118				

TABLE 4.13 Data Inventory per 1 ton of Sugarcane for SM FF Scenario—cont'd

Stage	Inputs	kg (kWh, Gcal)	Outputs	kg (kWh, Gcal)	Emissions	kg
	Juice	65.237				
	Fresh water	137.466				
	Electricity (kWh)	2.992				
	Steam (Gcal)	0.059				
Enzyme preparation	Glucose	3.423	Enzyme	16.651	NH_3	0.0006
	Corn steep liquor	0.198			CO_2	2.8652
	Ammonia	0.138			SO_2	1.77E−05
	Host nutrients	0.081				
	Sulfur dioxide	0.020				
	Electricity (kWh)	6.433				
	Steam (Gcal)	0.001				
Distillation	Beer from fermentation	368.138	Ethanol	24.203	NH_3	1.85E−07
	Fresh water	28.615	Lignin	56.972	CO_2	23.1071
	Electricity (kWh)	1.824	Filtrate	257.248	SO_2	1.85E−07
	Steam (Gcal)	0.035				
WWT	Filtrate	257.248	Biogas	4.388	NH_3	1.85E−07
	Caustic (as pure)	Sludge	5.749	CO_2	1.2922	
	Electricity (kWh)	24.379			SO_2	1.16E−05
					CH_4	0.0152
					NO_2	0.0002
Cogeneration	Biogas	4.3880	Total electricity (kWh)	165.545	NH_3	1.85E−07
	Sludge	5.7490	Steam (Gcal)	0.263	CO_2	290.2207
	Lignin	56.9723			NO_2	0.2683
	Bagasse	111.9999			CO	0.2683
	50% GHR	109.8885			SO_2	0.0023
	FGD lime	0.0384			Ash	13.2738
	Electricity	5.3268			Other emissions see at Appendix	

the SM FF scenario (sugar mill and feed production) calculated for 1 ton of sugarcane. As can be observed, the main environmental loads are associated with input chemicals such as lime, DAP, ammonia, nutrients, corn steep liquor, and with a consumption of utilities. The main output emissions are NH_3, SO_2, CH_4, NO_2, etc. It should be noticed that inventory data depend on the employed technologies and configuration of the plant.

Additionally, two important points should be highlighted. First, it is recommended to include heat and electricity in the LCI at each stage where they are consumed, even if the whole system is energy self-sufficient. Secondly, the combustion performance of different fuels at the CHP stage must be considered in the analysis. The challenge is that the Aspen model cannot calculate all possible emissions from simultaneous combustion of different fuels, and moreover such emissions are not present in the Ecoinvent database. To overcome such a limitation, an approximate evaluation of emissions performance can be done. The approach is to assume that combustion performance of a particular lignocellulose material can be calculated from known combustion data of other lignocellulose feedstock by adjusting it to certain combustion characteristics including level of combustion, low heat value and stream rate of fuel [16]. In the integrated sugarcane-based biorefinery under consideration, estimations of combustion emissions of bagasse and GHR fuels (Figs. 4.11 and 4.12) are based on the baseline combustion data of Brazil bagasse, which can be found in Appendix. Counterparts for biogas and lignin are methane and wood, whose combustion performances can be found in the Ecoinvent database.

The next stage in the LCA is an impacts analysis. This phase of the LCA quantifies selected environmental impacts based on the LCI flow results. It can be performed by using various databases and software (Chapter 2). In the present example the package SimaPro v8.0 and ReCiPe midpoint (E) methodology are employed as tools for the environmental impact assessment.

The last step is an interpretation. At this stage the results obtained are summarized and analyzed. Sets of recommendations, limitations, and conclusions of the whole LCA analysis are formulated. In addition, it is interesting to verify the obtained results through uncertainty analysis. This can be done using a Monte Carlo tool that is presented in the SimaPro software. In the Monte Carlo approach, every run computer takes a random variable for each value within the uncertainty range and recalculates the results. In this work, a lognormal distribution is assumed with a variance equal to 2, which corresponds to a relatively high level of uncertainty [42]. In the present work an uncertainty analysis comparing different biorefineries and associated RSs was done. This analysis calculates the number of comparison runs (in percent of total runs) in which values of absolute biorefinery impacts are lower than in corresponding RSs. The discussion of the results is presented in the next section.

4.4.7 Economic Analysis

In addition to the LCA, it is recommended to include an economic analysis in the process of biorefinery assessment. The four different biorefinery scenarios mentioned previously were compared in terms of their prospective economic performance value (PEP) calculated according to the value-based approach [7]. The details of the method are explained in Section 4.4.5. The value-based approach is based on evaluation and comparison of two crucial values of biorefinery feedstock: maximum purchasing price (MPP) and minimum selling price (MSP). MPP is the value of the feedstock calculated according to Eq. (4.6) at the final stage of the value-based analysis. It can be recognized as the highest price for feedstock that the biorefinery plant is ready to pay to be economically feasible. MSP is the lowest price of feedstock that a producer of biomass can assume. The PEP of the evaluated biorefinery is estimated as the difference between the MPP and MSP values (Eq. 4.7).

$$\text{PEP} = \text{MPP} - \text{MSP} \tag{4.7}$$

Therefore, the PEP value is an indicator of economic feasibility of the evaluated project—a higher PEP corresponds to higher economic prospects and therefore higher economic feasibility.

The estimations of capital investment, operation and fixed costs for various scenarios were made according to available literature data [21,31]. An exponential scaling expression was employed to calculate the costs of equipment for different scales (Eq. 4.8).

$$\text{NewCost} = \text{OriginalCost}\left(\frac{\text{NewSize}}{\text{OriginalSize}}\right)^n \tag{4.8}$$

The scaling factor n is 0.6 in the present work. Other parameters of the economic model as well as prices of products and feedstock (MSP) are presented in Table 4.12. It should be noticed that all calculations were made for the reference year 2010; however, the start year of the plant is assumed to be 2020. For work simplification, in this example economic calculations are based on the assumption that real prices have not changed from 2010 to 2020.

4.4.8 Results

The capabilities of the LCA and techno-economic assessment methodology presented here are tested through their application for four integrated sugarcane-based biorefinery scenarios recently presented in the paper [22]. All scenarios have the same capacity for the amount of feedstock (3.9 M ton) and work hours (7200 per year). However, they are different by level of integration and therefore by output products. The details of the scenarios are explained in Section 4.3.2.

The LCA of the starch-based biorefinery is presented in a brief summary of the work published in the paper [16] (Section 4.4.8.3).

4.4.8.1 Techno-Economic Analysis

The techno-economic analysis is performed according to the value-based approach explained in Sections 4.4.5 and 4.4.7. Table 4.14 shows the main results obtained by process simulation and economic analysis for each scenario. As can be seen, all scenarios show a positive PEP value and therefore a positive economic efficiency. However, the economic performance differs between various scenarios.

The results show that economic efficiency depends on the level of integration and therefore on the range and amount of issued products. Analyzing the results, it can be concluded that the best PEP value (16.27 $/ton SC) is provided by the ED OF scenario producing only ethanol and electricity for the market. The ED OF scenario represents the partial 1G-2G sugar mill

TABLE 4.14 Data and Results for Economic Analysis

Scenarios	ED FF	ED OF	SM FF	SM OF
Ethanol (ton/year)	320,070	359,063	94,392	133,235
C5 Syrup (ton/year)	106,400	0	106,229	0
Sugar (ton/year)	0	0	372,468	372,468
C molasses (ton/year)	0	0	402,239	402,239
Electricity (GWh/year)	53	97	192	394
TCI (M$/y)	794	798	750	760
OC (M$/y)	14	16	12	15
FOC (M$/y)	15	15	14	15
MPP ($/ton)	35.09	39.52	30.37	37.95
MSP ($/ton)	23.25	23.25	23.25	23.25
PEP ($/ton)	11.84	16.27	7.12	14.70

TCI, total capital investment; *OC*, operation cost; *FOC*, fixed operation cost; *MPP*, maximum purchasing price for feedstock (sugarcane); *MSP*, minimum selling prices for feedstock (sugarcane); *PEP*, prospective economic performance.

integration (see Section 4.3.2) and produces the largest amount of ethanol among all biorefinery scenarios (Table 4.14). In contrast, the scenario SM FF is a fully integrated biorefinery configuration that issues five various products; however, it represents the lowest PEP value (7.12 $/ton SC). Such a trend is largely explained by the product market prices that make ethanol production more profitable than other goods (Table 4.12). For example, scenarios with an integrated sugar mill (SM-) produce the same amount of sugar and C molasses, but the SM FF scenario issues C5 syrup at the expense of ethanol production, which leads to significant changes in economic performance: PEPs are 7.12 and 14.70 $/ton SC for SM FF and SM OF scenarios, respectively.

Despite the fact that the extent of ethanol-sugar mill integration is an important factor for assessment of economic performance of a biorefinery, it is not an obvious trend that high level of integration leads to worse economic efficiency. For example, the comparison of ED FF (partial integration) and SM OF (full integration) scenarios shows that a full integration configuration producing additional products like sugar and C molasses yields better PEP (11.84 and 14.7 $/ton SC for ED FF and SM OF scenarios, respectively).

The allocation factors of intermediate streams for selected biorefinery scenarios are presented in Table 4.15. As can be observed in the harvesting part, the main economic contribution corresponds to the sugarcane feedstock (96%) with near 4% attributed to GHR. At the milling and extraction stage the greatest economic importance is represented by the juice, in the case of ethanol distillery (ED-) scenarios, and sugar, in the case of sugar production (SM-) scenarios. At the distillation stage the ethanol shares the greatest part of the total economic value outputs. At the cogeneration stage, the contribution of electricity increases simultaneously with the amount of electricity production (see Table 4.8).

TABLE 4.15 Allocation Factors for Full Integration Biorefinery Process

		ED FF	ED OF	SM FF	SM OF
Harvesting	Sugarcane	0.9499	0.9548	0.9456	0.9536
	GHR	0.0501	0.0452	0.0544	0.0464
Milling, juice extraction	Bagasse CHP	0.0896	0.0716	0.0193	0.0199
	Bagasse 2G	0.0216	0.0985	0.0156	0.0666
	Juice	0.8889	0.8298	0.0849	0.0997
	Sugar	0.0000	0.0000	0.7842	0.7249
	Molasses	0.0000	0.0000	0.0961	0.0888
Hydrolysis and fermentation	Fermentation beer	0.9382	1.0000	0.7416	1.0000
	Syrup	0.0618	0.0000	0.2584	0.0000
Distillation and ethanol recovery	Ethanol	0.8834	0.8931	0.7456	0.8035
	Lignin	0.0207	0.0134	0.0105	0.0067
	Filtrate	0.0958	0.0934	0.2439	0.1899
WWT	Biogas	0.3348	0.4862	0.3190	0.5307
	Sludge	0.6652	0.5138	0.6810	0.4693
Cogeneration	Electricity	0.5822	0.6656	0.7205	0.8245
	Steam	0.4178	0.3344	0.2795	0.1755

4.4.8.2 *Life Cycle Analysis*

The evaluation of the environmental performance of the scenarios under consideration was made according to LCA methodology (see Section 4.4). Figs. 4.17 and 4.18 show the results of LCA in several impact categories. The results for Climate change impact is represented in kg CO_2 eq. (GHG emissions) per 1 ton of sugarcane (Fig. 4.17A). As can be observed, the environmental impacts provided by integrated biorefineries—105, 110, 77, and 90 kg CO_2 eq./ton SC for the ED FF, ED OF, SM FF, SM OF scenarios, respectively—are lower than GHG emissions obtained from associated RSs: 350, 385, 176, and 220 kg CO_2 eq./ton SC for RSs of ED FF, ED OF, SM FF, SM OF scenarios, respectively. Thus it can be concluded that the 1G-2G integration biorefinery in all configurations leads to significant mitigation of GHG in comparison with existing substituted systems.

Analyzing the results of absolute performances of all configurations, it can be seen that GHG emissions are increasing simultaneously with a decrease in the extent of sugar mill–ethanol production integration. In particular, the greatest absolute impact is provided by the ED OF scenario (110 kg CO_2 eq./ton SC) and the lowest by SM FF (77 kg CO_2 eq./ton SC). It can be explained by influence of ethanol production that makes the greatest contribution for all scenarios and has bigger production in ED- scenarios.

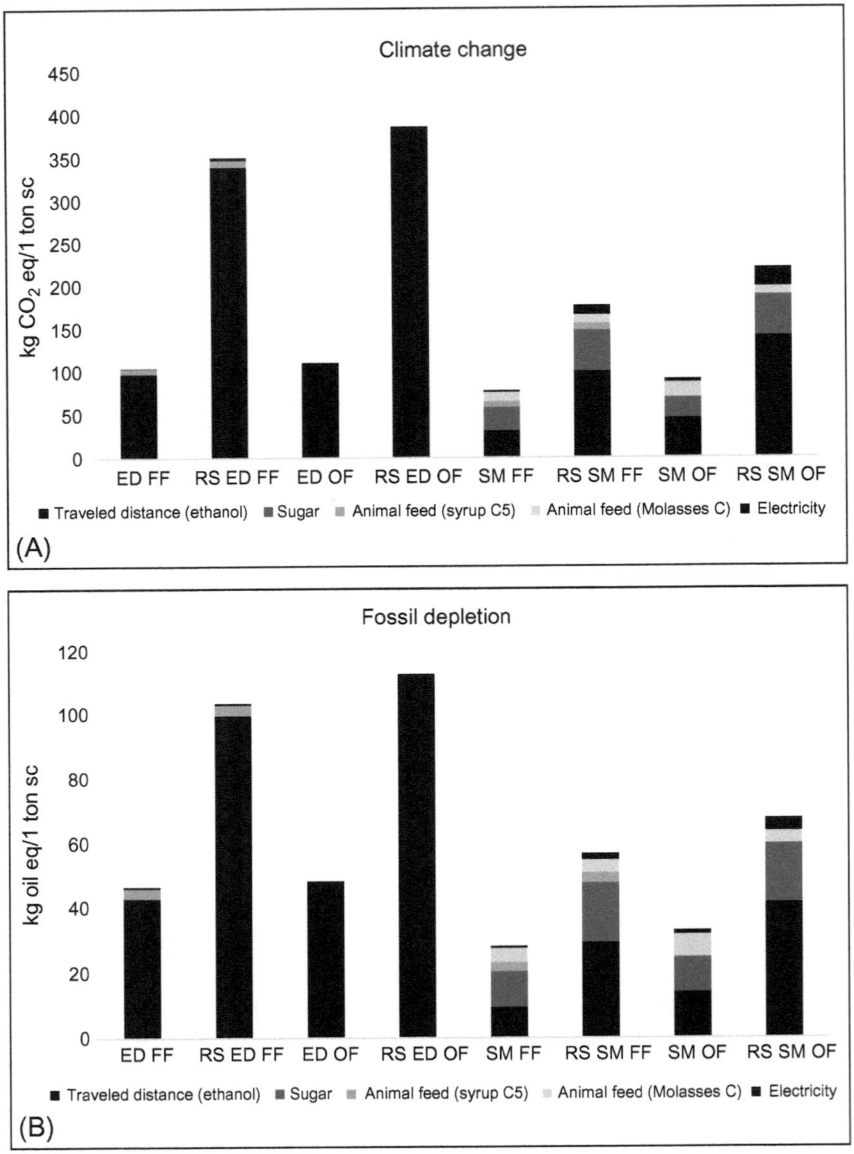

FIG. 4.17 Results of LCA for climate change (A) and fossil depletion (B) impacts.

The results of relative performances show at that extent impact increase or decrease when biorefinery substitutes the RS. Comparing such results, it can be noticed that SM FF and SM OF lead to less impact reduction (56% and 59% respectively) than ED FF and ED OF scenarios (70% and 71% respectively). Such behavior of scenarios is explained by higher proximity between the coproducts in the full integrated (SM-) scenarios and their counterparts in the RSs.

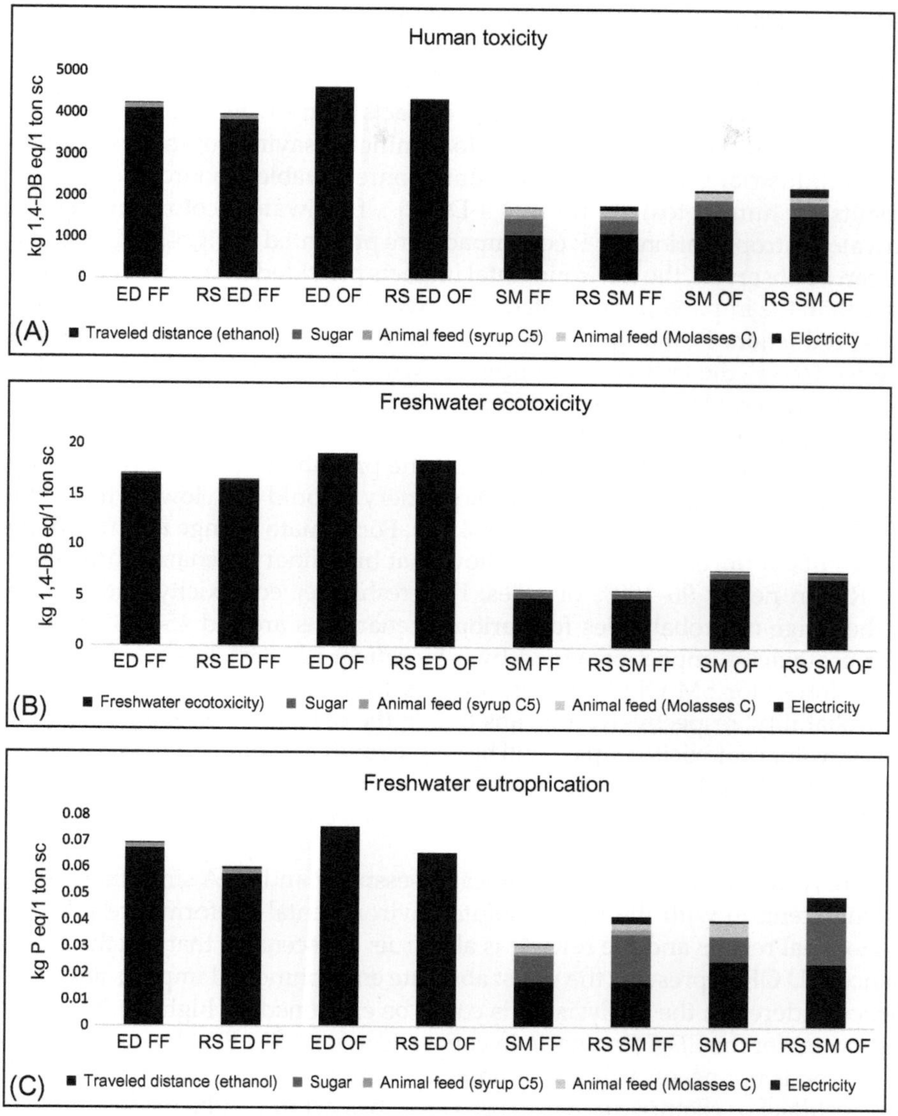

FIG. 4.18 Results of LCA for human toxicity (A), freshwater ecotoxicity (B), and freshwater eutrophication (C) impacts.

Fig. 4.17B shows the results for a fossil depletion environmental impact. The numerical results of absolute performances (in kg oil eq. per 1 ton SC) for considered biorefineries and associated RSs are following: ED OF: 48 and 112; ED FF: 46 and 103; SM OF: 33 and 67; SM FF: 28 and 57. The results for relative impact reduction are 57, 55, 51, and 51% for ED OF, ED FF, SM OF, and SM FF scenarios, respectively. The main conclusions are similar to those for climate change: all selected biorefineries provide lower impacts than those from associated RSs; the greatest contribution to total impact is associated with traveled distance service (ethanol production); the absolute impacts are higher for partially integrated ED FF and ED OF scenarios than for full integrated SM FF and SM OF configurations; and the latter scenarios provide lower relative reduction of impacts than ED FF and ED OF. These results prove that integrated biorefineries can lead to significant savings of fossil resources in comparison with industrial processes that consume nonrenewable resources.

The results for human toxicity (in kg 1,4-DB eq.), freshwater ecotoxicity (kg 1,4-DB eq.), and freshwater eutrophication (kg P eq.) impacts are presented in Fig. 4.18A, B, and C, respectively. As can be observed, the environmental impacts provided by a biorefinery are similar or even worse than the impacts of associated RSs, except the SM- scenarios that still show a positive impact reduction. Primarily this is explained by the fact that a biorefinery is based on sugarcane feedstock, the cultivation process of which requires high amounts of fertilizers, pesticides, and water. However, such effects of sugarcane cultivation apply less to the SM-scenarios where the agricultural phase of sugarcane is also included in the RS.

The results of uncertainty analysis represent the probability (in percent) that a particular environmental impact provided by a biorefinery would be lower than the impact corresponding to the associated RS (Section 4.4.6). For climate change and fuel depletion impacts, the results of uncertainty analysis show that biorefinery scenarios provide lower impact than RSs in nearly 96–100% of cases. For freshwater ecotoxicity and human toxicity impacts, the range of probabilities for various scenarios is around 45–55%. In contrast, the freshwater ecotoxicity impact provided by a biorefinery is 83% lower than the RS for SM FF and 77% lower for SM OF; however, impacts from ED FF and ED OF show only 40% and 38% probability, respectively. It means that in the case of ED- scenarios, it is more likely that the freshwater ecotoxicity impact will be worse for the biorefinery than for the RS. In general, it can be concluded that the results of the uncertainty analysis are consistent with the results of the LCA and, despite the high level of data uncertainty, the main findings are highly reliable.

Finally, analyzing the results of economical assessment and LCA simultaneously, it can be seen that the scenario with the best absolute environmental performance (SM FF) has the worst economical results and the reverse is also true: the scenario that has the best economic performance (ED OF) represents the worst absolute environmental impacts among all impact categories considered in the analysis. This could be explained by higher ethanol production (less integration) for the ED OF scenario in comparison with SM FF. However, the largest ethanol production is not necessarily a guarantee for better economic and worse absolute environmental results. For example, scenario SM OF shows better results than the ED FF scenario for economic and climate change indicators: PEP (14.7 for SM OF vs. 11.84 for ED FF, \$/ton SC) and absolute environmental impact (90 for SM OF vs. 104 for ED FF, kg CO_2 eq./ton SC).

In general, the results of the LCA prove that an integrated sugarcane- or starch-based biorefinery in various configurations can achieve good environmental benefits for climate change and fossil depletion environmental impact categories. However, the integrated biorefinery shows lower benefits for human toxicity, freshwater ecotoxicity and freshwater eutrophication. As was explained previously, this is not due to integrated biorefineries themselves but to the emissions associated with cultivation of sugarcane that are less present in the RS. Finally, it should be noted that such conclusions can be different for other conditions and contexts.

4.4.8.3 Results of Starch-Based Biorefinery

This section briefly presents the summary of the paper [16] which demonstrates the application of LCA to a starch-based biorefinery. The case study is based on a bioethanol plant consuming wheat as a feedstock and located in Switzerland. The capacity of the plant is 40,000 ton ethanol/year that corresponds to 134,000 ton/year of wheat consumption. The cultivation area required for such a scale is near 20,900 ha. In addition to fuel ethanol as a main product, the biorefinery system produces two coproducts: 48,600 ton/year of DDGS and 81,824 ton/year of wheat straw. The environmental performance of the biorefinery is evaluated according to net GHG emissions of bioethanol in a fuel blend consumed in car operation, and DDGS and wheat straw utilized in different services. The results of the biorefinery LCA were compared with the results of RSs that include gasoline car operation, previous land use and different services that could be substituted for DDGS and wheat straw. To simplify the calculations, the emissions associated with DDGS, wheat straw and LUC were transferred from the RS to the system under study with opposite sign, as is presented in Fig. 4.19. The functional unit assumed in the LCA study is 1 km of car operation.

In the analyzed paper [16], the sensitivity of bioethanol environmental performance to various methodological choices was tested on several scenarios. In general, the scenarios are varied by LUCs, substituted services for DDGS and wheat straw, allocation methods between coproducts, types of fuel blends consumed by car, and types of car performance. In total there are nine options of LUC (LUC1–LUC9), four allocations (A1–A4), four substitutions (S1–S4), three fuel blends (E5, E10, E85), and three fuel performances (1,2,3) which were tested in this work [16]. A summary of these options with short explanations is presented in Table 4.16.

Table 4.17 shows a summary of the results from the LCA obtained for selected scenarios [16]. The results are quantified as GHG emissions of fuel ethanol consumed in car operation, taking into account GHG emissions attributed to coproducts (Fig. 4.19), in kg CO_2 eq./km. The results from the RS are the GHG emissions of gasoline car operation, that is 0.237 kg CO_2 eq./km. The relative results are expressed as an index parameter in Table 4.17. The index is the ratio of biorefinery GHG to RSs GHG emissions, expressed in percent. It means if the index is less than 100%, the biorefinery scenario leads to reduction of GHG emissions. The default scenario consists of energy allocation (A1), set aside to cultivated land (LUC1), and fuel blend E5 with data of fuel performance obtained by actual vehicle tests (1 type).

As can be observed, the results of the LCA are very sensitive to the use of various methodological options. The first eight scenarios test different allocation methods and substitutes. The absolute results vary from 0.151 to −0.016 kg CO_2 eq./km. Such results show significant environmental reduction in comparison with the reference case; indexes vary from 63.8% to −6.7%. The best scenario which leads to a consumption of GHG from the atmosphere (negative value −0.016 kg CO_2 eq./km) corresponds to the S4 substitution option. This scenario employs straw and DDGS as fuels for generation of heat & power (CHP) instead of fossil fuels. It is worth noting that the worst scenario is related to substitution (S1) with DDGS and wheat straw used as animal feed.

Scenarios 9–18 investigated the effect of LUC on environmental performance. The results prove the importance of LUC to the GHG emissions balance. Absolute results vary from 0.047 kg CO_2 eq./km (grassland severely degraded to cultivated land, LUC5) to 0.249 kg CO_2 eq./km (forested land to cultivated land, LUC6). The indexes are changed from

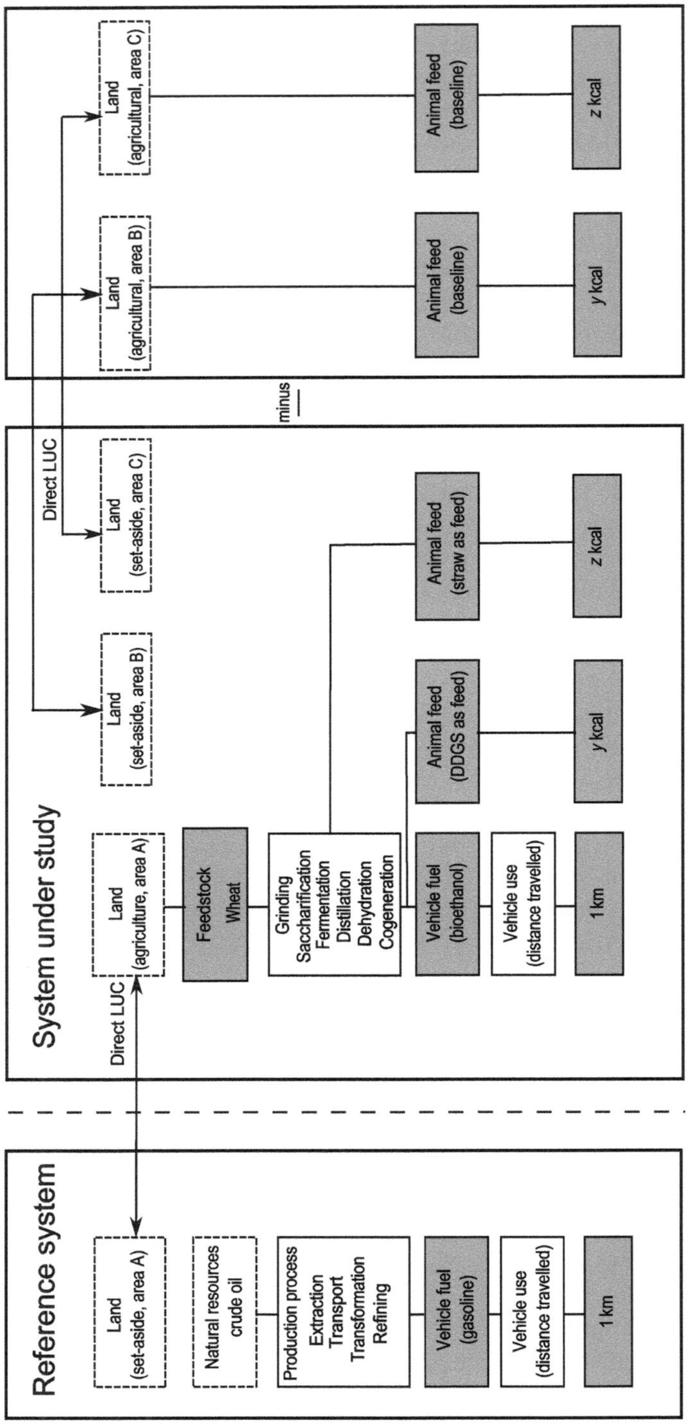

FIG. 4.19 System definition and boundaries (from reference system to system studied) for wheat-based biorefinery.

TABLE 4.16 Methodology Options Investigated in the Work

Ref.	Method	Key	Agricultural Stage		Industrial Stage	
			Wheat Grains	Wheat Straw	Bioethanol	Wheat DDGS
Allocation/substitution methods						
A1	Allocation	Energy content	15.1 MJth/kg	17.2 MJth/kg	26.8 MJth/kg	16.4 MJth/kg
A2	Allocation	Economic value	750 SFr/ton	100 SFr/ton	1139 SFr/ton	250 Sfr/ton
A3	Allocation	Carbon content	0.353 kg Ckg	0.365 kg C/kg	0.520 kg C/kg	0.321 kg C/kg
A4	Allocation	Dry mass	85 wt% d.m.	85 wt% d.m.	99.7 wt% d.m.	90 wt% d.m.
S1	Substitution		–	Animal feed	–	Animal feed
S2	Substitution		–	Fuel	–	Animal feed
S3	Substitution		–	Animal feed	–	Fuel
S4	Substitution		–	Fuel	–	Fuel

Ref.	From	To	Annual Soil Carbon Stock Change (tons of C/ha year)
Land-use change options and corresponding annual soil carbon stock changes			
LUC1	Set aside	Long-term cultivated, reduced tillage, medium inputs	−0.22
LUC2	Grassland, nondegraded		−1.07
LUC3	Grassland, improved		−1.74
LUC4	Grassland, moderately degraded		−0.84
LUC5	Grassland, severely degraded		0.35
LUC6	Native ecosystem (forested land)		−1.07
LUC7	Long-term cultivated, no tillage, medium inputs		−0.24
LUC8	Long-term cultivated, reduced tillage, medium inputs		–
LUC9	Long-term cultivated, full tillage, medium inputs		0.3

Ref.	Fuel	Basis	Variation of Fuel Consumption w.r.t. Gasoline (%)			Ethanol Component (MJth/km)
			(L/km)	(kg/km)	(MJth/km)	
Fuel blends and vehicle/fuel performance options						
E5-1	Ethanol, as E5	Actual tests	−1	−0.7	−2.7	1.413
E10-1	Ethanol, as E10	Actual tests	−4.3	−3.9	−7.5	1.174
E85-1	Ethanol, as E85	Actual tests	34.9	41.8	−2.5	2.485
E-2	Ethanol	Volume basis	0	–	–	1.703
E-3	Ethanol	Energy basis	–	–	0	2.564

TABLE 4.17 Results of Assessments of GHG Emissions for Selected Scenarios

Scenarios	Allocation	LUC	Fuel	WtT (kg CO_2 eq./MJth)		TtW MJth/km	TtW kg CO_2 eq./km		WtW (kg CO_2 eq./km)	Index (–)
	REF	REF	Gasoline	0.018	×	2564	+190	=	0.237	100
1	A1	LUC1	E5-1	0.047	×	1.413		=	0.066	27.9
2	A2	LUC1	E5-1	0.106	×	1.413		=	0.15	63.4
3	A3	LUC1	E5-1	0.048	×	1.413		=	0.068	28.5
4	A4	LUC1	E5-1	0.041	×	1.413		=	0.057	24.2
5	S1	LUC1	E5-1	0.107	×	1.413		=	0.151	63.8
6	S2	LUC1	E5-1	0.012	×	1.413		=	0.017	7
7	S3	LUC1	E5-1	0.084	×	1.413		=	0.119	50.1
8	S4	LUC1	E5-1	-0.011	×	1.413		=	-0.016	−6.7
9	A1	LUC1	E5-1	0.047	×	1.413		=	0.066	27.9
10	A1	LUC2	E5-1	0.068	×	1.413		=	0.095	40.2
11	A1	LUC3	E5-1	0.084	×	1.413		=	0.118	49.9
12	A1	LUC4	E5-1	0.062	×	1.413		=	0.087	36.9
13	A1	LUC5	E5-1	0.033	×	1.413		=	0.047	19.7
14	A1	LUC6	E5-1	0.177	×	1.413		=	0.249	104.9
15	A1	LUC7	E5-1	0.047	×	1.413		=	0.067	28.2
16	A1	LUC8	E5-1	0.042	×	1.413		=	0.059	24.7
17	A1	LUC9	E5-1	0.034	×	1.413		=	0.048	20.4
18	A1	LUC1	E5-1	0.047	×	1.413		=	0.066	27.9
19	A1	LUC1	E10-1	0.047	×	1.174		=	0.055	23.2
20	A1	LUC1	E85-1	0.047	×	2.485		=	0.116	49.1
21	A1	LUC1	E-2	0.047	×	1.703		=	0.08	33.7
22	A1	LUC1	E-3	0.047	×	2.564		=	0.12	50.7

19.7% to 104.9%. This means that the starch-based biorefinery would be less favorable than the conventional system in terms of GHG emissions in the unexpected case when production of wheat causes deforestation (LUC6).

Finally, the influence of fuel blends and car performance were investigated through the assessment of scenarios 19–22. All results show serious impact reductions and vary from 0.055 kg CO_2 eq./km in the case of the E10 blend and actual test data to 0.120 kg CO_2 eq./km in the case of E3—energy basis. This means that E10 is the most favorable type of fuel blend with respect to mitigation of GHGs according to the car test results. In general, the analyzed work [16] clearly illustrates that GHG emissions of the starch-based biorefinery strongly depend on the methodological choices made in the LCA analysis. Thus the benefits of a biorefinery with respect to gasoline could be both positive and negative depending on various scenarios. It is worth noting that the LUC effect provides the strongest influence on GHG emissions of such types of biorefineries.

4.5 CONCLUSIONS AND PERSPECTIVES

This chapter presents and analyzes different achievements and key issues in the LCA methodology that can be applied to evaluate environmental performance of sugar- or starch-based integrated biorefineries. Moreover, the chapter considers a comprehensive analysis of process design of a 1G-2G integrated sugarcane-based biorefinery with the possibility of issuing several products. In addition, an effective value-based approach for techno-economical assessment of such a type of biorefinery is presented.

Four distinct sugarcane-based biorefinery configurations were evaluated according to the proposed economic and LCA methodology. All scenarios selected show high-impact reductions for climate change and fossil depletion in comparison with conventional RSs. At the same time, scenarios that primarily focus on ethanol production (ED-) show less efficiency in human toxicity, freshwater ecotoxicity and freshwater eutrophication categories. All scenarios provide economic efficiency in the selected conditions. The best economic performance is provided by the ED OF scenario, which produces the largest amount of ethanol. The best absolute environmental performance is associated with the fully integrated scenario (SM FF) which issues five products. Moreover, it can be concluded that integration of residues based on 2G to 1G provides benefits regarding environmental impacts in comparison with 1G conventional industry; however, it proves to be more expensive. Finally, analysis of the study devoted to the LCA of a wheat-based biorefinery shows a high sensitivity of net GHG emissions to several methodological choices: allocation methods, RSs, type of fuel blend and type of LUC. Depending on the chosen assumptions, the GHG emissions associated with the starch-based biorefinery could be even higher than emissions provided by a fossil fuel system.

APPENDIX COMBUSTION PERFORMANCE OF 1 MJ OF LIGNOCELLULOSIC FEEDSTOCK (BAGASSE)

Emissions	Per 1 MJ of Feedstock	Units
Heat, waste	0.987	MJ
Carbon dioxide, biogenic	0.111877358	kg
Nitrogen oxides	0.000111571	kg
Particulates, <2.5 μm	5.10497E−05	kg

Continued

Emissions	Per 1 MJ of Feedstock	Units
Carbon monoxide, biogenic	7.95876E−06	kg
Methane, biogenic	4.93443E−07	kg
NMVOC, unspecified origin	6.93549E−07	kg
Sulfur dioxide	3.15696E−06	kg
Dinitrogen monoxide	2.91607E−06	kg
Acetaldehyde	7.73393E−08	kg
Hydrocarbons, aliphatic, alkanes, unspecified	1.03464E−06	kg
Hydrocarbons, aliphatic, unsaturated	3.52459E−06	kg
Arsenic	1.26786E−09	kg
Benzo(a)pyrene	5.68483E−10	kg
Benzene	1.03464E−06	kg
Bromine	7.60714E−08	kg
Calcium	7.41696E−06	kg
Cadmium	8.875E−10	kg
Chlorine	2.28214E−07	kg
Chromium	5.02071E−09	kg
Chromium VI	5.07143E−11	kg
Copper	2.78929E−08	kg
Dioxins, measured as 2,3,7,8-tetrachlorodibenzo-*p*-dioxin	3.52459E−14	kg
Benzene, ethyl-	3.4109E−08	kg
Fluorine	6.33929E−08	kg
Formaldehyde	1.47805E−07	kg
Benzene, hexachloro-	8.18615E−15	kg
Mercury	3.80357E−10	kg
Potassium	2.96679E−05	kg
Magnesium	4.57696E−07	kg
Manganese	2.16804E−07	kg
Sodium	1.64821E−06	kg
Ammonia	2.20607E−06	kg
Nickel	7.60714E−09	kg
Phosphorus	3.80357E−07	kg
PAH	1.25066E−08	kg

Emissions	Per 1 MJ of Feedstock	Units
Lead	3.15696E−08	kg
Phenol, pentachloro-	9.20942E−12	kg
Toluene	3.4109E−07	kg
m-Xylene	1.36436E−07	kg
Zinc	3.80357E−07	kg

NMVOC, nonmethane volatile organic compounds; *PAH*, polycyclic aromatic hydrocarbons.

References

[1] International energy outlook. Washington, DC: US Energy Information Administration; 2014.

[2] International energy outlook. Washington, DC: US Energy Information Administration; 2009.

[3] Karuppiah R, Peschel A, Grossmann IE, Martín M, Martinson W, Zullo L. Energy optimization for the design of corn-based ethanol plants. AIChE J 2008;54:1499–525.

[4] Seabra JEA, Macedo IC. Comparative analysis for power generation and ethanol production from sugarcane residue biomass in Brazil. Energy Policy 2011;39:421–8.

[5] Alvira P, Tomás-Pejó E, Ballesteros M, Negro MJ. Pretreatment technologies for an efficient bioethanol production process based on enzymatic hydrolysis: a review. Bioresour Technol 2010;101:4851–61.

[6] Čuček L, Martín M, Grossmann IE, Kravanja Z. Energy, water and process technologies integration for the simultaneous production of ethanol and food from the entire corn plant. Comput Chem Eng 2011;35:1547–57.

[7] Gnansounou E, Dauriat A. Techno-economic analysis of lignocellulosic ethanol: a review. Bioresour Technol 2010;101:4980–91.

[8] Furlan FF, Costa CBB, Fonseca GDC, Soares RDP, Secchi AR, Cruz AJGD, et al. Assessing the production of first and second generation bioethanol from sugarcane through the integration of global optimization and process detailed modeling. Comput Chem Eng 2012;43:1–9.

[9] Dias MOS, Ensinas AV, Nebra SA, Maciel R, Rossell CEV, Maciel MRW. Production of bioethanol and other bio-based materials from sugarcane bagasse: integration to conventional bioethanol production process. Chem Eng Res Des 2009;87:1206–16.

[10] Dias MO, Junqueira TL, Jesus CD, Rossell CE, Maciel Filho R, Bonomi A. Improving second generation ethanol production through optimization of first generation production process from sugarcane. Energy 2012;43:246–52.

[11] Dias MO, Junqueira TL, Cavalett O, Cunha MP, Jesus CD, Mantelatto PE, et al. Cogeneration in integrated first and second generation ethanol from sugarcane. Chem Eng Res Des 2013;91:1411–7.

[12] Macrelli S, Mogensen J, Zacchi G. Techno-economic evaluation of 2nd generation bioethanol production from sugar cane bagasse and leaves integrated with the sugar-based ethanol process. Biotechnol Biofuels 2012;5:1–18.

[13] Schaidle JA, Moline CJ, Savage PE. Biorefinery sustainability assessment. Environ Prog Sust Energ 2011;30:743–53.

[14] Lopes Silva DA, Delai I, Delgado Montes ML, Roberto Ometto A. Life cycle assessment of the sugarcane bagasse electricity generation in Brazil. Renew Sust Energ Rev 2014;32:532–47.

[15] Cherubini F, Strømman AH. Life cycle assessment of bioenergy systems: state of the art and future challenges. Bioresour Technol 2010;102:437–51.

[16] Gnansounou E, Dauriat A, Villegas J, Panichelli L. Life cycle assessment of biofuels: energy and greenhouse gas balances. Bioresour Technol 2009;100:4919–30.

[17] Fleming JS, Habibi S, MacLean HL. Investigating the sustainability of lignocellulose-derived fuels for light-duty vehicles. Transp Res Part D: Transp Environ 2006;11:146–59.

[18] González-García S, Gasol CM, Gabarrell X, Rieradevall J, Moreira M, Feijoo G. Environmental aspects of ethanol-based fuels from *Brassica carinata*: a case study of second generation ethanol. Renew Sust Energ Rev 2009;13:2613–20.

[19] Fu GZ, Chan AW, Minns DE. Life cycle assessment of bio-ethanol derived from cellulose. Int J Life Cycle Assess 2003;8:137–41.
[20] Papong S, Malakul P. Life-cycle energy and environmental analysis of bioethanol production from cassava in Thailand. Bioresour Technol 2010;101:112–8.
[21] Dias MOS, Junqueira TL, Cavalett O, Cunha MP, Jesus CDF, Rossell CEV, et al. Integrated versus stand-alone second generation ethanol production from sugarcane bagasse and trash. Bioresour Technol 2011;103:152–61.
[22] Gnansounou E, Vaskan P, Ruiz Pachón E. Comparative techno-economic assessment and LCA of selected integrated sugarcane-based biorefineries. Bioresour Technol 2015;196:364–75.
[23] Aden A, Ruth M, Ibsen K, Jechura J, Neeves K, Sheehan J, et al. Lignocellulosic biomass to ethanol process design and economics utilizing co-current dilute acid prehydrolysis and enzymatic hydrolysis for corn stover. National Renewable Energy Laboratory report, Seatlle/Golden, CO: National Renewable Energy Laboratory; 2002.
[24] Meyer J, Rein P, Turner P, Mathias K. Good management practices for the cane sugar industry. Bartnes; 2013.
[25] Gnansounou E, Dauriat A, Wyman CE. Refining sweet sorghum to ethanol and sugar: economic trade-offs in the context of North China. Bioresour Technol 2005;96:985–1002.
[26] Chen JC, Chou CC. Cane sugar handbook: a manual for cane sugar manufacturers and their chemists. 12th ed. New York, NY: John Wiley and Sons; 1993.
[27] Dias MO, da Cunha MP, Maciel Filho R, Bonomi A, Jesus CD, Rossell CE. Simulation of integrated first and second generation bioethanol production from sugarcane: comparison between different biomass pretreatment methods. J Ind Microbial Biotechnol 2011;38:955–66.
[28] Hugot E. Handbook of sugar engineering. Amsterdam: Elsevier Science Publishers; 1986.
[29] Rein P, Attard RG. Cane sugar engineering. Berlin: Bartens; 2007.
[30] Olbrich H. The molasses. Berlin: Biotechnologie-Kempe GmbH; 2006.
[31] Humbird D, Davis R, Tao L, Kinchin C, Hsu D, Aden A, et al. Process design and economics for biochemical conversion of lignocellulosic biomass to ethanol. National Renewable Energy Laboratory Technical Report, Golden, CO: NREL; 2011.
[32] Larsen J, Østergaard Petersen M, Thirup L, Wen Li H, Krogh Iversen F. The IBUS process—lignocellulosic bioethanol close to a commercial reality. Chem Eng Technol 2008;31:765–72.
[33] Larsen J, Haven MØ, Thirup L. Inbicon makes lignocellulosic ethanol a commercial reality. Biomass Bioenergy 2012;46:36–45.
[34] McAloon A, Taylor F, Yee W, Ibsen K, Wooley R. Determining the cost of producing ethanol from corn starch and lignocellulosic feedstocks. National Renewable Energy Laboratory Report, Seatlle/Golden CO: National Renewable Energy Laboratory; 2000.
[35] Kami Delivand M, Gnansounou E. Life cycle environmental impacts of a prospective palm-based biorefinery in Pará State-Brazil. Bioresour Technol 2013;150:438–46.
[36] Eggleston HS, Buendia L, Miwa K, Ngara T, Tanabe K. Agriculture, forestry and other land uses (AFOLU). 2006 IPCC/guidelines for national greenhouse gas inventories, 4. Hayama: IPCC/IGES; 2006.
[37] IPCC. Tool for estimation of changes in soil carbon stocks associated with management changes in cropland and grazing lands based on IPCC default data. Fort Collins, CO: Natural Resource Ecology Laboratory, Colorado State University; 2003.
[38] Dias Paes LA, Marin FR. Carbon storage in sugarcane fields of Brazilian South-Central region. Technical report, São Paulo: Centro de tecnologia canavieira; 2011.
[39] Carvalho JLN, Otto R, Franco HCJ, Trivelin PCO. Input of sugarcane post-harvest residues into the soil. Sci Agric 2013;70:336–44.
[40] CEPEA (Centre for Advanced Studies on Applied Economics). 2014. http://www.cepea.esalq.usp.br.
[41] USDA, Economic Research Service. Sugar and sweeteners yearbook tables, http://www.ers.usda.gov/data-products/sugar-and-sweeteners-yearbook-tables.aspx; 2014.
[42] Goedkoop M, Oele M, Leijting J, Ponsioen T, Meijer E. Introduction to LCA with SimaPro. The Netherlands: PRé Consultants; 2013.

CHAPTER

5

Life Cycle Assessment of Vetiver-Based Biorefinery With Production of Bioethanol and Furfural

J.K. Raman, E. Gnansounou†*

*Ecole Polytechnique Fédérale de Lausanne (EPFL), Lausanne, Switzerland

†Bioenergy and Energy Planning Research Group (BPE), Ecole Polytechnique Fédérale de Lausanne (EPFL), Lausanne, Switzerland

5.1 INTRODUCTION

Fossil energy has become a major part of human life for transport, shelter and life style. Though human dependence on fossil fuel has increased gradually over the years from its invention, in recent years it has increased drastically due to the human life-style improvements in developed and developing economies [1]. Contrary to the usage, the fossil fuel reserves are decreasing day by day, demanding several alternative energy sources to fulfill current and future energy needs of the growing population, economic and technology development. On the other hand, fossil-fuel burning is blamed for the greenhouse gas emission in the world leading to climate change, environmental pollution, and health problems [2]. The recent IPCC report on climate change warns that it is already late to tackle climate change and the use of fossil fuels should be phased out by 2100 [3]. Therefore it is necessary to address this energy crisis and the climate change issues through development of and use of alternative energy technologies.

Renewable energy such as solar, hydro, wind, etc. has been considered as alternative or additional choices for fossil energy sources. In recent years there has been a tremendous improvement in renewable energy technologies from research and development to commercialization, with new investment of atleast USD 249.4 billion in 2013, and renewable energy capacity of 1560 GW [4]. At present hydro power is the leading renewable energy source, followed by wind, solar, and biofuels. Though other forms of energy could be used in electricity generation, only biofuels can be used in the transport sector due to its properties that

Life-Cycle Assessment of Biorefineries
http://dx.doi.org/10.1016/B978-0-444-63585-3.00005-X

are similar to fossil gasoline and diesel. The total production of biofuels in 2013 was 113.5 billion liters and this is expected to increase in the future due to favorable biofuel policy changes in several countries. However, most of the biofuels (bioethanol and biodiesel) on the market now are produced from edible starch-, sugar- and oil-based feedstock that are blamed for food price hikes [5]. Therefore, nonfood-based biofuel feedstocks have recently been explored.

Nonfood-based biofuels include agricultural crop residues such as wheat straw, rice straw, corn cobs, corn stover, sugarcane leaves, and bagasse for bioethanol production, and jatropha, castor, cooking waste, and algae oil for biodiesel production. These feedstocks are waste products (wheat straw, rice straw, corn cobs, corn stover, and sugar cane leaves) often burnt in the field or unutilized, or they could be cultivated only for fuel production (jatropha, castor, and algae oil). Biofuel production from these feedstocks does not affect the food supply and in addition minimizes waste disposal problems and environmental pollution [6–11].

Agricultural residues, often called lignocellulosic, are composed of cellulose, hemicellulose, and lignin. Both cellulose and hemicellulose could be converted to their respective sugars using chemicals and enzymatic treatments and further could be fermented to produce bioethanol using several microorganisms [12–14]. However, unlike ethanol production from food-based feedstock, the process of converting cellulose and hemicellulose is complex, has high energy demands and lower efficiency, which makes the processes and product expensive. Nevertheless, if valuable added products are produced from hemicellulose along with ethanol from cellulose, the process cost could be balanced by the higher selling price of the value-added products [15,16]. Several value-added products such as sugar syrup for animal feed, furfural, acetic acid, levulinic acid, and lignin derivative chemicals could be produced from the hemicellulose and lignin fraction of biomass residues [17].

Furfural is used in several applications including fungicides, nematicides, specialist adhesives, flavor compounds, refinery lubricant recovery and as a precursor for 5-methyl furfural, furfuryl alcohol, tetrahydrofurfuryl alcohol, and tetrahydrofuran [18]. It is produced from a hemicellulosic fraction of lignocellulosic residues and is considered to be one of the high-value bio-based chemicals [19]. The total production of furfural in the world is around 430,000 tons per year and it is likely to increase in future due to the green chemicals initiative in various industries [20]. At present furfural is produced commercially from the hemicellulose present in biomass, while cellulose- and lignin-rich solid leftovers are used as boiler fuel. Since hemicellulose only makes a fraction of the total composition of lignocellulose, the sole production of furfural from biomass residues would be inefficient and costly and would generate more waste [21]. Therefore integrating bioethanol and furfural production from lignocellulosic residues would be an advantage for the biorefinery system to utilize the biomass fractions efficiently and economically [22].

The problem associated with traditional agriculture crop residues are that they are only grown in large quantities in developing counties such as China and India, and a significant amount of the residues are used as animal fodder. Apparently, when these residues are diverted to biofuel production or a biorefinery, there would be a scarcity of animal fodder. Hence, management of such crop residues and exploration of other nonfodder cellulosic residues that could grow in wastelands are very crucial for a specific geographical location to balance between animal feed and fuel production. There are several nonfodder crop residues such as vetiver and switch grass that grow well in wasteland with minimal water and nutrient requirements. In addition, they also have soil conservation with high carbon stocking [23,24] and phytoremediation properties [25,26]. Cultivation of such crops in wasteland and utilization for biofuels would be beneficial for biomass residue management, soil conservation and social economic status of a region.

Vetiver (*Chrysopogon zizanioides*), a perennial grass from the Poaceae family, is native to India and its cultivation system was developed by the World Bank in India in the mid-1980s. It is now being used in many countries for soil and water conservation, land stabilization and rehabilitation, sediment control, and phytoremediation. Vetiver leaves grow up to 1.5 m high and the roots are thick and grow fast and deep, binding strongly to the soil downward up to a depth of 2–4 m. It is a vegetative, low-maintenance, and drought-tolerant plant that could be used for reclamation of salt-affected soils [27–30]. Vetiver leaf biomass at present has only limited application, such as for boiler fuel, handicraft products, and mushroom cultivation [31,32]. Vetiver has the capacity to generate high biomass (leaves) in wasteland and its chemical composition is rich in cellulose, hemicellulose, and lignin content that could be used as a feedstock for bioethanol production [33] and biochemical production in a biorefinery in India, due to more wasteland availability.

India is highly dependent on fossil fuel for its energy needs, and is very keen on promoting biofuels in the transport sector with a policy framework to blend 20% biofuels by 2017 with conventional fossil fuels [34]. To meet the demand of the current bioethanol blend transport policy, several biomass residues such as wheat straw, rice straw, sugarcane tops and leaves, etc. are available. However, these residues are only generated in fertile zones and are mostly used as animal food and apparently could benefit only the socioeconomic development of these zones. On the other hand, to improve the soil and socioeconomic status of wasteland areas, and to satisfy the future's biomass residue demand for biofuels and biochemicals, nonconventional agriculture residues such as vetiver could be promoted. However, the environmental benefits of such a vetiver biorefinery system depends on the agricultural practices of the region, configuration of the integrated processes between biofuels and biochemicals, and transport systems of the raw materials and products that could be analyzed using life cycle assessment (LCA).

LCA is a tool to estimate the environmental benefits of any given process or product. Nowadays it is widely used in academia and industry to analyze the environmental benefits of an ongoing project or future projects, and most often the impacts (carbon dioxide and the fossil oil consumption) of product systems are reported. In this study the LCA of a vetiver biorefinery system in India to produce biofuel and a biochemical (furfural) is analyzed to determine the environmental benefits over the conventional system and to give suggestions on integration configuration improvements.

5.2 PROCESS DESCRIPTION

The scope of this work is to compare the processes and assess the life cycle environmental impact of bioethanol and furfural production from vetiver residues to a reference system composed of gasoline and furfural pathways. To assess the impacts, a clear understanding of the processes from vetiver cultivation to product production is important, and these are described in the following sections.

5.2.1 Bioethanol Production From Vetiver

Similar to other agricultural residues, the major composition of vetiver is cellulose, hemicellulose, and lignin. Bioethanol could be produced from the cellulosic and hemicellulosic fraction of vetiver, after converting to their respective sugars by a chemical or enzymatic process. The overall process of this production system is described here (Fig. 5.1). The vetiver plant is planted in the wasteland with minimal rainfall and fertilizer requirements. The leaves are harvested annually and are transported to the bioethanol production plant facility, where the leaves are milled to a desired size for the pretreatment process. There are several types of pretreatment processes reported in the literature. Among them, the most commonly used are acid, alkali and hot water treatment. Each pretreatment type has its own advantages and disadvantages. In the pretreatment process the milled vetiver leaves are treated with appropriate

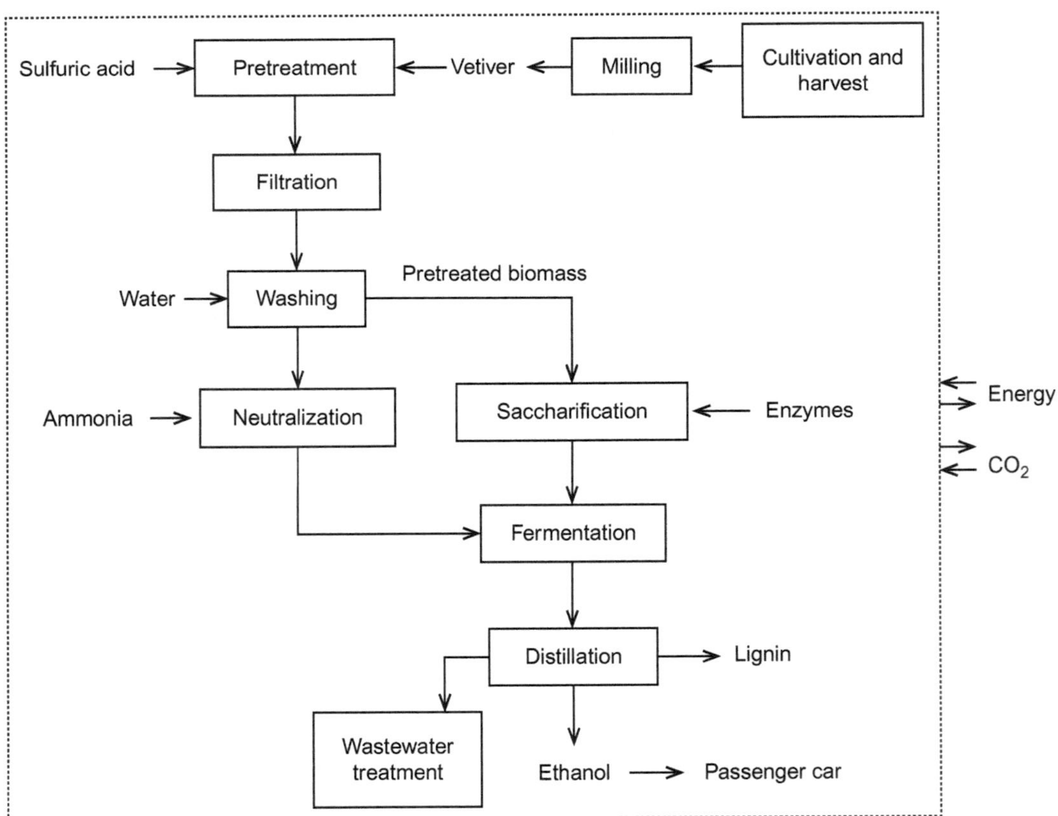

FIG. 5.1 Bioethanol production from vetiver.

chemicals (dilute acid in this study) at the desired temperature and time; later the solid fraction from the slurry rich in cellulose and lignin is separated from the hemicellulose rich liquid fraction using a filter and neutralized with water and alkali, and the solid fraction is further processed for bioethanol production. The solid fraction is subject to saccharification using cellulase and β-glucosidase enzyme where the enzyme hydrolysis converts the cellulose fraction present in the biomass into glucose. The hemicellulose or xylose solution obtained from pretreatment is mixed with a glucose solution and fermented with glucose and xylose fermenting microorganisms at the desired time and temperature to convert xylose and glucose to ethanol; later the broth is distilled to separate ethanol and lignin followed by molecular sieve separation to produce fuel grade ethanol. Lignin is sent to the combined heat and power (CHP) section to be used as a burning fuel to generate heat and electricity utilized in the process.

5.2.2 Bioethanol and Furfural Production From Vetiver

In this concept, unlike the previous ethanol production process, two products are produced from vetiver. With the cultivation, milling and pretreatment stage remaining the same as in the ethanol production process, the solid fraction from pretreatment, rich in cellulose, is neutralized with water and alkali and is saccharified using enzymes to convert the cellulose into glucose and is further fermented to ethanol using glucose fermenting microorganisms (Fig. 5.2). Later the broth is distilled to separate ethanol from lignin and is purified in molecular sieves. The lignin remaining after ethanol production is sent to the CHP section to be used as a burning fuel, whereas the liquid fraction with high pH is further subject to dehydration at the desired temperature and time. After dehydration, furfural in the mixture is removed with steam and condensed by cooling. Later, furfural from condensate is separated and purified by azeotropic distillation.

5.2.3 Gasoline Production From Crude Oil

This is the conventional system of gasoline production from crude oil in which the crude oil extracted from the well is transported to the refinery. The gasoline is separated by distillation and refining (Fig. 5.3).

5.2.4 Furfural Production From Vetiver

This system is similar to the conventional furfural production from cellulosic feedstock, in which the vetiver leaf biomass obtained from the cultivation site is milled and subjected to hydrolysis and dehydration in a dilute acid solution at the desired time and temperature in a reactor where the hemicellulose fraction in biomass is converted to xylose initially, and then further converted to furfural and other organic acids by the catalytic action of dilute sulfuric acid. Later, the furfural from the mixture is removed with steam and filtered to remove solid particles and condensed by cooling down. Furfural from the condensate is removed by azeotropic distillation (Fig. 5.4).

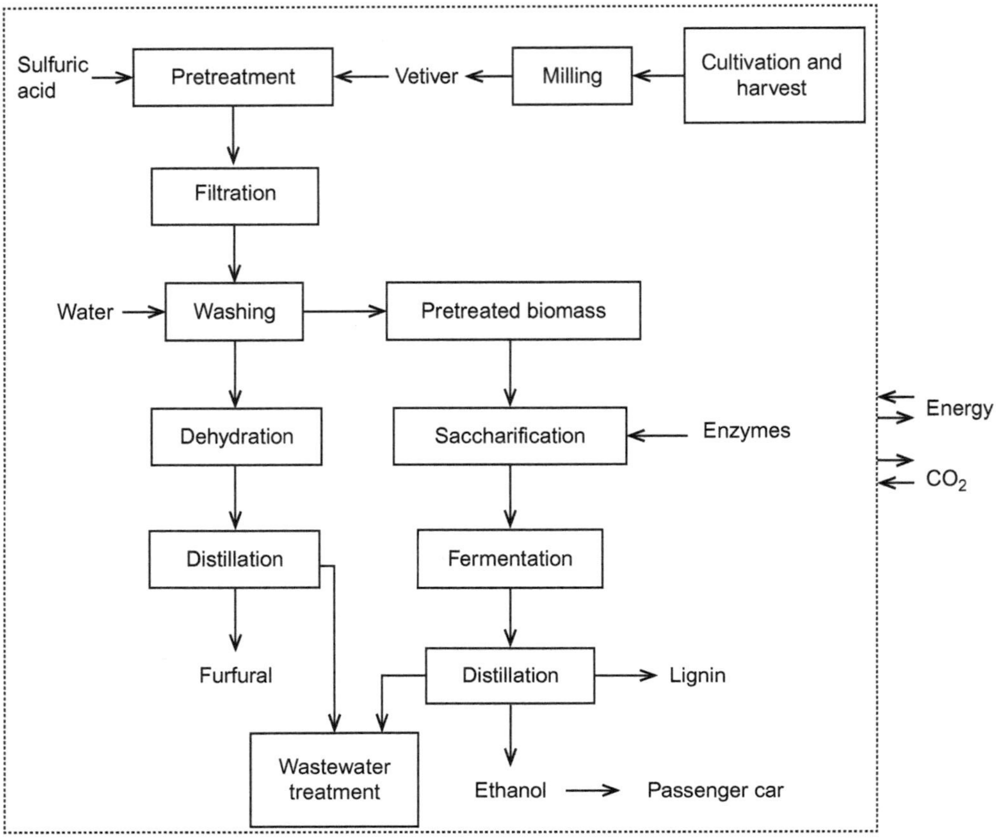

FIG. 5.2 Bioethanol and furfural production from vetiver.

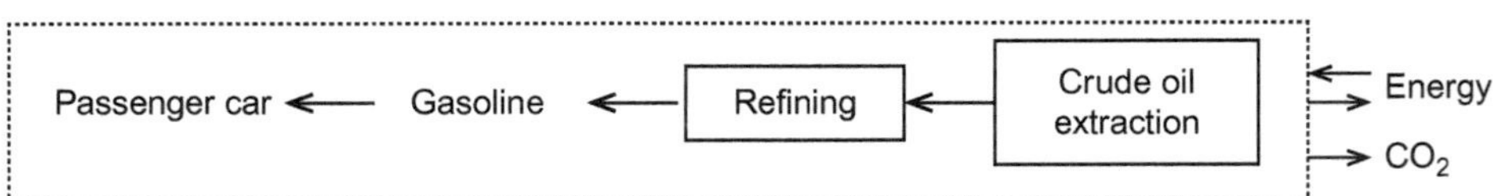

FIG. 5.3 Gasoline production from crude oil.

5.3 EXPERIMENTS AND DATA INVENTORY

5.3.1 Experiments

To access the life cycle impact of processes, data inventory is a very crucial part in which the specific data of each process is gathered from the literature and reported for the functional unit considered in the system being studied. In the case of biofuel and biorefineries, most of the data on product and processes are based on the composition of the biomass. Therefore, the compositional analysis of a vertiver leaf sample obtained from Haiti was done by conducting

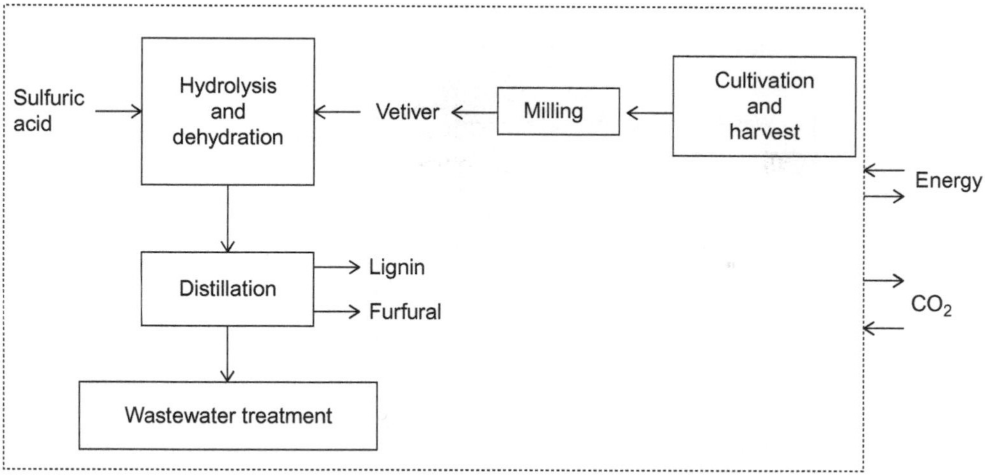

FIG. 5.4 Furfural production from vetiver biomass.

experiments according to NREL protocols [35,36]. High-performance liquid chromatography (HPLC) (Agilent Technologies, Germany) equipped with an Aminex HPX-87P, Aminex HPX-87H columns and refractive index detector was used to analyze the cellulose, hemicellulose and organic acid content in the biomass by comparing it with standards. Ash and lignin content were analyzed by gravimetrical methods. The heating value of vetiver leaf biomass was analyzed by bomb calorimeter using standard procedures (DIN 51900).

5.3.2 Functional Unit and System Boundary

Choosing an appropriate functional unit is very important to gather a relevant data inventory in any LCA study. There are several options for selection of functional units, such as energy basis, service basis, and land use basis. In biofuels and biorefineries, service from the products as the functional unit would be more appropriate in comparing it with the conventional products [37]. Therefore, service from products produced from 1 kg vetiver biomass was chosen in this study.

Two systems were compared in this LCA study with their respective conventional systems: (1) bioethanol production from vetiver leaves biomass (E-1) (system under study) was compared with conventional fossil gasoline production (P-1) (reference system); (2) bioethanol and furfural production from vetiver biomass (EF-1) was compared with conventional fossil gasoline and furfural production (PF-1). The system boundary for ethanol and gasoline is from well to wheel (WtW) as traveled distance and well to tank for furfural in kilograms (Figs. 5.5 and 5.6). For the system under study (bioethanol or biorefinery), the plant was assumed to be located in Cuddalore District in the South Indian state of Tamil Nadu. Vetiver leaf feedstock was assumed to be cultivated in wastelands available within the district and the neighboring district and was received in the biorefinery plant. A transport distance of 100 km each was assumed from vetiver cultivation farm to biorefinery, biorefinery to ethanol outlet and biorefinery to furfural outlet. To obtain the required data, a literature review was

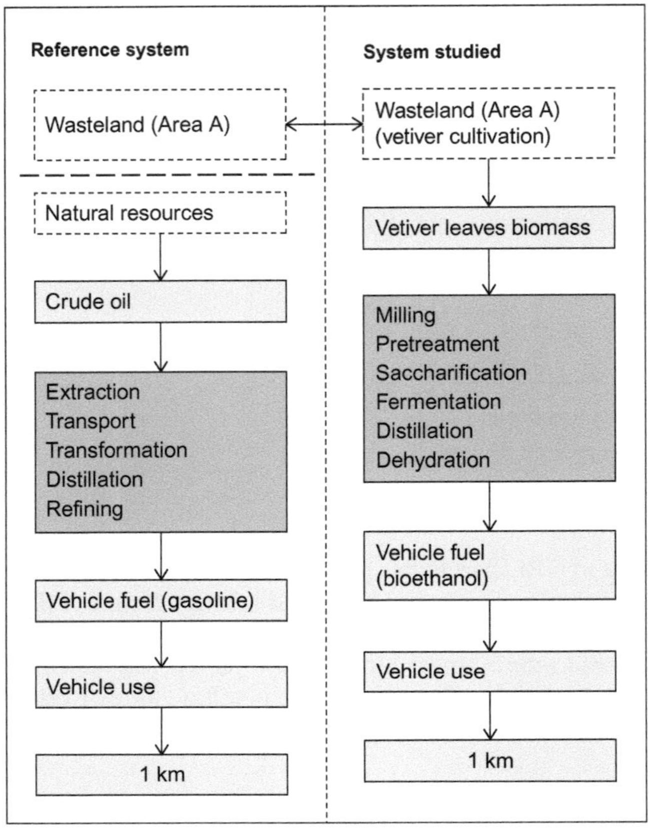

FIG. 5.5 System boundary of reference system (gasoline production) and studied system (bioethanol production) [38].

performed based on the process requirements for a large capacity plant (50 million liters per year) and then normalized for a functional unit of 1 kg vetiver leaf biomass (dry weight basis) (Table 5.1). For bioethanol WtW, the traveled distance was calculated using fuel efficiency and for furfural the amount in kilograms was calculated. Tables 5.2–5.4 show the data inventories of the system under study and the reference system. The carbon stocking due to land use change was also taken into account for the carbon dioxide emission calculation.

The fuel efficiency of bioethanol was calculated using Eq. (5.1) and the WtW carbon dioxide (GHG) emissions of bioethanol for 1 km traveled was calculated using Eq. (5.2), where biofuel, fuel blend, and fossil fuel represent bioethanol, bioethanol blend, and fossil gasoline per kilometer service [37]. An ethanol blend of 20% with gasoline was chosen for the impact assessment per kilometer with respect to the policy framework of the Indian government.

$$\text{Fuel efficiency}_{\text{biofuel}} = (\text{biofuel blend factor} \times \text{fuel efficiency}_{\text{fuel blend}})/1 - [(1 - \text{biofuel blend factor}) \times \text{fuel efficiency}_{\text{fuel blend}}/\text{fuel efficiency}_{\text{fossil fuel}}] \tag{5.1}$$

$$\text{GHG}_{\text{biofuel}}\text{WtW} = \text{GHG}_{\text{biofuel}}\text{WtT} \times \text{Fuel efficiency}_{\text{biofuel}} \tag{5.2}$$

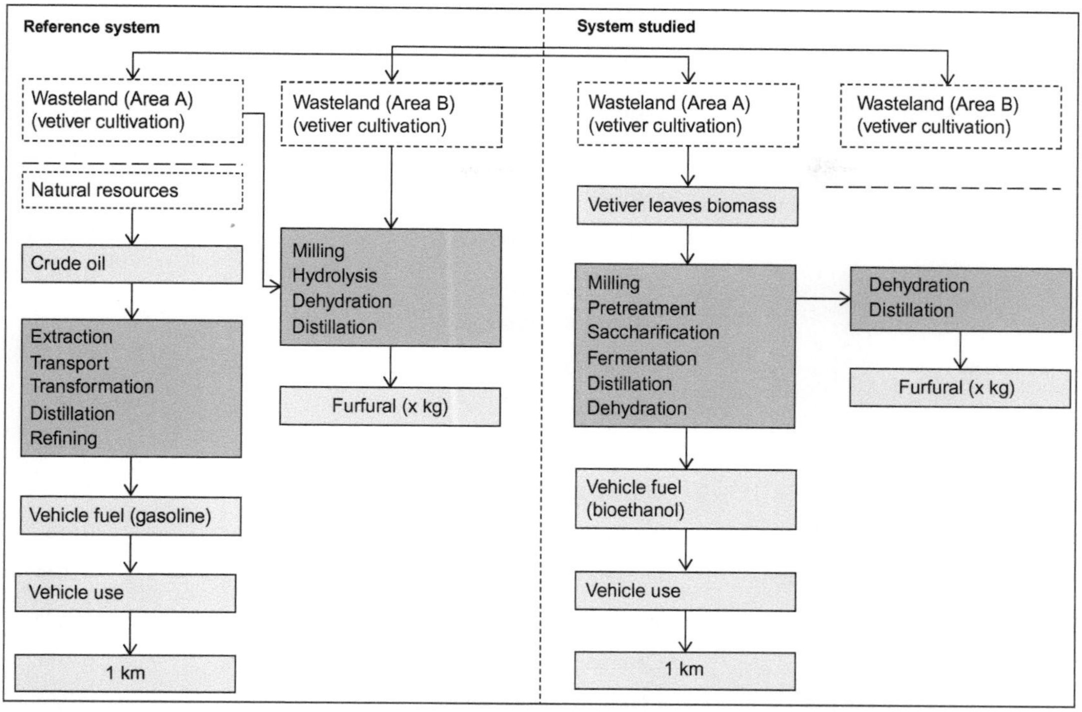

FIG. 5.6 System boundary of reference system (conventional refinery and furfural production) and studied system (bioethanol and furfural production).

TABLE 5.1 Conversion Efficiency of Ethanol and Furfural Production for System Under Study [38]

Cases	Conversion Compared to Theoretical Max (Dry Biomass)	Remarks
Ethanol production from C6 and C5	~80% and ~60%	Glucose and xylose loss in pretreatment, and washing Inhibition due to organic acids.
Ethanol production from C6	~82%	Glucose loss in pretreatment and washing
Furfural production from C5	~60%	Low process efficiency and mild side reactions
Furfural production from biomass	~47%	Low process efficiency and high side reactions

5.3.3 Vetiver Cultivation and Carbon Stock

The vetiver plant is assumed to be cultivated in sandy loam soil using only rainwater in south India through the propagation technique of splitting mature plants to produce root slips. The fertilizers and chemicals used for vetiver cultivation were assumed to be similar

TABLE 5.2 Life Cycle Inventory for Bioethanol Production from 1 kg Vetiver Leaf Biomass

Stages	Utilities/Raw Materials	Unit	Base Case
Cultivation	Fertilizer (N)	kg	0.0103
	Fertilizer (P_2O_5)	kg	0.0039
	Fertilizer (K_2O)	kg	0.0039
	Herbicide (Atrazine)	kg	8.62E−05
	Pesticide	kg	0.0043
Output	Vetiver leaf biomass	kg	1
Pretreatment	Vetiver leaf biomass	kg	1
	Sulfuric acid	kg	0.08
	Water	kg	3.9221
	Wash water	kg	40
	Ammonia	kg	0.014
Saccharification and fermentation	Enyzme	kg	0.0044
	Water	kg	1.7908
	Cornsteep liquor	kg	0.0151
	Diammonium phosphate	kg	0.002
	Heat	MJ	2.2117
	Electricity	kW h	0.0827
Output	Ethanol	kg	0.2433
Byproducts	Lignin	kg	0.1591
	Heat from lignin	MJ	1.6545
	Electricity from lignin	kW h	0.056
Transport	From farm to biorefinery	km	100
	From biorefinery to ethanol outlet	km	100

to the values [fertilizers (60 kg/ha of N, 22.5 kg/ha P_2O_5, and 22.5 kg/ha K_2O), herbicide (0.5 kg/ha) and pesticide (25 kg/ha)] reported in [39]. All the work related to cultivation and harvesting are considered done manually and the vetiver leaf biomass yield (5.8 ton/ha) was assumed to be according to the value reported in Ref. [33].

Carbon stock saving or loss due to land use change play important roles in carbon dioxide emissions [37,40]. Vetiver is well known for its carbon sequestration process in carbon dioxide mitigation [24,41–43]. In this study the carbon stock data of vetiver cultivation in wasteland (India) reported in Ref. [24], adjusted to the yield of 5.8 ton/year, were used in the LCA.

TABLE 5.3 Life Cycle Inventory for 1 kg Vetiver Biorefinery System (Ethanol and Furfural)

Stages	Utilities/Raw Materials	Unit	Base Case
Cultivation	Fertilizer (N)	kg	0.0103
	Fertilizer (P_2O_5)	kg	0.0039
	Fertilizer (K_2O)	kg	0.0039
	Herbicide (Atrazine)	kg	8.62E−05
	Pesticide	kg	0.0043
Output	Vetiver leaf biomass	kg	1
Pretreatment	Vetiver leaf biomass	kg	1
	Sulfuric acid	kg	0.0799
	Water	kg	3.9207
	Wash water	kg	40
Saccharification and fermentation	Enzyme	kg	0.0043
	Water	kg	1.7904
	Cornsteep liquor	kg	0.0078
	Diammonium phosphate	kg	0.0010
	Heat	MJ	1.3608
	Electricity	kW h	0.0509
Output	Ethanol	Kg	0.1497
	Lignin	Kg	0.159
	Heat from lignin	MJ	1.5269
	Electricity from lignin	kW h	0.0524
Dehydration	Water	kg	3.5045
	Sulfuric acid	kg	0.0451
	Sodium carbonate	kg	0.0008
	Lime	kg	0.0409
	Heat	MJ	6.3338
	Electricity		0.035
Output	Furfural	kg	0.1168
Transport	From farm to biorefinery	km	100
	From biorefinery to furfural outlet	km	100
	From biorefinery to ethanol outlet	km	100

TABLE 5.4 Life Cycle Inventory for Conventional Furfural Production From 1 kg Vetiver

Stages	Utilities/Raw Materials	Unit	Base Case
Cultivation	Fertilizer (N)	kg	0.0103
	Fertilizer (P_2O_5)	kg	0.0039
	Fertilizer (K_2O)	kg	0.0039
	Herbicide (Atrazine)	kg	8.62E−05
	Pesticide	kg	0.0043
Output	Vetiver leaves biomass	kg	1
Hydrolysis and dehydration	Vetiver leaves biomass	kg	1
	Cooling water	kg	3.893
	Sulfuric acid	kg	0.0524
	Sodium carbonate	kg	0.001
	Lime	kg	0.0486
	Heat	MJ	7.9132
	Electricity	kW h	0.0389
Output	Furfural	kg	0.0973
	Lignin	kg	0.3504
	Heat	MJ	2.989
	Electricity	kW h	0.096
Transport	From farm to biorefinery	km	100
	From biorefinery to furfural outlet	km	100

5.3.4 Energy Requirements and Enzyme Impact Data

Additional electricity energy required for the plant other than the electricity provided by the lignin residue was assumed to be drawn from the electricity grid, modeled similarly to the previous study [44]. Coal 44%, hydropower 21%, biomass and other renewables 22%, natural gas 9%, nuclear 2%, and diesel 2% formed the energy mix of the grid electricity [45], with an average transmission and distribution loss of 18% accounted for in the electricity supply.

In each system, the lignin remaining in the process was used as the primary source for heat and electricity. A CHP with 80% heat efficiency and 10% electricity efficiency was considered where the lignin was burnt. Heating value (12 MJ/kg) for lignin from a standalone furfural plant and heating value (10 MJ/kg) for lignin from a standalone bioethanol plant and biorefinery plant was assumed due to the process variations in the systems that generate different lignin quality. Additional heat energy requirements were produced using hard coal and the surplus electricity generated in the system was assumed to be sold to the grid. The reason for the higher heating value of lignin considered for the standalone furfural plant is due to a higher composition of cellulose in the residue as compared to other systems.

Enzymes play a major role in bioethanol production from lignocellulose. The cellulase enzyme is used in the saccharification for converting cellulose to glucose. Therefore, the impact of enzyme production has to be included in the assessment. LCA data of the enzyme with carbon dioxide emission 2.264 kg CO_2eq/kg and fossil oil consumption 24.8 MJ/kg of enzyme production reported in Ref. [46] were used in this study.

5.3.5 Furfural Production Data

Conventionally, furfural is produced through a single-stage process in which the C6 and C5 (cellulose and hemicellulose) are not separated by pretreatment, unlike the ethanol and furfural process considered in this study. A majority of the furfural is produced in China and the life cycle data are not available in the LCA database. However, a list of chemicals and energy data with a range is reported in [18]. Average data from the range were used in this LCA study for conventional furfural production. In the case of furfural production in a biorefinery system, lower energy and chemicals requirements were assumed compared to those of the conventional system. This is due to the fact that the processes are integrated with the ethanol production process and hence there are fewer requirements for furfural production. The energy and chemicals data considered for the conventional system and the biorefinery furfural system are shown in Table 5.5. Lignin remaining in the process is used in the CHP plant for heat and electricity generation used in the plant. As only xylose would be converted to furfural, the lignin remaining in the conventional furfural process was considered to be high compared to the biorefinery process.

5.3.6 Impact Assessment Method, Impact Categories, and Sensitivity Analysis

For all three scenarios, data inventories from the chosen system boundary were modeled in SimaPro 7.3.3 using the Ecoinvent database and ReCiPe Midpoint (E). The impact categories considered were carbon dioxide emission and fossil oil consumption. It is worth noting that in the impact assessment database there are no data available for Indian geography; therefore a European database was used for the assessment. A value-based economic allocation was made in order to know the life cycle impact of individual products within the biorefinery and reference system.

TABLE 5.5 Energy and Chemicals Data of Furfural Production Processes [38]

Energy and Chemicals (Per Ton Furfural)	Win [18]	Furfural Process (Conventional)	Furfural Process (Biorefinery)
Electricity (kW h)	200–600	400	300
Steam (ton)	25–35	30	20
Cooling water (ton)	30–50	40	30
Sulfuric acid (ton)	0.3–0.8	0.55	0.385
Sodium carbonate (ton)	0.01	0.01	0.007
Lime (ton)	~0.5	0.5	0.35

In a system study, the impact assessments are sensitive to various factors such as biomass yield, energy use in the process, transportation, etc. However, in this study the source of energy used played a crucial role in the impact assessment; therefore the sensitivity of the impacts when the conventional energy source (coal) was replaced by vetiver biomass energy was studied. The heating value of vetiver leaf biomass (17 MJ/kg) that was analyzed through standard procedures were used for the energy calculations.

5.4 LIFE CYCLE ASSESSMENT

The composition analysis results show that vetiver leaf biomass is rich in cellulose, hemicellulose, and lignin (Table 5.6), similar to that of several biomasses such as wheat straw and rice straw [47] that could be used as a feedstock for bioethanol production and biorefinery [33,48]. The lower heating value of vetiver leaf biomass was found to be 17 MJ/kg, which suggests vetiver biomass could also be used as a fuel for burning in boilers or CHP to generate process energy requirements similar to other crop residues [49]. Thus, the vetiver biomass system if implemented in the bioethanol or biorefinery system has the capacity to fulfill the energy need of its plant, like sugarcane and other biomass residues.

In relation to the carbon stock value of the soil, shoot and root of vetiver grown in sand soil reported in Ref. [24] for a shoot yield of 28.62 t/ha, the carbon stock of soil, shoot and root of vetiver for this study with a yield of 5.8 t/ha was found to be 0.112, 0.293, and 0.015 kg/m^2, respectively. The carbon stock value results conclude that vetiver has the highest carbon stocking capacity compared to other crops [24] in the soil, making vetiver cultivation or land use change to vetiver cultivation favorable to mitigate environmental impacts in terms of carbon dioxide emissions.

Figs. 5.7 and 5.8 show the carbon emission and fossil oil consumption results of a standalone ethanol plant of vetiver biomass (E-1) and the reference system (P-1R). The

TABLE 5.6 Composition of Vetiver Biomass [38]

Composition	% Dry Mass
Glucan (cellulose)	32.6
Xylan	22.6
Arabinan	7.1
Galactan	1.2
Acetic acid	3
HMF	0.3
Furfural	2.5
Acid soluble lignin	1.4
Acid insoluble lignin	15.9
Ash	2.8

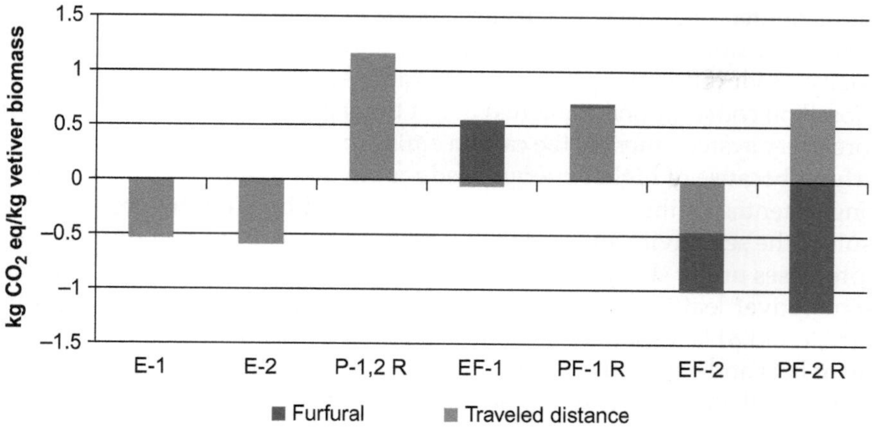

FIG. 5.7 Climate change impact comparison for different scenarios.

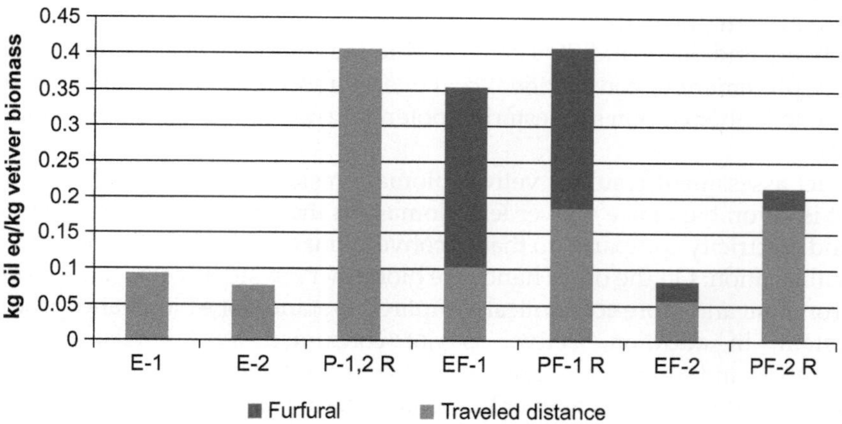

FIG. 5.8 Fossil oil consumption comparison for different scenarios.

reduction of carbon emission and fossil oil consumption of ethanol from vetiver leaf biomass (coal energy used for heat and electricity) compared to that of fossil gasoline are 145% and 77% respectively for the traveled distance in a car. These results show that production and use of bioethanol from vetiver is superior to fossil gasoline in environmental impact mitigation. Most of the carbon dioxide emission from the bioethanol system is attributed to the process energy requirements and enzyme production, though it was partially replaced by the lignin residues produced in the process and the carbon storage capacity of vetiver cultivation.

The carbon dioxide and fossil oil consumption results of bioethanol and furfural production (EF-1) from vetiver compared to that of gasoline and conventional furfural production (PF-1) are shown in Figs. 5.7 and 5.8. Compared to the reference system (PF-1), the reduction in carbon emission and fossil oil consumption of a vetiver leaf biomass biorefinery system was found to be 30% and 12% respectively. Most of the carbon emission in the reference system was attributed to the traveled distance using fossil gasoline, whereas the furfural production was responsible for fewer impacts due to the carbon sequestering potential and surplus electricity produced in the process due to more solid residue availability. On the other hand, the fossil oil consumption remained about the same for distance traveled and furfural. For the biorefinery system, most of the carbon emission and fossil oil consumption was attributed to furfural because of higher energy needs compared to that of ethanol, and the carbon sequestering potential of the vetiver biomass was shared by both the products.

The result of the sensitivity analysis with vetiver as the source for energy demand instead of coal in processes under study is also shown in Figs. 5.7 and 5.8 For traveled distance using ethanol from vetiver leaf biomass (E-2) the carbon emission and fossil consumption are reduced by 151% and 81% respectively compared to the reference system (P-1,2R) with vetiver energy use for heat and electricity. The impact reduction is increased by 6% and 4% compared to the coal-based ethanol production system for carbon emission and fossil consumption respectively. Whereas for an ethanol and furfural production system (EF-2), compared to that of gasoline and a conventional furfural production system (PF-2R), a similar trend is seen for the impacts when vetiver energy is used instead of conventional energy for the process energy requirement; however, the impact reduction values are significant. Compared to that of the reference system, the carbon emission and fossil consumption reductions are 100% and 62% respectively and 300% and 76% compared to that of a coal energy related system. This was due to replacement of conventional coal energy requirements with vetiver biomass energy, leading to higher carbon sequestering potential and surplus electricity generated in the EF-2 system.

The impact assessment result of vetiver biomass residue to bioethanol and furfural suggests that it is favorable to use vetiver leaf biomass as the energy source to generate the process heat and electricity compared to that of conventional coal energy that is used now in this geographical location. On the other hand, the biorefinery system seems to be more favorable to the environment and more economical compared to standalone bioethanol production for a vetiver system. In addition, vetiver is not consumed by animals; it is helpful in phytoremediation and soil regeneration and does not compete with animal feed crop residues in India. Due to these reasons, a vetiver biorefinery system could be considered as a potential alternative or additional source for conventional routes and the traditional standalone ethanol route in India for bioethanol and biochemical production.

5.5 CONCLUSIONS AND PERSPECTIVES

Among the developing countries in the world, India is highly dependent on fossil fuel sources for its energy needs and mainly for the transport sector. Biofuels and biochemicals from renewable sources are seen as an alternative to reduce fossil fuel dependency and also to mitigate environmental impacts leading to energy policy reforms in the transport sector. Lignocellulosic feedstocks are seen as a promising source for renewable energy in India. However, the conventional feedstocks have a demand in the animal feed sector that challenges the feedstock availability in future for biofuels and biochemicals. On the other hand, a vast area of wasteland is available in India that is not suitable for conventional crop cultivation.

Nonconventional feedstock production such as vetiver cultivated in available wasteland could be a win-win situation for future feedstock demand and regeneration of wasteland. However, the life cycle impact of such a system has to be assessed before implementation. Therefore, a prospective LCA was performed in this study on the vetiver system and compared with the conventional systems. The cellulose and hemicellulose composition of vetiver revealed that vetiver biomass could be a potential feedstock in bioethanol or a biorefinery for the production of bioethanol and furfural. The comparative LCA results of a vetiver leaf biomass system of different product configurations suggests that both bioethanol or bioethanol and furfural production from vetiver reduces carbon dioxide emissions and fossil oil consumption. Vetiver leaf biomass application in a biorefinery system to coproduce bioethanol and furfural could be a promising option in India to mitigate emission and partially fulfill the future feedstock demand. In future, more nonconventional feedstocks such as vetiver that could be cultivated in wastelands should be explored to cater to the future energy feedstock needs and to reduce environmental impacts in India.

References

[1] UNDP. World Energy Assessment: energy and the challenge of sustainability. New York, NY: United Nations Development Programme, Bureau for Development Policy, One United Nations Plaza; 2000.

[2] WHO. Climate change and health. Fact sheet No. 266, Geneva: World Health Organization; 2014.

[3] IPCC. Climate change 2014 synthesis report. Geneva: The Intergovernmental Panel on Climate Change (IPCC); 2014.

[4] REN21. Renewables 2014 global status report. Paris: Renewable energy policy network for the 21st century; 2014.

[5] Tenenbaum DJ. Food vs. fuel: diversion of crops could cause more hunger. Environ Health Perspect 2008;116: A254–7.

[6] Antizar-Ladislao B, Turrion-Gomez JL. Second-generation biofuels and local bioenergy systems. Biofuels Bioprod Biorefin 2008;2:455–69.

[7] IARI. Crop residues management with conservation agriculture: potential constraints and policy needs. New Delhi: Indian Agricultural Research Institute; 2012.

[8] Mirabella N, Castellani V, Sala S. Current options for the valorization of food manufacturing waste: a review. J Clean Prod 2014;65:28–41.

[9] Mohr A, Raman S. Lessons from first generation biofuels and implications for the sustainability appraisal of second generation biofuels. Energ Policy 2013;63:114–22.

[10] Sims REH, Mabee W, Saddler JN, Taylor M. An overview of second generation biofuel technologies. Bioresour Technol 2010;101:1570–80.

[11] Van Eijck J, Batidzirai B, Faaij A. Current and future economic performance of first and second generation biofuels in developing countries. Appl Energy 2014;135:115–41.

[12] Lin TH, Huang CF, Guo GL, Hwang WS, Huang SL. Pilot-scale ethanol production from rice straw hydrolysates using xylose-fermenting *Pichia stipitis*. Bioresour Technol 2012;116:314–9.

[13] Saha BC, Nichols NN, Qureshi N, Kennedy GJ, Iten LB, Cotta MA. Pilot scale conversion of wheat straw to ethanol via simultaneous saccharification and fermentation. Bioresour Technol 2015;175:17–22.

[14] Sun ZY, Tang YQ, Iwanaga T, Sho T, Kida K. Production of fuel ethanol from bamboo by concentrated sulfuric acid hydrolysis followed by continuous ethanol fermentation. Bioresour Technol 2011;102:10929–35.

[15] Bozell JJ. Feedstocks for the future—biorefinery production of chemicals from renewable carbon. Clean 2008;36:641–7.

[16] Holm-Nielsen JB, Ehimen EA. Biorefinery plant design, engineering and process optimisation. In: Waldron K, editor. Advances in biorefineries. London: Woodhead Publishing; 2014. p. 89–111.

[17] Cheng H, Wang L. Lignocelluloses feedstock biorefinery as petrorefinery substitutes. In: Matovic MD, editor. Biomass now—sustainable growth and use. Rijeka: InTech; 2013. p. 347–88.

[18] Win DT. Furfural—gold from garbage. AU J Technol 2005;8:185–90.

[19] NREL. Top value added chemicals from biomass: results of screening for potential candidates from sugars and synthesis gas. Oak Ridge: National Renewable Energy Laboratory. U.S. Department of Energy; 2004.

[20] DalinYebo. Past—present—future: global furfural market scenario. Kloof: DalinYebo Trading and Development (Pty) Ltd; 2014.

[21] Cai CM, Zhang T, Kumar R, Wyman CE. Integrated furfural production as a renewable fuel and chemical platform from lignocellulosic biomass. J Chem Technol Biotechnol 2014;89:2–10.

[22] IEA. Bio-based chemicals, value added products from biorefineries. IEA Bioenergy Task 42 Biorefinery; 2012.

[23] Cowie A, Smith P, Johnson D. Does soil carbon loss in biomass production systems negate the greenhouse benefits of bioenergy? Mitig Adapt Strateg Glob Chang 2006;11:979–1002.

[24] Singh M, Guleria N, Rao EVSP, Goswami P. A strategy for sustainable carbon sequestration using Vetiver (*Vetiveria zizanioides* (L.)): a quantitative assessment over India. Project Document CM PD-1101 CSIR, Bangalore: Centre for Mathematical Modelling and Computer Simulation; 2011.

[25] Anjum NA, Pereira ME, Ahmad I, Duarte AC, Umar S, Khan NA. Phytotechnologies: remediation of environmental contaminants. Boca Raton, FL: CRC Press; 2012.

[26] Chen BC, Lai HY, Juang KW. Model evaluation of plant metal content and biomass yield for the phytoextraction of heavy metals by switchgrass. Ecotoxicol Environ Saf 2012;80:393–400.

[27] World Bank. Vetiver Grass: the Hedge against erosion. Washington: The World Bank; 1987.

[28] Chaplot PC. Suitable techniques for establishment of vetiver grass (*Vetiveria zizaniodes*) in Rajasthan. Ann Agric Biol Res 1987;19:224–6.

[29] Truong P, Tran M, Pinners E. Vetiver system applications technical reference. San Antonio, TX: The Vetiver International Network; 2008.

[30] Yaseen M, Singh M, Ram D. Growth, yield and economics of vetiver (*Vetiveria zizanioides* L. Nash) under intercropping system. Ind Crop Prod 2014;61:417–21.

[31] Balasankar D, Vanilarasu K, Selva Preetha P, Rajeswari S, Umadevi M, Bhowmik D. Traditional and medicinal uses of vetiver. J Med Plants Stud 2013;1:191–200.

[32] Danh LT, Truong P, Mammucari R, Pu Y, Foster NR. Phytoremediation of soils contaminated by heavy metals, metalloids, and radioactive materials using vetiver grass, *Chrysopogon zizanioides*. In: Anjum NA, Pereira ME, Ahmad I, Duarte AC, Umar S, Khan NA, editors. Phytotechnologies: remediation of environmental contaminants. London: CRC Press; 2012. p. 255–79.

[33] Wongwatanapaiboon J, Kangvansaichol K, Burapatana V, Inochanon R, Winayanuwattikun P, Yongvanich T, et al. The potential of cellulosic ethanol production from grasses in Thailand. J Biomed Biotechnol 2012. Article ID 303748.

[34] National policy on biofuels. New Delhi: Ministry of New and Renewable Energy, Government of India; 2009.

[35] NREL. Standard procedures for biomass compositional analysis. Golden, CO: National Renewable Energy Laboratory; 2014.

[36] Sluiter JB, Ruiz RO, Scarlata CJ, Sluiter AD, Templeton DW. Compositional analysis of lignocellulosic feedstocks. 1. Review and description of methods. J Agric Food Chem 2010;58:9043–53.

[37] Gnansounou E, Dauriat A, Villegas J, Panichelli L. Life cycle assessment of biofuels: energy and greenhouse gas balances. Bioresour Technol 2009;100:4919–30.

[38] Raman JK, Gnansounou E. LCA of bioethanol and furfural production from vetiver. Bioresour Technol 2015;185:202–10.

[39] Mohideen MK, Shakila A, Anburani A. Production technology of medicinal and aromatic crops. New Delhi: Agrobios; 2011.

[40] Searchinger T, Heimlich R, Houghton RA, Dong F, Elobeid A, Fabiosa J, et al. Use of U.S. croplands for biofuels increases greenhouse gases through emissions from land-use change. Science 2008;319:1238–40.

[41] FAO. Food security and agricultural mitigation in developing countries: options for capturing synergies. Rome: Food and Agriculture Organization of the United Nations; 2009.

[42] Lavania UC, Lavania S. Sequestration of atmospheric carbon into subsoil horizons through deep-rooted grasses—vetiver grass model. Curr Sci 2009;97:618–9.

[43] Taranet P, Wattanaprapat K, Meesing I, Nopmalai P. Carbon sequestration and carbon dioxide emission in vetiver grass cultivation areas in Thailand. Bangkok: Land Development Department; 2012.

[44] Ajayebi A, Gnansounou E, Raman JK. Comparative life cycle assessment of biodiesel from algae and jatropha: a case study of India. Bioresour Technol 2013;150:429–37.

[45] Energy Statistics. New Delhi: Central Statistics Office, National Statistical Organization, Ministry of statistics and programme implementation, Government of India; 2012.
[46] MacLean HL, Spatari S. The contribution of enzymes and process chemicals to the life cycle of ethanol. Environ Res Lett 2009;4:014001.
[47] Saratale G, Chen SD, Lo YC, Saratale RGS, Chang JS. Outlook of biohydrogen production from lignocellulosic feedstock using dark fermentation—a review. J Sci Ind Res 2008;67:962–79.
[48] Zhao F, Yang W, Zeng Z, Li H, Yang X, He Z, et al. Nutrient removal efficiency and biomass production of different bioenergy plants in hypereutrophic water. Biomass Bioenergy 2012;42:212–8.
[49] Koopmans A, Koppejan J. Agricultural and forest residues—generation, utilization and availability. In: Paper presented at the Regional Consultation on Modern Applications of Biomass Energy, 6-10 January; Kuala Lumpur; 1997.

CHAPTER

6

Life Cycle Assessment of Thermochemical Conversion of Empty Fruit Bunch of Oil Palm to Bio-Methane

P. Montafia, E. Gnansounou†*

*Ecole Polytechnique Fédérale de Lausanne (EPFL), Lausanne, Switzerland

†Bioenergy and Energy Planning Research Group (BPE), Ecole Polytechnique Fédérale de Lausanne (EPFL), Lausanne, Switzerland

6.1 INTRODUCTION

The increase of internal and foreign demand for oil palm is leading Brazil to encourage expansion of oil palm plantations. Biofuels and the food industry are driving this request. The state of Pará, the largest oil palm producer, is the main player in this game. Oil palm production generates a huge amount of biowastes, such as fibers, kernel shells, and empty fruit bunches (EFBs). For several decades, fibers and shells have been combusted on the oil palm mill site, obtaining process steam and power; EFB is usually collected and spread out on the plantation as mulch, as its high water content makes it a bad fuel. By the way, energetic valorization of EFB has been investigated in several studies in recent years. However, the use change could cause an environmental impact on the biosphere and it has to be evaluated.

More generally, a considerable literature about green fuels compares the world energy demand and the biomass energy potential. In 2012 the total nonrenewable energy supply was 11,565.915 Mtoe (484.24 EJ). The biomass availability worldwide has been estimated to be between 200 and 500 EJ per year [1]. Thus, biomass looks to be able to cover a large amount of the global energy demand. However, the environmental performance of biofuels may be even worse than that of fossil fuels in certain conditions; for instance, deforestation, long transportation, and intensive culture produce high emissions. And we need to find out just how renewable biofuels are: indeed the total impact may be negative if their exploitation rate is

Life-Cycle Assessment of Biorefineries
http://dx.doi.org/10.1016/B978-0-444-63585-3.00006-1

larger than biomass growth. Improving the biofuels yield and blend could be an encouraging strategy to reduce the global impact. Sharing the environmental effects of common processes, such as cultivation and raw material transportation, causes a decrease in environmental effects. That is why the energy valorization of residues is advisable. Thus, smart and sustainable development has to promote low-impact technologies that can energetically value all the feedstock. Because of this, multifunctional plants can efficiently reach the goal. Moreover, other criteria to sort out the most promising processes are: economic sustainability, reliability, ease of control, and biomass availability. Life cycle assessment is presenting as the most promising tool for environmental impact evaluation.

It allows one to estimate the process effects on the environment: it takes into account each main contribution and the analysis is led from the cradle to the grave. Each significant step of the process has to be considered, from primary source extraction to waste disposal. The procedure to do an LCA is fixed and it is explained in the second chapter of this book. In any case, it consists of several steps: preliminary, defining of the functional unit, scope, and system boundaries are determined; then the emissions inventory is carried out, followed by the impact evaluation; finally, the interpretation ends the analysis. In this study, allocation of coproducts is essential in order to correctly attribute a share of the impact to each of them. Another key point in the LCA procedure is the reference system choice: it has to be representative of the scenario which it is compared with. The studied system and the reference systems must provide the same services. Moreover, the worst-case scenario should not be used as a reference system, in order to avoid any overestimation of real benefits that an alternative system may provide. This is valid especially when comparing new renewable energy facilities with traditional ones. Concerning the LCA of bioenergy (energy from biomass), Delivand and Gnansounou [2] and Johnson [3] highlight the importance of land use change when carrying out an LCA of organic matters: thus, the previous land use has to be considered in the assessment. This study embraces this approach and particular emphasis is given to this aspect in the following section.

This chapter focuses on an LCA of the thermochemical conversion of oil palm EFB. Among several choices of technology, this purpose is carried out by hydrothermal gasification of the feedstock which, considering the feedstock being studied, is supposed to have the best performance compared with other gasification techniques. It consists of a nontraditional thermal biomass degradation: the thermal process occurs in a hydrothermal condition, which means at very high pressure and relatively low temperature in an aqueous environment. This gasification state is characterized by 400°C and 300 bar of temperature and pressure, respectively, and it enhances the methane production; higher temperature shifts the equilibrium towards higher hydrogen yield. The process ends with gas-water separation and biogas purification. Finally, the gas can be delivered to the final user and burnt in a furnace. A coproduct is delivered along with bio-methane: inorganics held in biomass are collected and separated in a brine that can be used as fertilizer. The present level of development is a pilot plant at the laboratory scale. Few plants (at laboratory scale) exist around the world and more research work must be carried out. An overall description of the technology is provided in the next section, as well as the model that leads the analysis.

This work intends to evaluate the environmental performance of the whole process valorizing the EFB. The system is operated in Pará State, Brazil. It is supposed to operate for several years in a projected oil palm biorefinery. This situation is inspired by a real project proposed in 2011 by the ADM company (Archer Daniels Midland Company) [4]. The scenario proposes to install a biorefinery on the mill site. It treats the EFB, gathering bio-methane and fertilizing salt brine. The new plantation will be settled in degraded lands [4], complying with the local law. The study proposes to determine by means of LCA the impacts of such a system. It also compares it with traditional ways to provide fertilizers and heat.

6.2 HYDROTHERMAL GASIFICATION: PROCESS DESIGN

6.2.1 General Considerations of Hydrothermal Gasification

Conversion of biomass into methane is mainly divided into two categories: biochemical and thermochemical processes. The former transforms biomass into biogas by means of bacteria, while the latter yields bio-methane with thermal degradation of feedstock. Biochemical digestion is carried out by groups of bacteria which can be sorted by preferred life temperature: thermophilic (above 50°C), mesophilic (between 20 and 45°C) and psychrophilic (under 20°C) digestions. The last category is not really used because of the low efficiency. The thermochemical process produces bio-methane, avoiding high temperatures. The process can be classified by means of the heat supply system (direct combustion or indirect, by an external source), or the gasification temperature level and gasification support (fixed or fluidized bed). Hydrothermal gasification is a particular case of thermochemical treatment of the biomass. At the end of the 1970s Professor Modell introduced hydrothermal gasification at the Massachusetts Institute of Technology [5], demonstrating its high potential to yield synthetic natural gas (SNG) from wet biomass; traditional processes cannot manage such types of feedstock efficiently, given that the drying process and hydrolysis in conventional thermochemical conversion are high energy-consuming.

Hydrothermal gasification is a thermochemical process that aims to produce SNG. It occurs in a pressurized water environment, it allows particular biomasses to decompose that nowadays are not exploited because of technological constraints. The process is made up of five main steps: first the sludge is pumped up to the operating pressure, then the hydrolysis takes place; afterwards the liquid mixture is heated up and the inorganics are separated. After that, the catalytic gasification is carried out. Finally, the mixture is cooled down, bio-methane, and water are separated and the gas is purified.

Hydrothermal conditions mean that the temperature and the pressure are near or above the critical point of water (22.064 MPa and 374°C). In the so-called supercritical phase, water behaves intermediately like vapor and liquid. These two phases are not observable separately. This phenomenon is due to the fact that, when temperature and pressure grow, both liquid and vapor tend to the same thermodynamic condition. Increasing the temperature, liquid density decreases, whereas when the pressure grows, the gas density augments. Over the supercritical point water vaporization is avoided and vaporization heat injection as well. The supercritical state is characterized by the absence of surface tension. No liquid-gas bond exists, so the solubility increases with the pressure and it decreases with the temperature in the vicinity of the critical point; hence, the solubility rises again at higher temperature. All the supercritical fluids are completely miscible and they produce monophase mixtures; the pseudo-critical point (it is the critical point of a mixture) is computable by the arithmetic mean of each critical temperature and pressure. A better accuracy can be reached using equations of state such as the Peng-Robinson equation [6]. The first experiments of hydrothermal gasification have investigated the most efficient process conditions. Many authors have published results of their studies and have stated that a supercritical environment is more promising than a near-critical environment for hydrothermal purposes. Here, the water becomes at the same time the reaction medium and the reactant. Reactions of hydrolysis and gasification can occur in the same batch with short residence times; polar fluids at subcritical conditions become nonpolar and high temperature and density increase water solvation skill. The process converts all the fed biomass into bio-methane [7]. In order to better understand why hydrothermal water can convert lignocellulosic biomass, a short overview about these polymers is given. Cellulose, hemicellulose, and lignin are natural polymeric compounds, very hard to be thermally decomposed. The cellulose is a linear polymer, and ether bonds connect many molecules of glucose; amorphous and crystalline zones are made by glucose chains connected together. Hemicellulose is composed of amorphous and branched chains of hexoses, pentoses, and glucuronic acid. Lignin is a mixture of cross-linked polymers which contain a quota of amorphous aromatic polymers. The great thermal stability of lignin comes from this amorphous part contained in a three-dimensional network [8]. The hydrothermal decomposition of lignocellulosic biomass achieves high standards: indeed its transformation is faster and complete. Cellulose is dissolved by hydrolysis in glucose and other oligomers. In supercritical condition, this process works on the entire cellulose crystal rather than only on the surface, as in the subcritical condition. Moreover, the transformation is homogeneous whereas in a subcritical environment it is heterogeneous. Glucose and oligomers produced by the hydrolysis react in water. The glucose production rate is higher than the dissolution one at high temperature. Furthermore, glucose produces 5-HMF, alcohols, carboxylic acids, aldehydes, and ketones. Also, supercritical water easily hydrolyzes the hemicellulose by producing several monosaccharides that react onto products mentioned previously. Lignin is thermally stable. However, because of its high solvation behavior, supercritical water destroys ether bonds, creating phenolic compounds.

A brief summary of the state of the art in hydrothermal gasification and a literature review are useful. In 1985 Modell [9] found that in hydrothermal conditions wooden biomass can be gasified without char and tar formation; he operated in batch reactors. This finding started off research on the topic, investigating the biomass-to-gas efficiency, the gas yields at the outlet, the char, and tar formation, the required residence time, the catalyst efficiency, and its deactivation [10–13]. The process was first studied by means of model compounds, such as glycerol or glucose, which allowed the reaction path to be studied and general issues which occur with real biomass (reactor plugging) to be avoided. In addition, at Pacific Northwest National Laboratory (PNNL), Washington, United States, and the Paul Scherrer Institut (PSI), Villigen, Switzerland with Vogel and Waldner [10], researchers have carried out significant experimental work on catalytic gasification. It takes place at relatively low temperature, in the vicinity of the supercritical point; due to energy requirements, it needs catalysts to reach an interesting speed and an adequate selectivity. Swiss scientists have found nickel (Ni) and ruthenium (Ru) to be the most effective catalysts [10,12]. Vogel's research has addressed catalyst selection. Skeletal nickel shows deactivation after 100 h of running, whereas the Ru/C is stable and active also after a 120 h test; hence, experiments have shown that the Ru catalyst has better performance and its use is advisable. Furthermore, at the PSI, researchers have simultaneously studied the reaction path through model compounds: notably they have investigated tar and char formation and they have proven what Modell first stated: the process avoids formation of tar and char. Also gasification of real biomasses has been studied. The achievements from these works have been as follows: suitable catalysts, best gasification temperature, and pressure have been found, and they have highlighted gasifier issues such as reactor plugging and catalyst deactivation, due to salt deposition. In the frame of the TEES project (Thermochemical Environment Energy System) of PNNL, Sealock Jr. has achieved one of the first supercritical wood gasifications, obtaining a yield of 0.33 g_{CH4}/g_{wood} [14]. After this step, particularly at the PSI, a continuous process has been developed [15], focused on the investigation of a reliable salt separation able to fulfill severe requirements on the purification from inorganics. At the PSI, Vogel and his colleagues continue their work about catalysts and salt separation, running one of the first continuous hydrothermal gasifiers. In 2010 they founded a start-up (Hydromethane AG) in order to enhance the scale-up process. They aim to design a commercial demonstrative plant [7,16]. Nowadays, salt separation remains the main issue that scientists are focused on, although other research topics have been opened in recent years. Among them we can mention process integration with biomass production or with solar concentration technologies [15,17,18]. For this aim, we can cite the project SunCHem developed with the involvement of the PSI and EPFL (École Polytechnique Fédérale de Lausanne) [19,20]. Finally, as confirmed by many efforts, particularly the Swiss ones made at the PSI [10,11] and at EPFL [21,22], hydrothermal gasification has a great potential for the gasification of lignocellulosic and wet biomasses.

In the following, just a little hint is given about a similar process, supercritical water oxidation (SCWO). It has been developed simultaneously but it is already at the commercial stage. The German professor E. U. Franck has started studying this technology and the research has been pursued by ETH Zurich. The process aims to destroy hazardous waste by hydrothermal oxidation. Supercritical water, at very high temperature (above 500°C, usually at 600°C) can decompose 99% of every waste. It transforms the wastes into a gas made up of carbon dioxide, oxygen, and nitrogen, and it reduces NO_x production. The issue of salt separation is almost completely avoided, since it does not use any catalyst. However, SCWO reactors have to withstand high corrosive conditions and plugging and fouling caused by supercritical water and inorganics, respectively. Japan and the United States are at the forefront in this field. They are already running several waste treatment plants using this technique [23].

6.2.2 Hydrothermal Gasification

Details on the continuous hydrothermal gasification process are reported in this chapter. This is the most suitable configuration for industrial purposes: a batch reactor, although it has an easier design, provides a pulsed stream of bio-methane and has issues concerning catalyst deactivation. Moreover, all the heat and power generators, and cogenerators as well, need a continuous flow of fuel, which can be achieved with a storage tank in the batch mode.

The first step is the biomass treatment to produce a sludge. The water and the solid (biomass) content highly affect the performance and the pump reliability. Indeed, the solid fraction has to be in the range of 15–30%. Under the lower bound the methane yield is small, whereas above the higher bound the pumping can have some issues. If the hydrogen yield is high, regulations force the use of special pipes and engines. At the state of the art, hydrothermal gasification is used to produce methane, although a few studies are related to hydrogen production. Solid feedstocks have to be crushed in order to produce very small chips, a few millimeters long. This procedure enhances the overall process efficiency since it increases the hydrolysis effectiveness: the smaller the biomass chips, the easier the hydrolytic decomposition. On the other hand, when the feedstock is aqueous (manure, sewage), dewatering is mandatory to increase the organic fraction.

Afterwards, the pumping takes place. The stream is pumped up to 25–30 MPa. The multiphase nature of the stream makes the operation tricky, as pipes and pump can be plugged. Moreover, the pressure to be reached is very high and such a level does not allow rotating pumps. The most promising technology is the Triplex pump, an alternative piston pump typically made with three pistons. With an efficient control, this system ensures an almost perfect continuity of flow rate and pressure stability.

In the hydrolyzer, the first reactions take place. This is carried out in a heated reactor. In a batch experiment this is the first step of the whole transformation. The hydrolysis occurs between 250 and 350°C with a peak at 320°C, as Kruse and Gawlik showed [24]. The hydrolysis decomposes the organic matter into simpler compounds, gaseous and liquid; in this condition pyrolysis is slower, so it does not happen. As many authors have stated, fast hydrolysis of holocellulose produces glucose in a near-supercritical environment and it breaks up quickly into polyphenols and 5-HMF; they rapidly become carboxylic acids, aldehydes, and alcohols. The main products of lignin decomposition are phenols and aromatic compounds. The literature affirms that the reaction paths of such a hydrolysis are very similar and the common schema shown by Vogel and Waldner is valid.

Fig. 6.1 also includes the gasification step. Although it is a simple one, it is still too complex to be easily used in a model. Thus, in the following the hydrolytic decomposition is presented

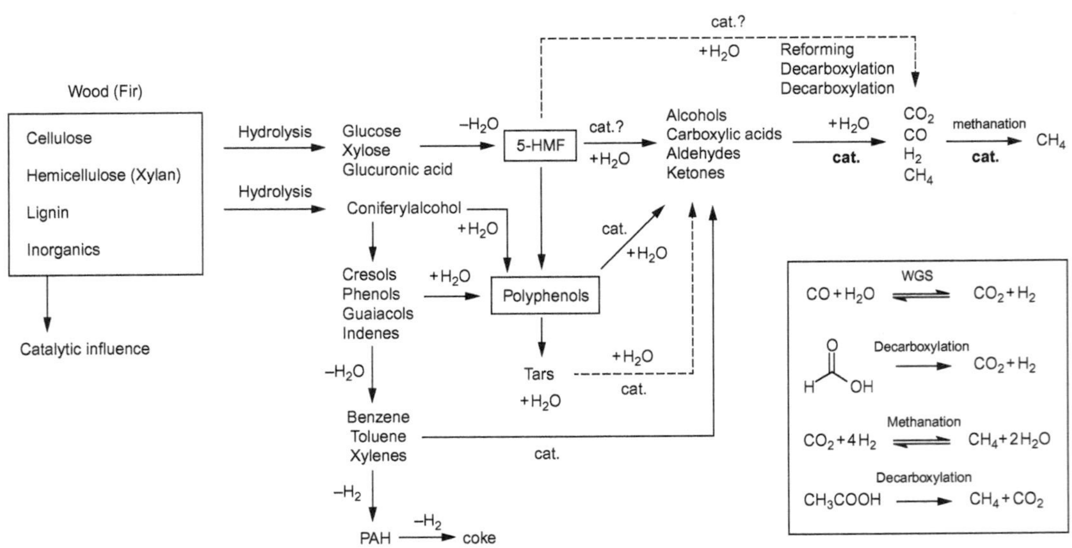

FIG. 6.1 Common hydrolysis reaction path [10].

along with its implementation as proposed by Gassner [21] and slightly modified in this study. First, two main compounds are identified in the feedstock, the lignin and the holocellulose, as shown in the following:

$$\text{Biomass} \rightarrow \text{holocellulose} + \text{lignin}$$

i.e.:

$$CH_{r_{bm,H}}O_{r_{bm,O}} \rightarrow (1 - c_{lignin})CH_{r_{cel,H}}O_{r_{cel,O}} + c_{lignin}CH_{r_{lig,H}}O_{r_{lig,O}} \tag{6.1}$$

assuming the biomass and lignin molar ratios r (notably, biomass and lignin molar composition) and the lignin mass fraction c in the feedstock known. To compute the cellulose molar ratio and the lignin molar fraction, molar balances are proposed:

$$r_{cell,H} = \frac{r_{bm,H} - r_{lig,H}}{1 - c_{lignin}} + r_{lig,H} \tag{6.2}$$

$$r_{cell,O} = \frac{r_{bm,O} - r_{lig,O}}{1 - c_{lignin}} + r_{lig,O} \tag{6.3}$$

$$c_{lignin} = \frac{M_{biomass}}{M_{lignin}} C_{lignin} \tag{6.4}$$

with M and C the molar fractions (of biomass and lignin) and mass fraction of lignin in the feedstock, respectively, and r the hydrogen and oxygen molar ratios referred to each compound.

During the hydrolysis, each of these two compounds reacts in a specific and simplified path, reported previously. The model considers only the most abundant and representative substances observed throughout several experiments. The holocellulose mainly dissolves in acetic and formic acid, acetaldehyde, and methanol:

$$\text{holocellulose} + \text{water} \rightarrow \text{acetic acid} + \text{formic acid} + \text{acetaldehyde} + \text{methanol}$$

i.e.:

$$\begin{aligned} CH_{r_{cel,H}}O_{r_{cel,O}} + a_{H_2O}H_2O \rightarrow a_{acetic\,acid}CH_3COOH + a_{formic\,acid}HCOOH \\ + a_{acetaldehyde}CH_3CHO + a_{methanol}CH_3O \end{aligned} \tag{6.5}$$

Stoichiometric coefficients are indicated by the letter a.

The lignin decomposition generates phenol, furfural, and glycerol:

$$\text{lignin} \rightarrow \text{phenol} + \text{furfural} + \text{glycerol}$$

i.e.:

$$CH_{r_{lig,H}}O_{r_{lig,O}} \rightarrow a_{phenol}\,C_6H_6O + a_{furfural}\,C_5H_4O_2 + a_{glycerol}\,C_3H_8O_3 \tag{6.6}$$

Eqs. (6.7), (6.8) are introduced in order to find the stoichiometric coefficients for the holocellulose decomposition. Indeed, carbon, hydrogen, and oxygen balances are not enough. They are based on experimental data provided by Waldner and Vogel and Kruse and Gawlik:

$$R_{acids} = \frac{a_{acetec\ acid}}{a_{formic\ acid}} = \frac{c_{acetec\ acid}}{c_{formic\ acid}} = \frac{c_{acetec\ acid + glucose}}{c_{formic\ acid}} \quad (6.7)$$

$$R_{alcohol/aldehyde} = \frac{a_{methanol}}{a_{acetaldehyde}} = \frac{c_{methanol}}{c_{acetaldehyde}} \quad (6.8)$$

with R as concentration ratios between the most abundant compounds resulting from hydrolysis.

The following interesting findings were obtained experimentally by Kruse and Gawlik [24]:

- The concentration of all kinds of phenols slightly decreases when temperature increases, between 320 and 410°C, whereas it grows passing from 40 to 50 MPa.
- Organic acid concentration falls down over the critical point of the water, but acetic acid has a maximum at 370°C and the formic one slightly decreases with the temperature.
- Furfurals disappear at very high temperature.

Waldner observed, without a catalyst, that the addition of alkali salts enhances biomass liquefaction. Unfortunately, some of them could be highly corrosive for Ni-based alloys.

In the continuous process, the inorganic matter has to be sequestered by dedicated equipment. Indeed, the presence of salts in the gasifier reactor can cause plugging, corrosion (production inside the reactor of H_2S gas) and catalyst deactivation in a very short time. Furthermore, the salts which can be recovered have an economic value as fertilizer since the feedstock is rich in nitrogen and phosphor; they are widely used in agriculture or in the integrated process of the feedstock production, like microalgae [19]. In this way, the cycle of these nutrients is closed. In particular, the phosphor recovery is extremely interesting. Nowadays phosphate fertilizers are obtained from rock phosphate which is an unrenewable source. It should be remarked that the salt brine mainly consists of water (80%) from which salts are separated. Salts are held in long polymer chains of lignin and holocellulose and they are released during the hydrolysis and biomass liquefaction. When the salts are released, they dissolve in the subcritical water since their solubility is high. It may be possible to thermally separate and collect them in a concentrated salt brine. By the way, the brine produced is a mixture of extracted salts and dissolved organic compounds, because a perfect separation is infeasible. Thermal salt separation happens because of the supercritical water features. As Leusbrock [25] writes, at this state many water properties change quite rapidly and among them is the solvation of inorganic compounds. This behavior is shown well by the dielectric constant evolution, as depicted in Fig. 6.2. The water is a polar fluid in subcritical conditions, but it becomes a nonpolar one. The dielectric constant is a good indicator for understanding the solvation behavior. Changes in hydrogen bonds and solvation behavior cause a decrease of salt solubility. It reaches a new equilibrium and produces a solid phase, which provokes severe issues in industrial applications. When the process is kept under control, this behavior promotes the inorganic separation and sequestration.

The state of the art of the technology relies on this process to separate and sequestrate the salts throughout the continuous hydrothermal gasification. The component is typically a heated vessel which collects a concentrated salt brine on the bottom; then the flow is thrown away. The stream, coming from the hydrolyzer, is heated up to over 450°C and overpasses the supercritical conditions; the salt solubility decreases quickly and the mixture is gathered. It is made by inorganics, water and a few percent of the products from the hydrolysis. The vessel design is still under investigation because its effectiveness is sometimes weak. In particular,

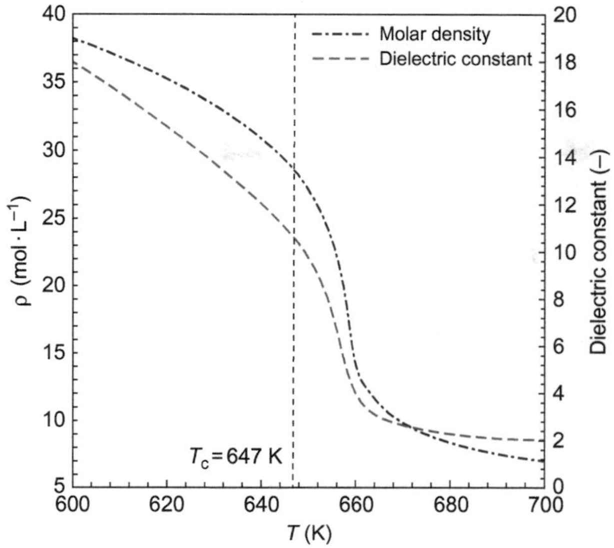

FIG. 6.2 Dielectric constant at 25 MPa [25].

Peterson et al. [17] highlighted in several experiments some issues that can occur when the design or the operating conditions change: the temperature distribution, the salt type, the biomass flow rate, the length of the entrance dip tube. In any case, blockages in the vessel can lead to shorter residence times and to an ineffective separation. In Fig. 6.3 the vessel tested by Peterson and his colleagues and by Schubert is shown.

Once the flow is salt free, the gasification can take place in a dedicated vessel. It is almost energetically balanced. The transformation needs a catalyst in order to be carried out in a reasonable time. Research on the most suitable catalyst has been and is still intense. American and Swiss scientists have spent much time on this subject. They found the most promising metals are Raney nickel and ruthenium, as Vogel, Waldner and Elliot confirm [10,12,26]. Between them, the latter looks more reliable and performs better.

This process consists of several reactions, among which the water gas shift and methanation are the most relevant. Those reactions also drive the gas composition at equilibrium. Due to the water-gas shift (WGS) reaction, the maximum yield of methane is reached at low temperature (around 400°C) [10], whereas the maximum yield of hydrogen is reached at high temperature, above 600°C. Also the catalyst choice can influence the gas composition at the outlet [12]. A simplified reaction path is proposed by Vogel et al. [26] for wood gasification, made by the gasification, methane, CO_2, and hydrogen reactions (Eqs. 6.9–6.11, respectively); v and s in brackets mean vapor and solid states, respectively.

$$CH_{1.49}O_{0.67(s)} + 0.33H_2O_{(v)} \rightarrow CO + 1.08H_2 \tag{6.9}$$

$$0.52CO + 1.56H_2 \rightarrow 0.52CH_4 + 0.52H_2O_{(v)} \tag{6.10}$$

$$0.48CO + 0.48H_2O_{(v)} \rightarrow 0.48CO_2 + 0.48H_2 \tag{6.11}$$

$$CH_{1.49}O_{0.67(s)} + 0.29H_2O_{(v)} \rightarrow 0.52CH_4 + 0.48CO_2 \tag{6.12}$$

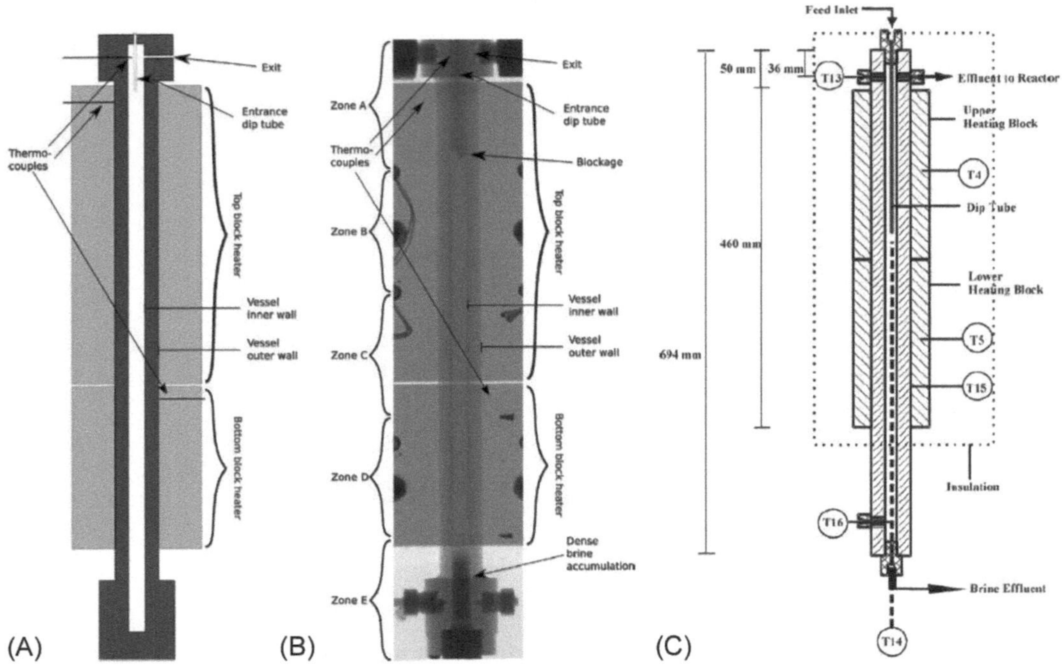

FIG. 6.3 (A) Scheme of the salt separator. (B) Image of the vessel acquired with the neutron radiography technique [17]. (C) Scheme of continuous salt separator studied by Schubert [15].

Eq. (6.12) provides the overall reaction which makes carbon dioxide and methane available from wood (slightly exothermic). These reactions are specific for the wood composition assumed in Eqs. (6.9), (6.12) but the same model can be generalized for every kind of biomass. A more general and precise reaction path is provided by Waldner: all the simple compounds that come from the hydrolysis can be gasified in the following pattern: acetic and formic acid undergo the decarboxylation (Eqs. 6.13, 6.14), whereas acetaldehyde, phenol, furfural, glycerol, and methanol undergo reforming reactions (Eq. 6.15). Actually, acetic acid is also partially decomposed by reforming:

$$CH_3COOH \leftrightarrow CH_4 + CO_2 \tag{6.13}$$

$$HCOOH \leftrightarrow CO_2 + H_2 \tag{6.14}$$

$$C_nH_mO_k + (2n-k)H_2O \leftrightarrow nCO_2 + \left[2n + \frac{m}{2} - k\right] H_2 \tag{6.15}$$

The steam reforming reaction in Eq. (6.15) is the total one and it can be split into a partial reforming (Eq. 6.16) and the water gas shift reaction (Eq. 6.17):

$$C_nH_mO_k + (n-k)\,H_2O \leftrightarrow nCO + \left[n + \frac{m}{2} - k\right] H_2 \tag{6.16}$$

$$CO + H_2O \leftrightarrow CO_2 + H_2 \tag{6.17}$$

Water-gas shift reaction (WGSR) is fundamental in industrial processes for hydrogen production. Biomass and coal gasification is based on this reaction which allows methanation afterwards. The precise reaction is:

$$CO + H_2O \leftrightarrow HCOOH \leftrightarrow CO_2 + H_2 \tag{6.18}$$

High density of water enhances the reaction [27] but Sato et al. [28] stated the reaction kinetic is influenced only by the CO concentration when the partial pressure of water is below 30 MPa, according to:

$$-\frac{d[CO]}{dt} = k[CO] \tag{6.19}$$

It is also proven that alkali salts accelerate the reaction as well as the temperature increase and a metal catalyst. Neither the salts nor the temperature increase technique is practically feasible, however. Alkali salts suppress methanation whereas the temperature must be chosen to improve the yield of methane, produced at lower temperature.

Finally, the methanation reaction (Eq. 6.20) occurs in the gasifier:

$$CO + 3H_2 \leftrightarrow CH_4 + H_2O \tag{6.20}$$

In batch reactor experiments, the equilibrium condition is often achieved by varying the residence time at fixed temperature and pressure [26]. In a continuous reactor the residence time is fixed when the flow rate and the size are chosen: the equilibrium can be reached by varying temperature and pressure. In any case, it is advisable to keep a suitable trade-off between the methane production and the thermodynamic overall efficiency.

In the last step, energy recovery has to be performed to make the process energetically and environmentally sustainable. Then, the produced SNG has to be separated from the liquid phase by means of two flash separators, and finally purified from CO_2 in order to achieve the natural gas grid standards (96%mol of methane, for Swiss standards). Gassner et al. [22] proposes chemical methods of purification. He has noticed the different solubility in water of methane and carbon dioxide when the pressure decreases. This procedure and the membrane filtration can provide SNG at grid quality. The progress in membrane separation systems allows this technology to be used directly for the gas purification. ENEA, the Italian national agency for new technologies, energy and sustainable development, states that high purity standards can be reached by using a set of membranes [29]; purification level depends on the layout of filters as shown in Fig. 6.4.

These devices cause a big pressure drop to occur, which can be suitable to decrease the pressure at the gas grid level. The literature review reveals few membranes can operate at such a pressure. The rapid development which characterizes this technical field allows the possibility that in a few years the technology will be widely available for a more extended range of pressure and temperature.

Energy recovery is performed, especially for the thermal part: both gas and liquid have to be cooled down and a regenerating heat exchanger network has been designed. Wasted streams contain a more-than-negligible amount of hydrogen and methane: the energy is recovered by means of combustion. Otherwise, when the temperature is too high (salt separation),

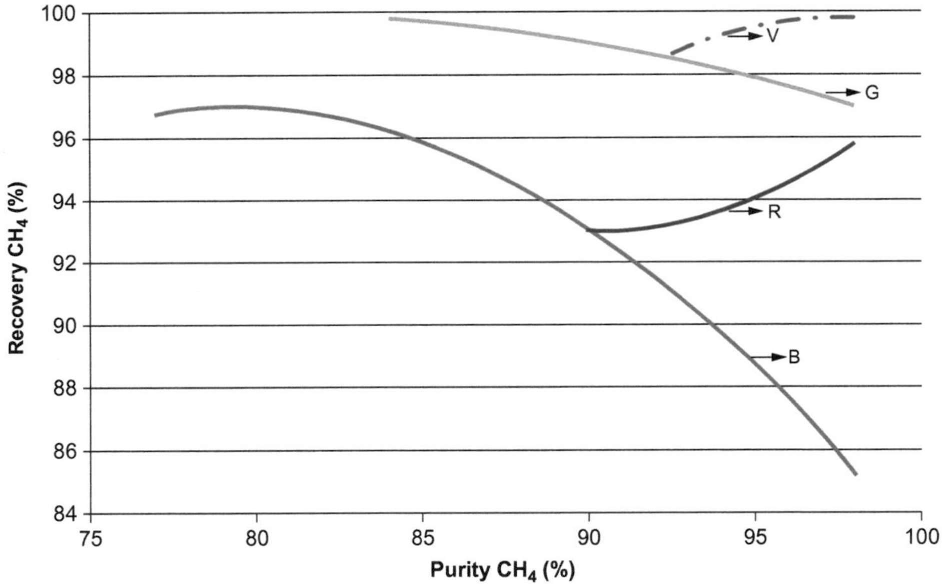

FIG. 6.4 Different layouts' performance of membrane separation system evaluated by ENEA. *Blue line/B*: single stage separation; *red line/R*: double stage separation with second retentate recirculation but without second power recovery system (gas turbine); *green line/G*: double stage separation system without recirculation and with second power recovery equipment; *violet line/V*: double stage separation system with recirculation and energy recovery system; the last one is the chosen one [30].

external sources or a portion of the SNG are used. The energy recovery cannot supply the whole heat demand; thus external sources or a part of the biogas are employed to fulfill this lack.

Some possible application of the biogas can be pointed out: injection into the natural gas grid, direct combustion in boilers or furnaces, direct use in internal combustion engines and direct combustion in gas turbines. Currently, this last technology is coupled with a Rankine cycle in a combined gas turbine-Rankine cycle facility, which is the most efficient technique to produce power. The last two options can be run in cogenerating mode, getting heat and power simultaneously.

6.2.3 Model Implementation

Here we have a look at the model used in this study and at the results which it has given. The model has been made on the Aspen Plus© platform. This is chemical process simulation software, developed especially for the oil industry. By means of a user-friendly layout, the software allows the building of a representation of the plant, offering a library of blocks which covers almost all the components requested in a chemical plant. For each block, the physics and chemistry is already implemented. In addition, a dedicated tool permits the user to customize additional components. Moreover, it has an extended library of thermodynamic models with which the analysis can be led; it includes also a tool to perform sensitivity analysis. The software can manage utilities too: it can simulate the power, heat, and cold supply by heat carries. The most significant elements of the model implementation are reported in the following paragraphs.

The biomass elemental composition has been provided by laboratory standardized analysis, whereas the composition of lignin and holocellulose has been taken from the literature [31]. The hydrolyzer has been modeled by blocks Ryield, which compute the equilibrium composition by means of yields of incoming compounds. It has been customized by the implementation of an automatic Microsoft Excel© datasheet (the software allows communication with a Fortran compiler) of the hydrolysis reaction path. Furthermore, the transformation has been split into 11 steps to better evaluate the thermal power-temperature curve. The model provided by Leusbrock et al. [32] to calculate the concentration of salts in supercritical water has been used to assess the salt separation. The results have been implemented in a traditional separator provided by Aspen Plus. Since the final gas composition was unknown, the gasifier has been reproduced with RGibbs reactors. They compute the equilibrium composition by means of Gibbs free energy minimization. The process has been split into two steps: the reforming and the WGS and methanation. All the reactions have been applied in a dedicated section of blocks. Finally, the gas separation and purification are handled by two flash separators and two polymeric membranes, respectively. Filters have been designed in a recirculating configuration. Since Aspen Plus has no block representing a membrane filter, calculations have been performed by hand and implemented in a traditional separator later on.

Simulations have given interesting results. As the sensitivity analysis has shown, the most suitable gasification temperature and pressure are 400°C and 300 bar, respectively. Also, the solid fraction in the incoming sludge significantly affects the gas yield: it has to be included between 15% and 30% in order to avoid pumping issues. In the best conditions the gasification provides 15.68 MW of biogas, equal to 0.323 kg/s with a lower heating value of 48.48 MJ/kg. The plant is fed by 20 MW of dry EFB (based on the biomass LHV). Thus the gross efficiency is 76.14%. As already seen, the process is not self-sustainable and it requires external heat and power. The minimum thermal energy requirement has been calculated with a pinch analysis: the sum of this quantity and the demanded power gives the minimum energy requirement (MER). Accounting also for the MER, the net efficiency becomes 55.20% and the net energy content in the bio-methane is 11.04 MW.

Finally, an interesting result is introduced that the analysis has gathered from a second sensitivity analysis (20 different biomass elemental compositions have been used). It highlights the relation between the methane yield in the biogas and the biomass elemental composition. The ratio oxygen/hydrogen is linearly related to the methane molar fraction in the gas. Indeed, the lower the O/H ratio, the higher the molar fraction, as shown in Fig. 6.5. This plot may be useful for the biomass selection. It is the most significant parameter because the proportion of oxygen and hydrogen moves the equilibrium between the methane and carbon dioxide yield.

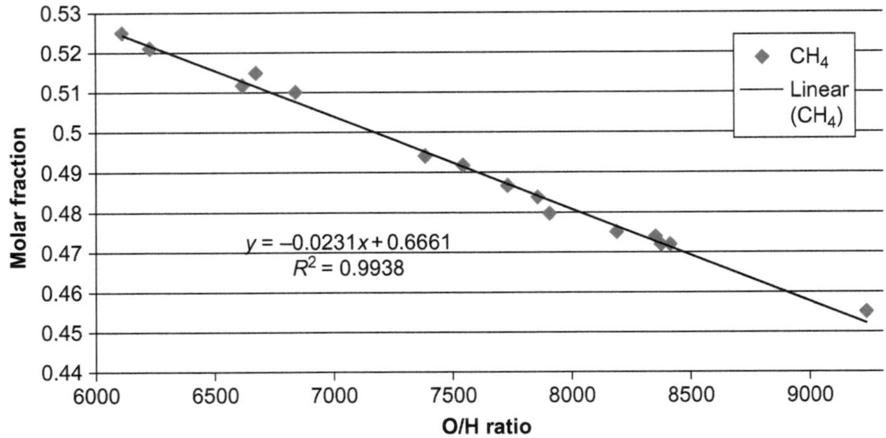

FIG. 6.5 Relation between O/H ratio in the biomass and methane molar fraction in biogas.

6.3 LIFE CYCLE INVENTORY

6.3.1 Goal, Functional Unit, and Scope of the Study

The use of biomasses for energy is an alternative to fossil fuels. Since we are facing global problems like fossil resources depletion and climate changes, biofuels seem a possible substitute, or at least might be integrated with current energy supply carriers. But the real environmental benefit of such fuels has to be proven; their production and transportation can involve issues such as deforestation and intensive land use. Petroleum oil and natural gas and coal are still the most used fuels worldwide, even though the production and trade of biofuels can greatly reduce climate change and fossil resource depletion. The LCA performed in this chapter has two main goals, which can be carried out simultaneously. First, it aims to assess the major environmental impact caused by processing Brazilian EFB using hydrothermal gasification. Second, it intends to compare this impact with that associated with the current supply procedure: the two main gasification products, bio-methane and salt brine, are compared with natural gas and traditional fertilizers. Moreover, the effects of land use change and the plantation end of life have been evaluated in detail. Finally, different scenarios of the hydrothermal MER fulfillment have been compared among them.

Since the comparison occurs between two products that have different origins, the functional unit concerns the main product: the heat produced by the gas combustion. The functional unit that has been chosen is 1 MWh of heat. The amount of the coproduct "salt brine" which is obtained throughout the hydrothermal process is directly related to the bio-methane production: thus, the salt brine availability is 142 kg. Depending on the scenario, several raw materials had different previous uses. Replacement products have been accounted for according to the main functional unit.

Finally, the scope includes all the bio-methane production steps, from the fresh-fruit bunch cultivation at the oil palm plantation to the hydrothermal gasification and the gas combustion. Transportation of raw material and final products has been taken into account. It does not consider all the oil palm mill chain but only the first three stages, until the fruit threshing: here, the EFB is produced along with the fruits. Thus, the thermal and electrical energy consumed in this phase and produced by oil palm mill wastes (palm fibers and shells) is allowed for an external source. The construction and decommissioning of buildings and equipment has not been taken into account. The direct land use change has been strongly considered: the work aims to highlight the great importance it has on the overall environmental impact of biofuels.

6.3.2 System Definition

This section gives a detailed overview of the system; the main inputs are presented.

The system location is assumed to be in Brazil, in the Pará State. Here it is assumed that a new biodiesel production site will be built, in an area almost 100 miles from Belém. The feedstock supply will be ensured by a large network of smaller producers. FFB will be purchased from almost 600 family farms which manage about 6000 ha of degraded land. The total required land for oil palm plantation will be 12,000 ha, spread around the municipality of São Domingos do Capim and Mãe do Rio.

Pará State is the largest oil palm producer in Brazil because of its location and climate: oil palm (*Elaeis guineensis*) is a tree crop requiring a tropical climate. The maximum temperature must range between 24 and 28°C, whereas the minimum must be higher than 18°C. Temperatures must be stable for all the year. Moreover, the yearly rainfall should be higher than 1800–2000 mm [33], in order to achieve optimal growth. The annual productivity of oil palm varies between 7 and 25 tons of FFB per hectare, with an average of 18 tons/ha during the whole plantation life span. The typical economic life span of such a plantation is 25 years. These data are reported in [2].

The system includes the biomass production, which starts from the plantation seeding which is carried out by using poly plastic bags $38 \times 45\ cm^2$. On average, 300 seeds are required for 143 plants/ha [2]. Throughout the plantation economic life, the cultivation needs fertilizers because of the lack of nutrients of the Amazonia soil. Usually, the suggested amount of fertilizers is 2.6 kg/palm/y of urea (nitrogen), 3.7 kg/palm/y of SSP (phosphorus), 2 kg/palm/y of MOP (potassium), 0.5 kg/palm/y of magnesium sulfate (magnesium), and 0.1 kg/palm/y of borax (boron) [34]. Also, pesticides and herbicides have to be used, as well as mulch; all these products are assumed to be purchased in the regional storehouse. They are made outside the region and it is assumed they are carried by train for 300 km and finally 100 km (return trip) by lorry from the rail logistical platform to the storehouse. Typically, mulching raw material consists of the EFB, which is provided by the oil mill. Here, since the EFB is the energetic feedstock, mulch will be purchased and it is made from wasted wood chips. Although the yearly rainfall is very high, the plantation needs an irrigation system, especially during the drier period (summer). Indeed, oil palms require 200 L/day on average [34]. Fresh-fruit bunches are harvested each 10–12 days and this operation is carried out manually and FFB are collected in tractors (3.5–7 tons). In order to prevent fruit acidification, the fruit has to be processed within 24 h from the harvesting. Thus, the processing mill location has to be close to the plantation.

The transportation to the mill is carried out by small trucks which collect the tractors' load. Trucks can carry between 7.5 and 16 tons. Since the new Brazilian transportation regulation is as restrictive as the EURO 5 regulation in Europe, this emission category has been taken into account. The mean distance (return trip) between the palm field and the mill has been guessed to be about 100 km.

This system considers just the first three stages of FFB processing, namely until the EFB-fruit separation. Therefore, FFB charge, sterilization, and threshing are inside the system boundaries. All the downstream steps have not been taken into account because they primarily belong to another process (oil palm production) and they do not actively occur in the gasification process. The FFB comes to the mill and it is charged and washed on a conveyor belt. Then it is sterilized with low pressure steam (60–90 min at 120–130°C [35]) in order to deactivate the free fatty acid production and to simplify the stripping process. Afterwards, the threshing takes place in a rotating thresher; the machine consists of a metal cage where sterilized bunches are slightly spun. The threshing outputs are oil palm fruits and EFB and, consequently, the first allocation is done. Downstream steps produce crude oil palm (200 kg/ton_{FFB}), kernel oil palm (50 kg/ton_{FFB}) and several wastes: POME (palm oil mill effluents) (670 kg/ton_{FFB}), fibers (140 kg/ton_{FFB}), and shells (60 kg/ton_{FFB}) [36]. Nowadays the POME is treated in a wastewater plant or it is digested in biogas. Fibers and shells (LHV: 9.2 and 16.9 MJ/kg, respectively [37]) are burnt onsite in biomass furnaces. They produce high pressure steam which runs back to a pressure turbine (power production) and the discharged steam is delivered to sterilizer tanks and digestion equipment. Usually, all the fibers and 40% of the shells are enough to fulfill the heat and power demand of the mill; 60% shell excess is burnt and the generated steam runs a turbine. Thus, the heat and the power demand of the first three milling stages is supplied from outside. Moreover, this process needs tap water (160 kg/ton_{FFB}), decarbonized water as heat carrier (1 kg/ton_{FFB}), lubricant and diesel to start the furnace. Transportation of raw material is always taken into account.

After the threshing, the EFB is carried to the hydrothermal gasifier. The system boundaries include all the gasification process, from the sludge preparation to the gas purification. It proceeds according to the scheme shown in the previous section. The energy and material requirements are based on the model results. Decarbonized water and thermal oil are needed as heat carriers: in order to compute their replacement rate, a general assumption has been made, looking at existing supercritical processes. The plant also consumes lubricant for its rotating machine. The model also provides the amount of catalyst (ruthenium) which has to be substituted because of the salts. In this section the replacement mulch is accounted for. Raw material transportation is also considered. The electrical energy is supplied by the grid, whereas two different scenarios have been describe for the thermal energy supply. As shown in Fig. 6.6, the former uses the residual amount of shells to MER fulfillment and it also requires diesel for the boiler start-up. The latter takes a part (Fig. 6.7) of the produced biogas to satisfy the MER. The gas boiler does not need diesel starter. Moreover, since the shell exploitation prevents the extra power production at the oil mill, Brazilian power facilities have to cover this lack. Thus, this extra power generation is accounted for in the hydrothermal gasification inventory. That happens because the new system causes a change in the use of the remaining shells.

The gasification produces 106.9 Nm^3 of bio-methane and 142 kg of salt brine in the first case and 106.9 Nm^3 of bio-methane and 260 kg of salt brine in the second case. The second delivers more salt brine because it requires much more EFB to fulfill the biogas production

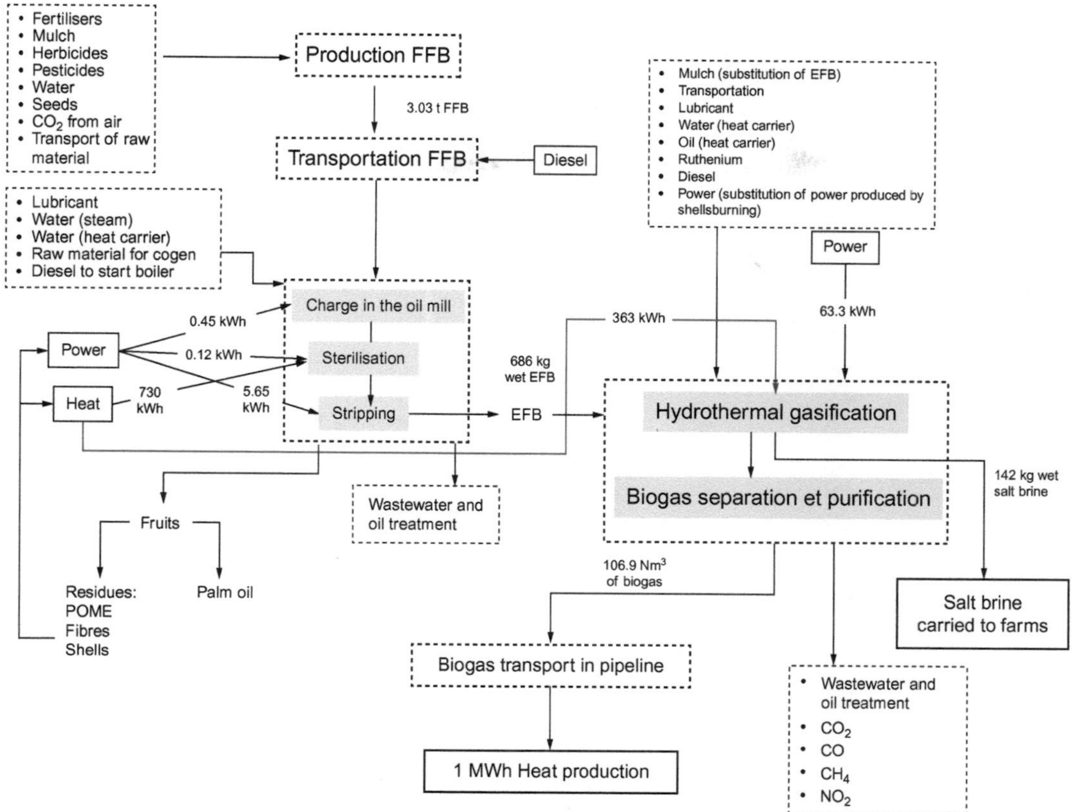

FIG. 6.6 System pattern when the hydrothermal gasification MER is provided by residual shells.

goal; a relevant amount of gas is burnt directly in the plant. At this level, the second allocation is performed. Since the process burns some fuels (waste gases and biogas or shells, depending on the case), it releases GHG into the atmosphere, such as biogenic CO_2, CO, CH_4, and NO_x. Furthermore, industrial wastewater and oil treatments have been considered.

Finally, the bio-methane that is available at the grid pressure of 70 bar is delivered to a generic final user by means of a pipeline. A travel distance of 100 km has been estimated. No compression station has been considered, due to the high available pressure and the short distance to be covered.

Bio-methane combustion happens in an industrial furnace with a 90% net efficiency. Heat production with hydrothermal biogas generates GHG emissions. Only the most abundant ones in methane combustion have been taken into account: biogenic carbon dioxide, biogenic carbon monoxide, and NO_x. No sulfur oxide has been allowed for, since the gas stream is free from sulfur.

6.3.3 Reference System Definition

The chosen reference system includes all the main services the system provides: the heat production, the fertilizer production, mulching and the land use. A service is considered in

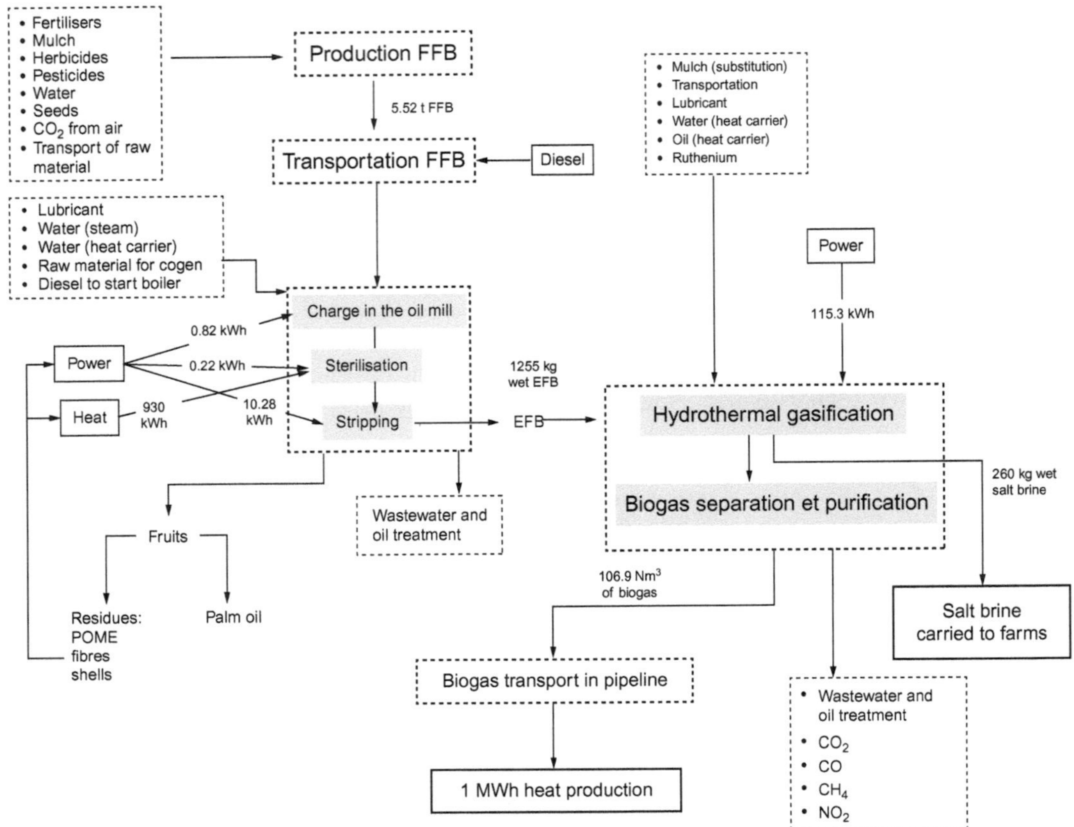

FIG. 6.7 System pattern when the hydrothermal gasification MER is provided by biogas spillage.

the reference system only if a difference occurs between the studied system and the reference or if the change in exploitation is economically significant. Thus, three different reference systems have been built, distinguishing the general case, the "residual shells" case (case where the hydrothermal MER is supplied by residual shells) and the "EFB as waste" case (case in which the EFB is previously considered as a waste at the oil mill).

Concerning the heat production from a gaseous fuel, the study has referred to the natural gas heat supply. As all the following three figures show, the heat production occurs by natural gas combustion in an industrial furnace (90% thermal efficiency). The gas is carried by pipelines for a distance of 2000 km (from Bolivia to Pará State). The reference also includes the natural gas extraction which takes place in Bolivia. It refers to associated Bolivian natural gas and not to the Brazilian shale gas extracted in the south.

The fertilizer reference consists of a mix of nitrates, sulfates, phosphates, and magnesium, having the same nutrient content of salt brine. Thus, the basis of comparison is nutrients instead of the weight. The reference system accounts for fertilizer production, transportation, and storage at the regional storehouse and final transportation to the farm.

Mulching has always been considered except when the EFB has been allocated as a waste (0% allocation to EFB, 100% allocation to oil fruits). Traditional mulching practice joins the reference system, which means the EFB is used as mulch. Typically EFB is not burnt because of its high water and inorganic content; thus it is collected on the mill site and carried back to farms, where it is put under palms as a soil conditioner. The reference system includes the EFB production for mulching purposes and its transportation to the farm. Transportation has been assumed to occur by small trucks (up to 16 tons) and tractors (up to 7 tons); the travel distance is 40 km (return trip).

The land use change strongly affects the environmental performance of biofuels and it cannot be neglected since the carbon stock and carbon emissions depend heavily on land management [3]. Brazilian regulation is very strict concerning new crop fields: no deforestation can be carried out [38] and the State government is involved in degraded land upgrading projects. Thus, the new plantation is supposed to take place in degraded lands such as pastures (grasslands). The approach which has been used is proposed by Delivand and Gnansounou [2]. The carbon stock change can be divided into two main portions: the stock in vegetation and the stock in soil, according to Eq. (6.21).

$$\Delta C_{LU} = \Delta C_{veg} + \Delta C_{soil} \tag{6.21}$$

The carbon stock change in vegetation is also distinguished in aboveground and underground vegetation. The difference in soil stock depends on the soil type and management: the carbon stock change, carbon losses equivalent to N_2O and CO_2. Vegetation stock data rely on data provided by a literature review done by RSPO (Roundtable in Sustainable Palm Oil) because of their more conservative approach. The authors stated IPCC (International Panel of Climate Change) data are too optimistic compared to those of other valid authors [39]. Those values are not specifically referred to a region because the stock in biomass depends on the vegetal species and not on the country. An IPCC tool has been used to calculated the carbon stock in soil: the institution makes software available which calculates old and new carbon stock. This computation is based on land exploitation (grade of tillage, fertilizing, and use), country and soil type. With 25 years of economic life span and 18 ton_{FFB}/ha yr of yearly FFB yield, carbon stock amounts are the following: 0.078 kg_C/kg_{FFB} in new plantation vegetation, 0.026 kg_C/kg_{FFB} in the existing degraded grasslands and −0.024 kg_C/kg_{FFB} (net emission) of carbon stock in land use change. Thus, the total modification in carbon stock is positive: 0.028 kg_C/kg_{FFB}. As shown in the following section, the land use change strongly affects the biofuel impact assessment. However, spatial and temporal system boundaries determine the choice of land use emission/stock. Since at the end of the economic life span, the plantation could be cut and burnt, vegetal carbon stock is temporary. On the other hand, if the palm wood is saved and recycled for further long-term use, the carbon stock can be longer and significant for GHG savings. Therefore, it is clear the biomass cannot be carbon neutral. This study aims to point out the importance of land use change in the biofuel impact assessment; however, it depends on the hypotheses and scenarios which are considered.

Finally, the previous use of palm shell has been considered only when the hydrothermal gasification thermal requirement is fulfilled by the residual palm shell. Indeed, its exploitation does not change in all the other scenarios and it has not been taken into account. Thus, the reference system includes the power production, which is generally carried out by means of residual shell combustion (60% of the shell in the oil mill is burnt and it produces power which is sold to distributors). The system consists of a biomass furnace which produces pressurized steam, the electrical generation with a steam turbine and its delivery into the power grid. The global efficiency has been assumed to be 25% since the furnace and the turbine efficiencies are usually low (70% and 35%, respectively). Figs. 6.8 and 6.9 summarize the main features of the reference systems.

6.3.4 Allocations

Throughout the assessment two allocations have been made: the first one involves palm fruit and EFB, the second one the hydrothermal gasification coproducts heat and salt brine.

The latter is an economic allocation and it depends on the gasification heat supply. When the process is fed by a portion of hydrothermal biogas, the biomass requirement is higher and the salt brine production as well. Therefore, the allocation percentage changes. Since no data are available on hydrothermal biogas and salt brine prices (no commercial plants are ready to produce), maximum purchase prices of competing products have been taken into account. Indeed, those prices represent the maxima amounts customers can pay for those facilities. If the production cost is considered, especially at this stage of development of hydrothermal gasification, the economic allocation can be twisted. Utility prices of 2013 have been used and they have been assumed to stay constant in the future. The natural gas price in Pará State was 38.18 $/MWh [40] whereas the calculated biogas cost is 66 $/MWh. The difference means a governmental subsidy is needed. The salt brine price has been computed as the price of a

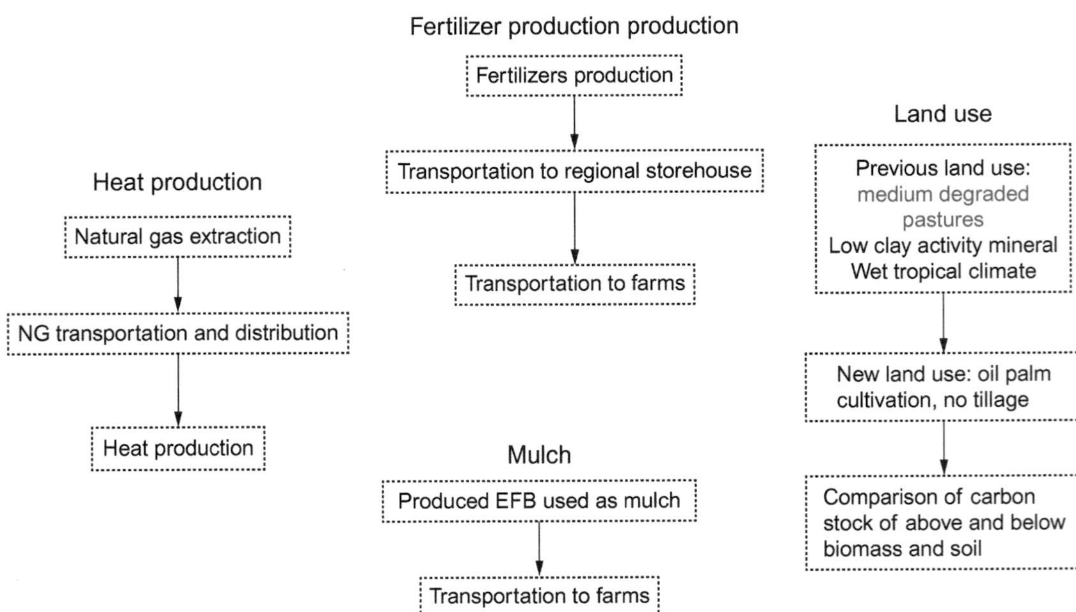

FIG. 6.8 Scheme of general reference system.

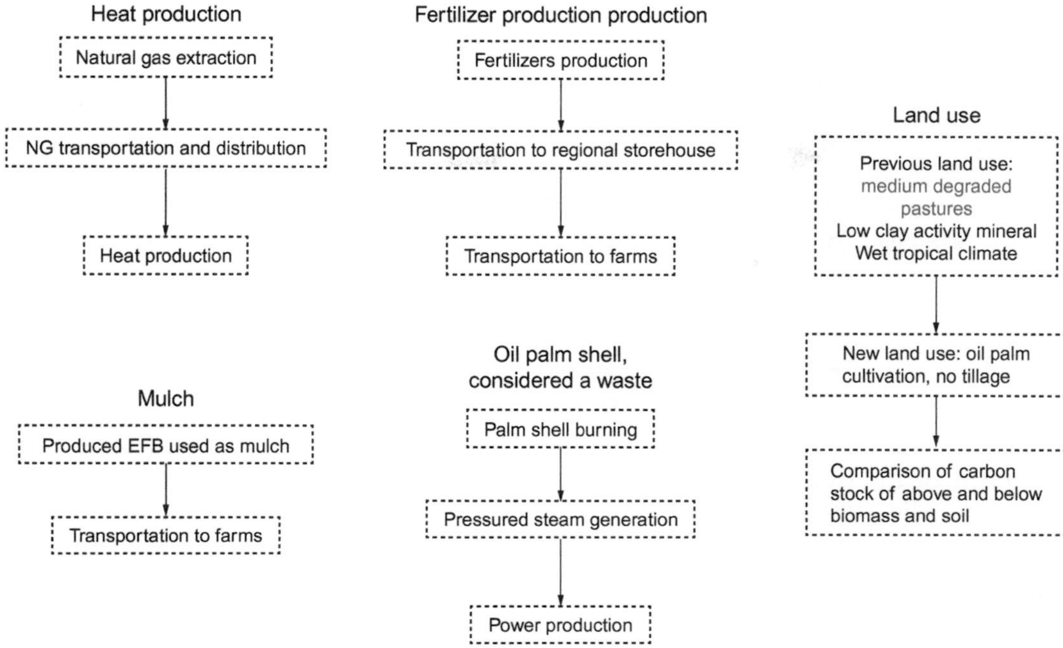

FIG. 6.9 Reference system of the hydrothermal MER case provided by residual shell.

fertilizer mixture with equivalent nutrient content; it is 8.61 $ for 142 kg of salt brine (amount produced when the process MER is fulfilled by palm shells). Thus, the allocation is 81.6% for the heat and 18.4% for the salt brine. If the gasification additional heat is provided by hydrothermal biogas spillage, the produced salt brine is 260 kg and its total price is 1566 $. Since the gas price is always 38.18 $/MWh, the allocation becomes 70.9% for the heat and 29.1% for the salt brine.

The first allocation depends on the EFB uses: three different ones have been considered. First, the economic allocation has been performed. Since the EFB can be used as mulch and it is used directly in palm plantations (no prices are available), common mulch price has been used. A mulch type has been taken into account with the same nutrient content and soil conditioner ability. The palm fruit price is also unknown because it is not traded: its price has been computed through the FFB price. The fruit price is the sum of the FFB price and the manufacturing costs to obtain it (transportation, charge, sterilization, and threshing). The FFB maximum price is provided by [41]. The prices are 0.012 $ and 0.91 $ for 1 kg of wet EFB and 3.39 kg of palm fruits, respectively. The allocation which follows is: 1.35% and 98.65% for EFB and fruits, respectively.

Another allocation criteria is the energetic one. The parameter of the comparison is the energy content (Lower Heating Value) of each coproduct. The total LHV of palm fruits is 31.5 MJ/kg whereas the LHV of wet EFB is 14 MJ/kg [42]. Per each kg of EFB, 3.3965 kg of fruits are produced: thus, the EFB energy content is 14 MJ and that one of fruits is 107 MJ. Therefore, the energetic allocation becomes 11.57% and 88.43% for EFB and fruits, respectively.

A third EFB reference exploitation has been highlighted and a third allocation method has been used in order to evaluate the impact of hydrothermal gasification if the EFB is thrown away as a waste. Indeed, in a Brazilian oil palm mill a common practice consists of recycling half of the EFB and disposing of the rest. The former is used as mulch; the latter is just a waste. Thus, only this second half can be addressed toward the gasification. Since it is a waste, all the allocation is charged on fruits: 0% for EFB and 100% for fruits. Furthermore, as already stated in the previous section, no modification in mulch production occurs between the new and the reference systems. Thus, for this scenario this aspect has not been considered in both systems.

6.4 IMPACT ASSESSMENT

The following section shows the analysis histograms of the assessment. Results are based on the last version of the Ecoinvent database, developed by a Swiss association with the involvement of EPFL, ETHZ (Eidgenössische Technische Hochschule Zürich), PSI, EMPA (Eidgenössische Materialprüfungs- und Forschungsanstalt), and Agroscope (Helvetic Confederation research group). The database is already implemented in the software SimaPro8.

The system requires 3.03 tons of FFB. At the end of the threshing process the available EFB is 689 kg and it is delivered to the hydrothermal gasification. Here, 106.3 Nm^3 of biogas are produced along with 142 kg of salt brine. Thus, the process yields 35.1 Nm^3/ton_{FFB} and 46.9 kg/ton_{FFB}. These values take into account all the losses along the whole production chain. At the same time, the biogas yield is lower if the MER of hydrothermal gasification is supplied by a part of the biogas; indeed, EFB feeding has to fulfill the process thermal energy requirement. So, the system needs 5.5 tons of FFB which guarantee 1255 kg of EFB. The salt brine amount rises up to 260 kg; biogas yield becomes 19.3 Nm^3/ton_{FFB} whereas the salt brine yield stays constant. That's why the brine amount depends on EFB feeding.

The most affecting polluting items are the same among different scenarios, although the percentages which are given back by simulations slightly change. Except for the EFB-waste scenario, palm cultivation (especially fertilizers, pesticides, and herbicides use) and FFB transportation account for about 11% and 19%. However, the highest contribution is made by the power consumption. Although the Brazilian electricity mix has a low carbon footprint, it is the main pollution factor, accounting for about 40%. Indeed, the electricity is the most used energy source throughout the whole process. When the EFB is considered a waste, EFB cultivation is not accounted for; thus, emissions from raw material transportation and electricity generation increase significantly.

The carbon stock difference in vegetation has been taken into account. The last section is devoted to comparing the effects of considering different scenarios of carbon stock.

6.4.1 Assessment With Economic Allocation of EFB Fruits

The climate change gas emissions have been analyzed separately for the heat and the fertilizer. As the histogram shows (Fig. 6.10), the main contribution is due to the heat production because of the allocation. The total emissions reduction is significant compared with references; the comparison between two methods of gasification MER supply highlights

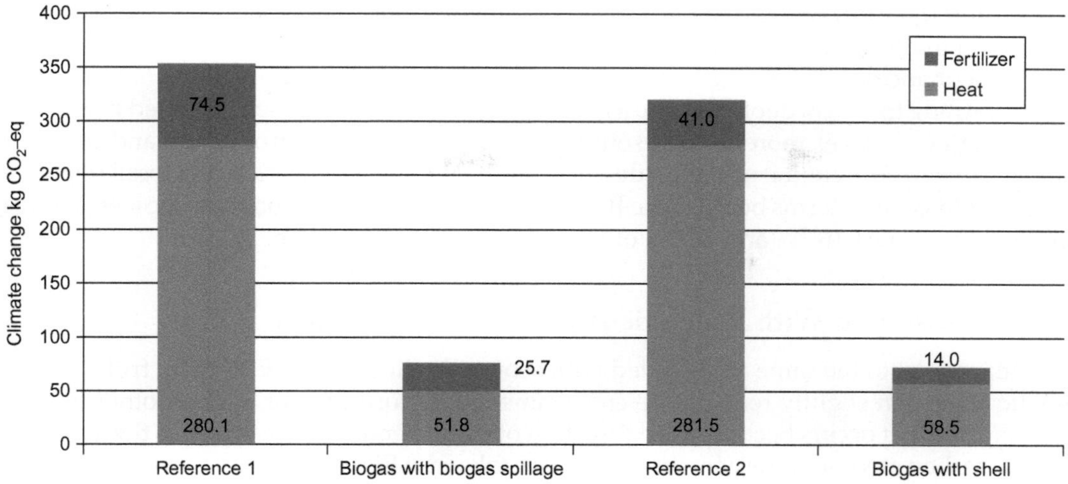

FIG. 6.10 Climate change histogram with economic allocation; Reference 1 refers to the biogas spillage scenario, Reference 2 refers to the shell combustion scenario.

a small difference. The biogas spillage case gives 14.4 g CO_2-eq/MJ_{biogas} (instead of 77.8 g CO_2-eq/MJ_{biogas}) whereas the shell use case emits 16.25 g CO_2-eq/MJ_{biogas} (reference: 78.1 g CO_2-eq/MJ_{biogas}). Fertilizer-specific emissions are based on traditional manure amounts (29.4 and 16.06 kg, respectively). In the first case the value is 874 g CO_2-eq/$kg_{traditional\ manure}$ (reference: 2534 g CO_2-eq/$kg_{traditional\ manure}$) and in the second one it is 872 g CO_2-eq/$kg_{traditional\ manure}$ (reference: 2553 g CO_2-eq/$kg_{traditional\ manure}$). Fig. 6.11 helps to identify the best scenario because it shows the percent climate change footprint. They differ because of the palm shell use change (power production method varies and the LCA takes it into account) and the way to feed the gasification which changes the EFB requirement.

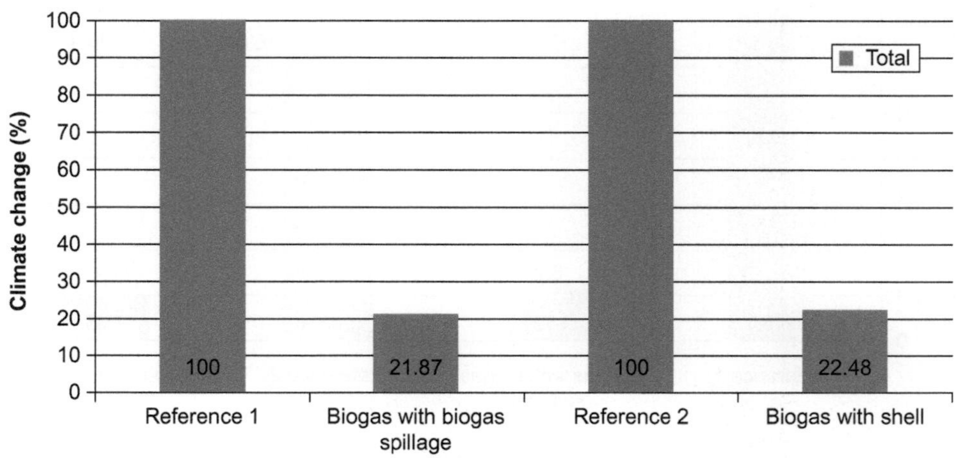

FIG. 6.11 The percent comparison between two new and reference systems. It compares their global climate change gas emissions.

The carbon footprint of Brazilian electricity is comparable to that of biomass power generation. Since the production of FFB stores more CO_2 than its cultivation releases, higher use of biomass enhances the environmental performance of the first scenario.

Concerning the fossil depletion, the assessment gives the result which is plotted in Fig. 6.12. The second case saves more fossil resources because the biomass production and transportation is lower. Although it also includes an additional power generation, the fossil depletion is smaller in general terms because the Brazilian power mix is based on hydropower. The fossil depletion is just 16.7% and 14.6% of the corresponding reference.

6.4.2 Assessment With Energetic Allocation of EFB Fruits

With respect to the same analysis led with economic allocation of EFB-palm fruits, the energetic allocation slightly reduces the environmental performance of the hydrothermal process (Fig. 6.13). It occurs because the cultivation of carbon emissions overpasses the growth of vegetal carbon stock and the energetic allocation takes the EFB production more into account. Thus, the thermal energy production emits 15.14 g CO_2-eq/MJ_{biogas} in the first case and 16.7 g CO_2-eq/MJ_{biogas} in the second one. Concerning the fertilizer production, climate change equivalent emission becomes 911.6 and 890 g CO_2-eq/$kg_{traditional\ manure}$ for the first and second scenario, respectively. GHG emission is cut down by 77.07% and 76.94% The fossil depletion grows because of the allocation which charges more on the EFB and makes it environmentally worse. Fig. 6.14 well represents this: the biogas production uses 7.9 g oil-eq/MJ_{biogas} in the first scenario and 6.4 g oil-eq/MJ_{biogas} while the references burn 32.8 and 32.9 g oil-eq/MJ_{biogas}, respectively. The fertilizer production has the following fossil depletion indexes: 449 and 361 g oil-eq/$kg_{traditional\ manure}$. References utilize 990 and 996 g oil-eq/$kg_{traditional\ manure}$, respectively. New systems save 72.7% and 78.4%, respectively, of fossil resources.

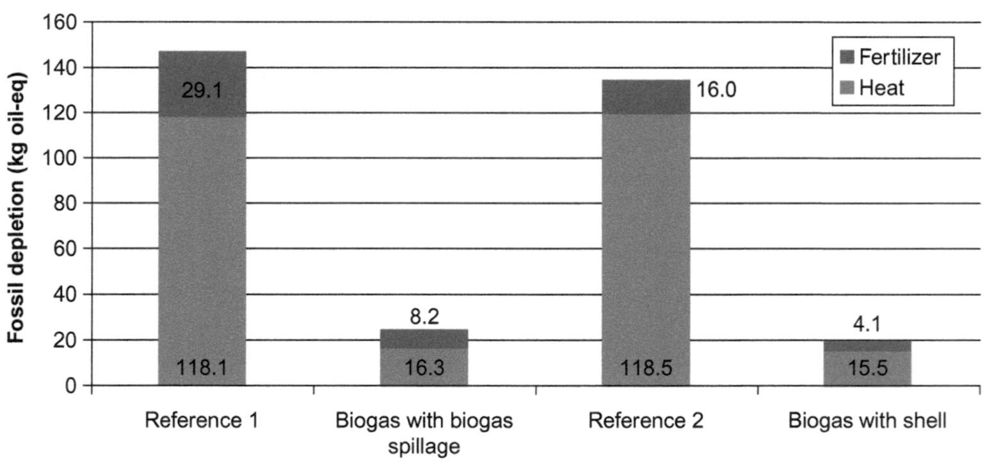

FIG. 6.12 Fossil depletion histogram.

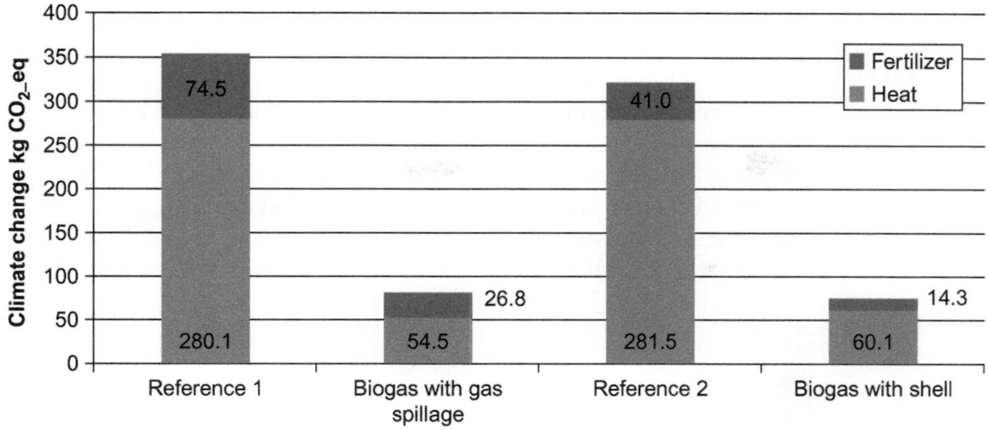

FIG. 6.13 Equivalent carbon dioxide emissions of two cases and comparison with reference systems.

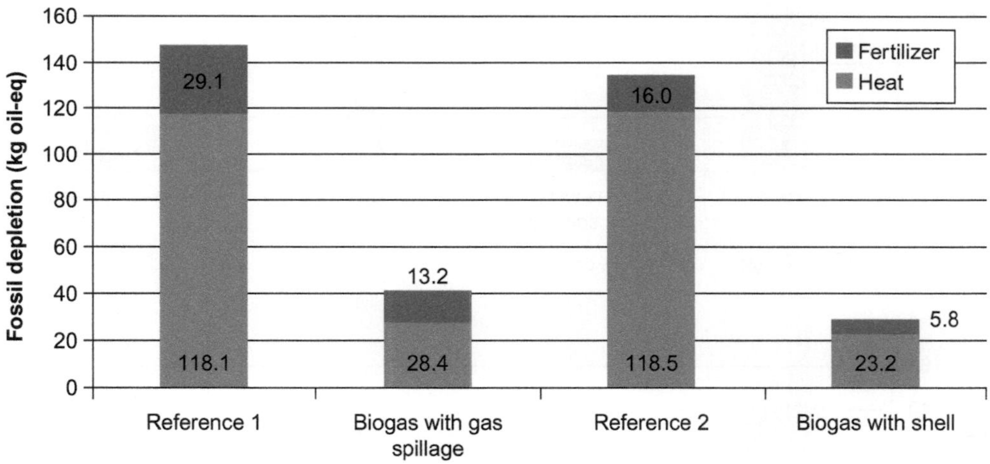

FIG. 6.14 Fossil depletion histogram, obtained by means of the energetic allocation between EFB and palm fruits.

6.4.3 Assessment With EFB Considered as a Waste

As already mentioned, if this hypothesis is made, the EFB has no economic value. This is valid when the mulching purpose is fulfilled by half of all the available EFB. Thus, the allocation percentage is 0 and environmental effects of the biomass cultivation and treatment are charged to the palm fruits. This reason, along with the lack of mulch replacement, explains the decrease of climate change gas emission and fossil depletion.

Specific emissions of two scenario are 6.08 and 9.8 g CO_2-eq/MJ_{biogas}, respectively, whereas those of the references are 74.5 and 74.9 g CO_2-eq/MJ_{biogas} (Fig. 6.15). Concerning the bio-fertilizer, specific values are 306 and 511 g CO_2-eq/$kg_{traditional\ manure}$. The reduction is clear and it is due to a very favorable allocation. Global decreases of GHG release with respect to the references are 90.85% and 85.9%, respectively.

At the same time, the fossil depletion specific values also fall down (see Fig. 6.16). The heat production by hydrothermal biogas just exploits 1.1 and 1.7 g oil-eq/MJ_{biogas}, respectively, whereas the salt brine production uses 54 and 112 g oil-eq/$kg_{traditional\ manure}$, respectively. With respect to references, the overall reduction of fossil depletion is 96.2% and 93.9%.

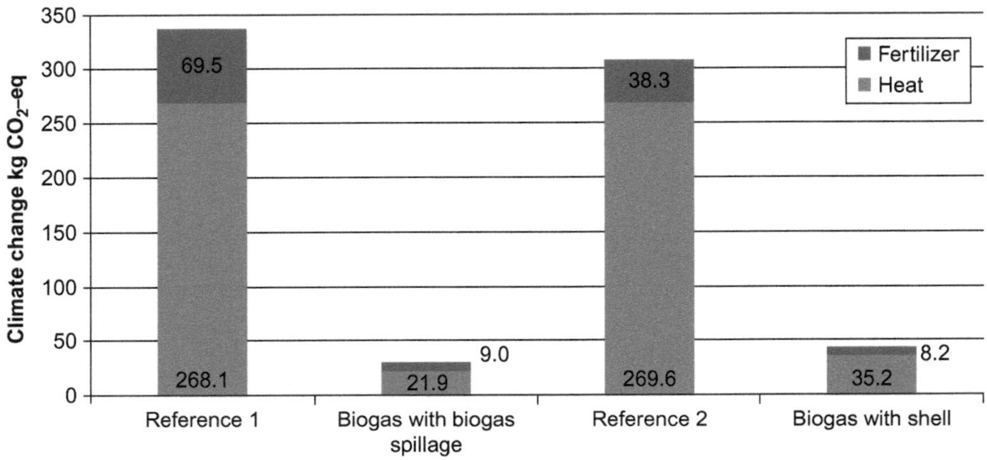

FIG. 6.15 Histogram of climate change gas emissions.

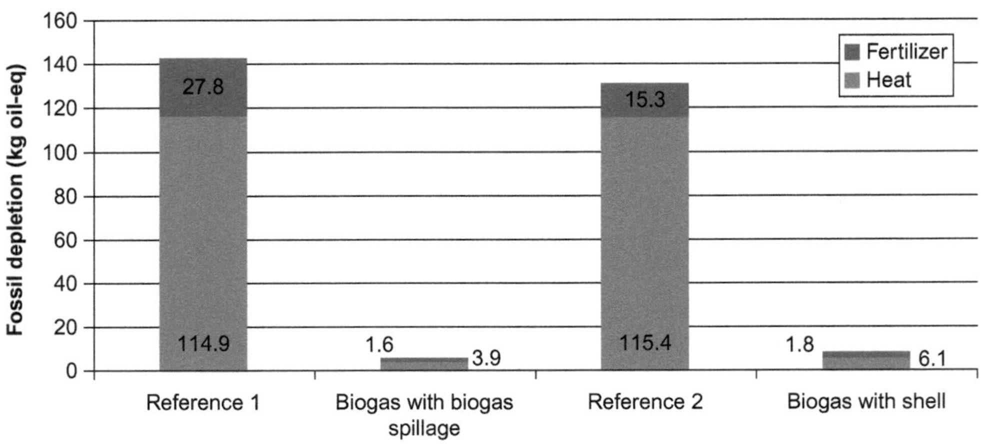

FIG. 6.16 Fossil depletion histogram for 0% allocation scenario.

6.4.4 Comparison Between Different Final End Use Scenarios of Oil Palm

Land use change has to be evaluated, as already shown in Section 6.3.3. The difference of the carbon stock in soil is usually provoked by soil management (tillage, fertilizing). The difference of vegetal carbon stock depends on the kinds of previous and new biomasses. In the scenario investigated, the soil stock change must be always taken into account, because an untouched land (degraded grassland) becomes a palm plantation. This cultivation requires work that moves the ground and the stored carbon. On the other land, the vegetal carbon stock can be considered or not. When the biomass is burnt, their emissions commonly are zero because that carbon is "biogenic" and directly coming from the atmosphere; but all the residual stock life (in roots, branches, logs, and leaves) depends on the plantation end of life. Oil palm plantations have an economic life span lower than the real plant life, because the productivity decreases after 25 years. Thus, the vegetal carbon is stored at least for this time, which is longer than the grass stock (prone to faster replacement and decomposition which quickly close the carbon cycle). In any case, at span end the biomass is treated and the vegetal carbon stock depends on it. The biomass can be cut and burnt or it can be used again in wood manufactories. Therefore, the choice among several tree final uses leads to different vegetal carbon stock evaluations. Indeed, carbon stock extension allows the vegetal carbon stock to be kept in the input inventory; if the plantation is burnt, the stock span is relatively short and the vegetal carbon stock is finally absent for the system under investigation. Clearly, GHG global emissions are highly affected by these hypotheses. Figs. 6.17 and 6.18 show the different

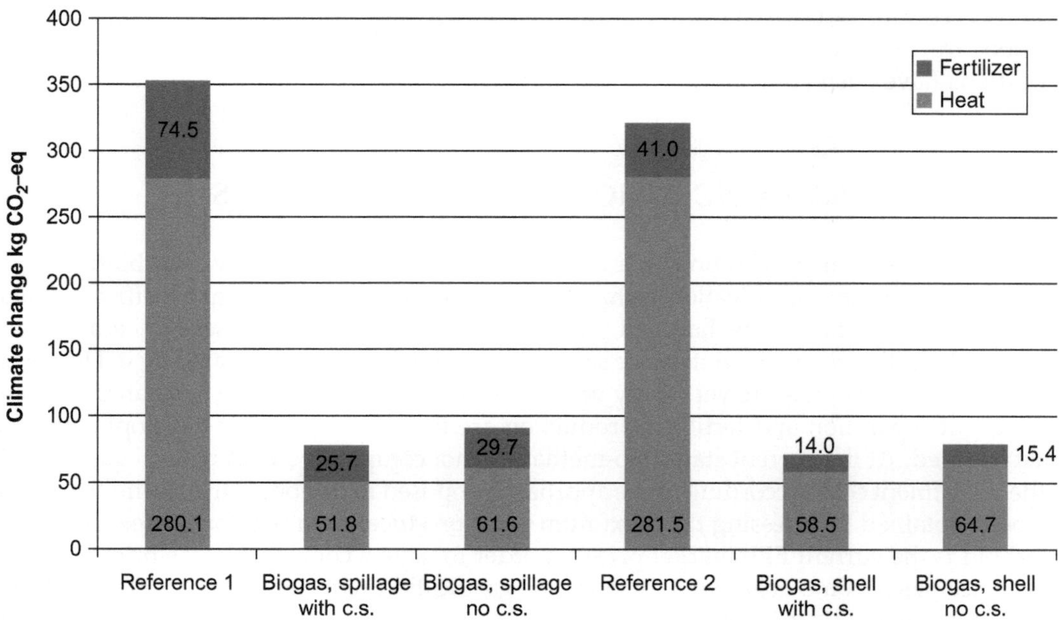

FIG. 6.17 Comparison of different vegetal carbon stock evaluations with economic allocation. C.S. means vegetal carbon stock.

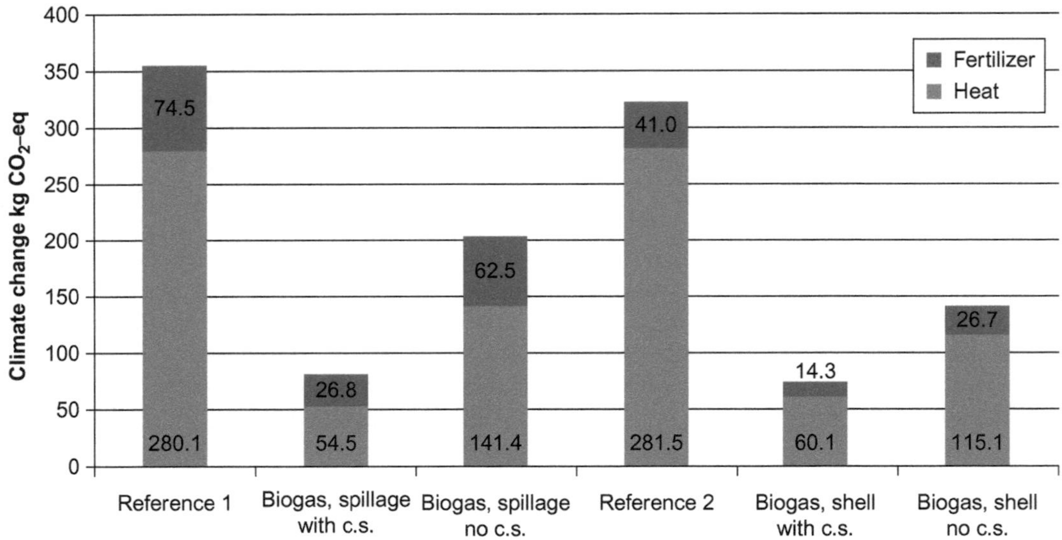

FIG. 6.18 Comparison of different vegetal carbon stock evaluations with energetic allocation. C.S. means vegetal carbon stock.

characterization of climate change performance, varying the vegetal carbon stock evaluation. The comparison has been carried out between two of the three assessments: if the EFB production accounts for 0%, the carbon stock as well as emissions throughout the biomass cultivation are not included. Thus, only evaluations made with economic and energetic allocations have been compared.

6.5 CONCLUSIONS AND PERSPECTIVES

The hydrothermal gasification is an advisable process in the innovative biofuel field. It overcomes all other gasification techniques because of the transformation efficiency and the biomass types that can be handled. Indeed, its net efficiency surpasses 50% if a suitable process integration is carried out; at the same time, wet biomasses are treated by it. However, the technology is not mature yet. Many drawbacks and weaknesses have to be analyzed and solved: salt separation and fertilizer production are the most important developments that must be faced. At the current stage, bio-methane is not competitive with natural gas because of the investment cost. According to the approach proposed in this book, the bio-methane cost has been obtained by guessing the maximum price producers can pay for the feedstock (the threshold is the current EFB market price) in order to have a comparable gas price. The biomass price has become negative, which means gasification requires subsidies. The environmental performance is remarkable, as Section 6.4 shows. According to the selected allocations and gasification energy feeding methods, climate change gas emission and fossil depletion indexes change. The best scenario occurs when the EFB has been previously considered as a waste. Here, EFB replacement carbon cost, which is the item that most affects feedstock production, is not taken into account. Furthermore, other allocations give a comparable result since the carbon footprint of EFB production is almost zero when the vegetal carbon stock is considered. Actually it is slightly emitting, thus energetic allocation makes the climate change gas emission index worse. Concerning fossil depletion, the carbon stock effect disappears and biogas spillage and energetic allocation make the index worse. Thus, the biogas spillage is a better MER supply method from the GHG emission point of view and a worse one concerning the fossil depletion. In this specific case, a larger feedstock requirement (especially mulch replacement and its transportation) negatively affects the fossil fuel consumption index. The comparison has to be carried out by means of emission and depletion percent reductions referring to the respective reference system: indeed, configurations have little differences which make systems directly incomparable. Moreover, the allocation type highly influences the assessment result, leading to omitting or overrating some chain steps. If the EFB was considered a waste, it does not bring any contribution to the hydrothermal gasification. All environmental charges of the palm cultivation go to the main product which hails from (oil palm or biodiesel). In this way, the analysis refers to a previous use assumption, whereas such an assessment should be based on known future biomass employment. That is why energetic allocation is actually the most significant. According to the energetic final use of the whole product range, allocation of feedstock components is performed by means of an objective factor: the total mean energetic content. This hypothesis spreads environmental impacts of biomass cultivation and treatment onto the energetic potential of each feedstock part. Furthermore, assumptions about the life end have to be carefully evaluated and discussed. Indeed, the medium vegetal carbon stock can last for the crop life span (25 years) or it can be extended by a proper use of the residual biomass. With respect to traditional rainforest carbon stock, economic life of palm plantation is a smaller interval as well as degraded grasslands. Thus, this storage can be effectively considered if it is kept beyond. In the case of tree cutting and onsite burning, the carbon stock is lost and it is accounted for in the analysis. As Section 6.4.4 shows, final plantation management highly affects the total climate change emission of the biogas. This effect is highlighted when the feedstock production accounts for a larger percentage (energetic allocation between EFB and palm grains). If the vegetal carbon stock is not considered, the net EFB production emissions become highly positive. Thus, climate change impact of the biogas spillage technique surpasses that of the shell technique. As a consequence, the latter is more advisable than the former. Assumptions made for the end of life should be improved in the future, after more experience in Brazil. Finally, this study has characterized the hydrothermal gasification environmental impact according to impact categories: climate change gas emission and fossil depletion. Although the performance is mainly affected by allocation and land use change hypotheses, the carbon footprint and the fossil resources exploitation significantly enhance the performance with respect to the reference scenario. The mean reduction of GHG emissions and fossil depletion is around 80% and 85%, respectively.

Acknowledgments

The authors would like to thank the Hydrothermal Gasification team (PSI institute) and Alberto Mian from LENI laboratory (Ecole Polytechnique Fédérale de Lausanne) for providing information. We are grateful to lamented Prof. Emilio Cafaro (Politecnico di Torino) for crucial support on thermodynamic aspects.

References

[1] Dornburg V, Faaij A, Verweij P, Langeveld H, van de Ven G, Wester F, et al. Biomass assessment: global biomass potentials and their links to food, water, biodiversity, energy demand and economy, main report (climate change scientific assessment and policy analysis). Bilthoven: The Netherlands Environmental Assessment Agency; 2007.

[2] Delivand MK, Gnansounou E. Life cycle environmental impacts of prospective palm-based biorefinery in Parà State-Brazil. Bioresour Technol 2013;150:438–46.

[3] Johnson E. Goodbye to carbon neutral: getting biomass footprints right. Environ Impact Assess Rev 2009;29:165–8.

[4] ADM. ADM to invest in sustainable palm production in Brazil. Decatur: ADM Media Relations; 2011.

[5] Modell M, Reid RC, Amin SI. Gasification process. USA Patent 4113446; 1978.

[6] Clifford AA. Calculation of thermodynamic properties of carbon dioxide using Peng-Robinson equation of state. North Yorkshire: Critical Processes; 2007.

[7] CCEM. 2nd generation biogas, new pathways to efficient use of biomass. Villigen: Competence Center Energy and Mobility; 2010.

[8] Molino A. Stato dell'arte industriale della gassificazione con acqua in condizioni supercritiche e potenzialita' di applicazioni per il territorio nazionale, con individuazione delle principali materie convertibili. Rome: ENEA; 2012.

[9] Modell M. Gasification and liquefaction of forest products in supercritical water. In: Overend RP, Milne TA, Mudge LK, editors. Fundamentals of thermochemical biomass conversion. Amsterdam: Springer; 1985. p. 937–50.

[10] Waldner MH, Vogel F. Renewable production of methane from woody biomass by catalytic hydrothermal gasification. Ind Eng Chem Res 2005;44:4543–51.

[11] Waldner MH. Gasification of biomass for the production of sythetic natural gas. Zurich: ETH; 2007.

[12] Elliot DC. Catalytic hydrothermal gasifi cation of biomass. Biofuels Bioprod Biorefin 2008;2:254–65.

[13] Fang Z, Minowa T, Fang C, Smith RLJ, Inomata H, Kozinski JA. Catalytic hydrothermal gasification of cellulose and glucose. Int J Hydrogen Energy 2008;33:981–90.

[14] Sealock Jr, Elliot DC. Method for the catalytic conversion of lignocellulosic materials. US Patent 5.019.135; 1991.

[15] Schubert M. Catalytic hydrothermal gasification of biomass—salts recovery and continuous gasification of glycerol solutions. Zurich: ETH; 2010.

[16] Département fédéral de l'environnement, des transports, de l'énergie et de la communication DETEC: la biomasse aqueuse a un potentiel énergétique. Office fédéral de l'énergie OFEN, 2012.

[17] Peterson AA, Vontobel P, Vogel F, Tester JW. In situ visualization of the performance of a supercritical-water salt separator using neutron radiography. J Supercrit Fluids 2008;43:490–9.

[18] Mian A, Ensinas AA, Ambrosetti G, Maréchal F. Optimal design of solar assisted hydrothermal gasification for microalgae to synthetic natural gas conversion. Chem Eng Trans 2013;35:1009–14.

[19] Haiduc AG, Brandenberger M, Suquet S, Vogel F, Bernier-Latmani R, Ludwig C. SunCHem: an integrated process for the hydrothermal production of methane from microalgae and CO_2 mitigation. J Appl Phycol 2009;21:529–41.

[20] Mian A, Ensinas AV, Marechal F. Multi-objective optimization of SNG production through hydrothermal gasification from microalgae. In: 23rd European symposium on computer aided process engineering, Lappeenranta, Finland; 2013.

[21] Gassner M. Process design methodology for thermochemical production of fuels from biomass: application to the production of synthetic natural gas from lignocellulosic resources. Lausanne: EPFL; 2010.

[22] Gassner M, Vogel F, Heyen G, Maréchal F. Optimal process design for the polygeneration of SNG, power and heat by hydrothermal gasification of waste biomass: thermo-economic process modelling and integration. Energy Environ Sci 2011;4:1726–41.

[23] Gidner AV, Stenmark LB, Carlsson KM. Treatment of different wastes by supercritical water oxidation. In: IT3 conference, Philadelphia; 2001.

[24] Kruse A, Gawlik A. Biomass conversion in water at 330–410°C and 30–50 MPa: identification of key compounds for indicating different chemical reaction pathways. Ind Eng Chem Res 2002;42:267–79.

[25] Leusbrock I. Removal of inorganic compounds via supercritical water: fundamentals and applications. Groningen: University of Groningen; 2011.

[26] Vogel F, Waldner MH, Rouff AA, Rabe S. Synthetic natural gas from biomass by catalytic conversion in supercritical water. Green Chem 2007;9:616–9.

[27] Yoshida T, Oshima Y. Partial oxidative and catalytic biomass gasification in supercritical water. Ind Eng Chem Res 2004;43:4097–104.

[28] Sato T, Kurosawa S, Smith R, Adschiri T, Arai K. Water gas shift reaction kinetics under noncatalytic conditions in supercritical water. J Supercrit Fluids 2004;29:113–9.

[29] Monteleone G, Carewska M, Galli S, Naticchioni V. Stato dell'arte dei processi di up-grading del biogas. Rome: ENEA; 2011.

[30] Braccio G, de Bari I, Molino A. Sviluppo di un processo in membrane polimeriche per la purificazione del biogas da CO_2 e H_2S. Rome: ENEA; 2011.

[31] Abdullah N, Sulaiman F, Gerhauser H. Characterization of oil palm empty fruit bunches for fuel application. J Phys Sci 2011;22:1–24.

[32] Leusbrock I, Metz SJ, Rexwinkel G, Versteeg GF. Quantitative approaches for the description of solubilities of inorganic compounds in near-critical and supercritical water. J Supercrit Fluids 2008;47:117–27.

[33] Cavalcanti de Moraes B, Nogueira da Costa JM, Lôla da Costa AC, Costa MH. Variação espacial e temporal da precipitação no estado do Pará. Acta Amazon 2007;2:207–14.

[34] Prasad MV, Arulraj S, Mounika B. Oil palm cultivation practices. Pedavegi: Directorate of Oil Palm Research; 2012.

[35] Wai-Lin S. Palm oil milling process. Malaysian Palm Oil Board, 2011. Available from: http://lipidlibrary.aocs.org/processing/palmoil/index.htm [accessed September 2014].

[36] Elbersen W, Meesters K, Bakker R. Valorazation of palm oil (mill) residues. Croeselaan: Netherlands Programmes Sustainable Biomass; 2013.

[37] Husain Z, Zainal Z, Abdullah MZ. Analysis of biomass-residue-based cogeneration system in palm oil mills. Biomass Bioenergy 2003;24:117–24.

[38] Departamento De Gestão De Risco Rural. Portaria no 83, Cooerdeaçao-General de Zonamento Agropecuario. Ministério da Agricultura; Pecuària e Abastecimento; 2011.

[39] Klaarenbeeksingel F. Greenhouse gas emissions from palm oil production: literature review and proposals from RSPO working group on greenhouse gases. Hoevelaken: Brinkmann Consultancy; 2009.

[40] Gomes I. Brazil: country of the future or has its time come for natural gas? Oxford: Oxford Institute for Energy Studies; 2014.

[41] Agropalma. Fòrmula para càlculo do preço de acordo com o contrato: preço CFF. São Paulo: Agropalma; 2015.

[42] Fehrenbach H. Greenhouse gas balances for the german biofuels quota legislation. In: GBEP task force meeting GHG methodology, Washington, DC; 2007.

CHAPTER

7

Life Cycle Assessment of Algal Biorefinery

E. Gnansounou, J.K. Raman[†]*

*Bioenergy and Energy Planning Research Group (BPE), Ecole Polytechnique Fédérale de Lausanne (EPFL), Lausanne, Switzerland

[†]Ecole Polytechnique Fédérale de Lausanne (EPFL), Lausanne, Switzerland

7.1 INTRODUCTION

Energy security and climate change were among the crucial global issues in the last few years due to fossil fuel resource depletion and higher usage on one hand, and carbon dioxide emission due to fossil fuel use on the other hand. Renewable energy is seen as an alternative to help overcome these issues and there has been a significant improvement in renewable energy technologies, such as solar, wind, geothermal, and biofuels. However, only biofuels such as biodiesel and bioethanol could replace fossil-based fuels in the transport sector, which consumes nearly 30% of the world's total energy, including 50% of the global oil production [1] and is responsible for approximately one-quarter of energy-related global greenhouse gas (GHG) emissions [2]. The properties of biofuels are similar to those of fossil-based transport fuels, which allows blending of the fuels in various ratios. Development of biofuel technologies is on the rise and currently most of the biofuels (biodiesel and bioethanol) are produced from food-based feedstocks such as sugarcane, corn, soybean oil, palm oil, etc. Biofuel production from such feedstocks is often blamed for food price hikes in recent years [3]. Therefore, biofuels production from nonfood-based feedstocks is currently being heavily researched.

Recent advancements in biofuel technologies show that lignocellulosic feedstock could be a promising source for bioethanol, whereas algae and jatropha could be promising feedstocks for biodiesel production. However, the main issue with biofuel production from these feedstocks is the process costs, which are high compared to fossil-based fuel [4]. Thus only fuel production from these feedstocks are uneconomical; however, any added value product that could be produced with the same feedstock in addition to fuel production may be a viable option for overcoming the cost issue.

http://dx.doi.org/10.1016/B978-0-444-63585-3.00007-3

Soybean, corn oil, palm or algal oil could be used for biodiesel production. However, algae oil seems to be more attractive because of the higher oil yield capacity when compared to other crop oils and the fact that it does not require arable land and does not compete with the food supply, and it has the possibility to produce coproducts along with biodiesel [5]. Recent research is being concentrated on algal biorefineries capable of producing ethanol and other value-added products such as animal feed, heath care products, cosmetic products and biochemicals [6,7] to make the algae system economically viable. However, the environmental impact of such an algal biorefinery system has to be assessed to select the appropriate biorefinery configuration.

India is the fourth-largest fossil fuel consumer in the world with nearly 3.7 million barrels consumed per day; most of India's transport fuel demand is for motor gasoline and diesel. On a purchasing power parity basis, India is the third-largest economy and its dynamic economic growth and modernization will increase the energy demand significantly in the future [8]. With increasing pollution and rising energy demand being main concerns, India's energy sector has to explore alternative energy sources to provide energy and reach the renewable policy (20% biofuels blending by 2017 in the transport sector) goal set by the government [9]. In the case of bioethanol production and blending with gasoline, there are agricultural residues such as rice straw, wheat straw, sugarcane tops and leaves, bamboo, etc., available in India to produce bioethanol. However, there are no appropriate nonfood-based energy sources other than jatropha to produce biodiesel and blend with diesel. Ajayebi et al. suggested algae biomass could be a potential feedstock for biodiesel in India owing to the coastal line, industry and the technology availability [10]. However, as mentioned earlier, producing biodiesel alone from algae could not be economically favorable and other coproducts such as animal feed and biochemicals from algae have to be produced in order to utilize the full potential of algae nutrient composition [11].

With 4.11% of the total gross domestic product (GDP) during 2012–13 [12], livestock production is a significant industry in India. As an exporter of soybean meal, any additional protein meal available would be beneficial to the export market in the country, improving the economic status. Algae contain up to 50% protein and could be a potential source for animal feed production; in addition it could also provide several health and marketing benefits in the livestock industry [13–15]. Therefore, protein production from algae would be a profitable option for an algal biorefinery system.

Succinic acid has its application in industries such as 1,4-butanediol, polyurethane, plasticizers, resins, coatings, dyes and inks, pharmaceuticals, food, cosmetics, solvents, lubricants, and de-icing solutions [16,17] and it is identified as one of the 30 top-valued chemicals. At present, succinic acid is produced from fossil-based resources and there is high demand for this product due to its growing application industry in India and China. On the other hand, the long-term perspectives of increase in cost of petroleum-based products and the high environmental burden demand a bio-based succinic acid on the market [16,18]. Bio-based succinic acid is produced by the fermentation of carbohydrate in the biomass. Up to 15% of algae is carbohydrate that could be a potential source for bio-based succinic acid production [13].

Lipid, protein, and carbohydrate are the major components in algae. When biodiesel is produced, only the lipid is utilized and the remaining protein and carbohydrate are burnt as fuel in the boiler, underestimating their potential commercial value. If all the components could be used, other value-added products such as protein for animal feed and biochemicals such as succinic acid or levulinic acid from carbohydrate could be produced along with biodiesel, and as the other two products have a high value in the market, the overall economics of the processes would be improved significantly. In addition, it would be environmentally favorable if the coproducts from algae replace the existing fossil-based products.

Life cycle assessment (LCA) is a tool to evaluate the environmental impact of any process or a product. In literature, a few of the previous scientific attempts are available that evaluated production of transportation fuels from algae. The net energy ratios and the GHG emissions range from 0.1 to 4.2 and 35 to 140 gCO_2eq respectively for a functional unit of 1 MJ of produced biodiesel. Some studies claimed that biodiesel from algae could be sustainable and superior compared to fossil diesel [10,19–22]. Some others indicated a low difference between algal biodiesel and fossil diesel [23–27], while a few claimed poor performance with negligible advantages [28–33]. The significant variance in the studies could be due to different system boundaries, functional unit and the reference system, which are all very important for impact assessment studies [21,34]. However, in the present situation, producing only biodiesel from algae is not favorable; therefore, the future trend would be towards an algae biorefinery system capable of producing multiproducts [13,35]. Thus the focus of this prospective LCA study is to compare an algae biorefinery system with a conventional system in India considering the production of biodiesel, animal feed protein, and succinic acid.

7.2 PROCESS DESCRIPTION

In this section the process details of the products from algae (Figs. 7.1 and 7.2) and the reference system (Fig. 7.3) are given.

7.2.1 Microalgae Cultivation and Harvesting

Various methods have been proposed for prospective cultivation of microalgae for biofuels, ranging from floating containers on the surface of the oceans [37] to cocultivation with agricultural crops on paddy fields [38]. The most studied methods for production of algal biomass so far are raceway ponds and closed bioreactors. Protection from contaminants, which provides the option of high-value added coproducts, higher productivity, water, CO_2 and nutrients savings, are the advantages of closed bioreactors [39,40]. However, higher maintenance costs and technical difficulties have made this method uneconomical [27,41]. On the other hand, raceway ponds are relatively economical with easier maintenance and energy-efficient operation [42,43] that would suit for countries such as India. As a result, only raceway ponds were considered in this study for microalgae cultivation. A few studies have modeled cultivation of algae in raceway ponds. A well-established numerical method was used by Stephenson et al. considering an energy input of 7.2 MJ/kg of algal biodiesel [27]. Lardon et al. also modeled a raceway pond system where the energy input was 20.8% lower compared to the latter [26]. In this study, the model described by Frank et al. was considered as a base, which was adjustable based on conditions in this study that characterized the raceway ponds [25]. Microalgae can be cultivated in seawater or freshwater depending on the strain; however, only cultivation of the algae strain *Chlorella vulgaris* in seawater was assumed in this study, considering the capacity of a large-scale biodiesel production and the scarcity of

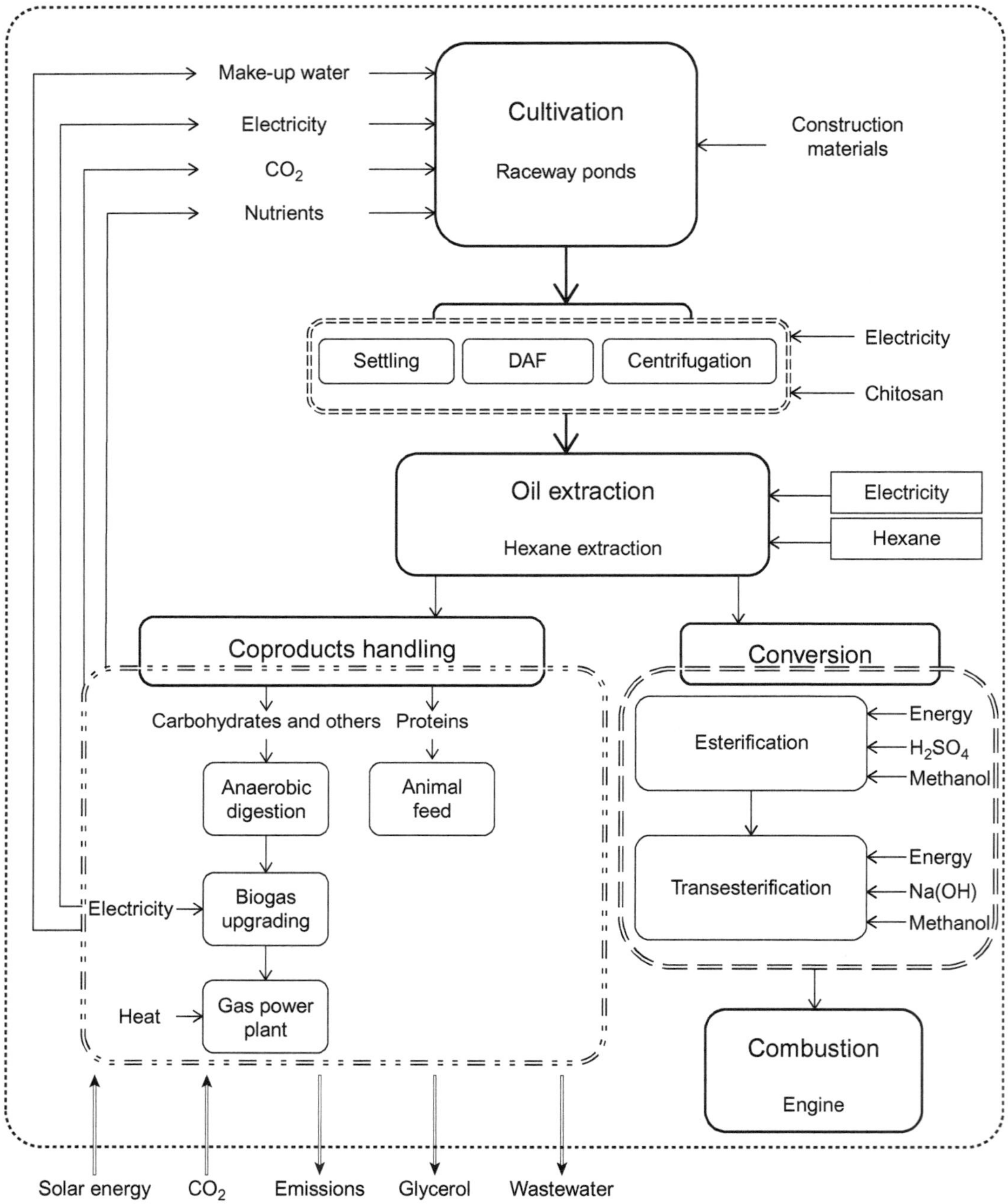

FIG. 7.1 System boundary of the algal biodiesel and animal feed production [36].

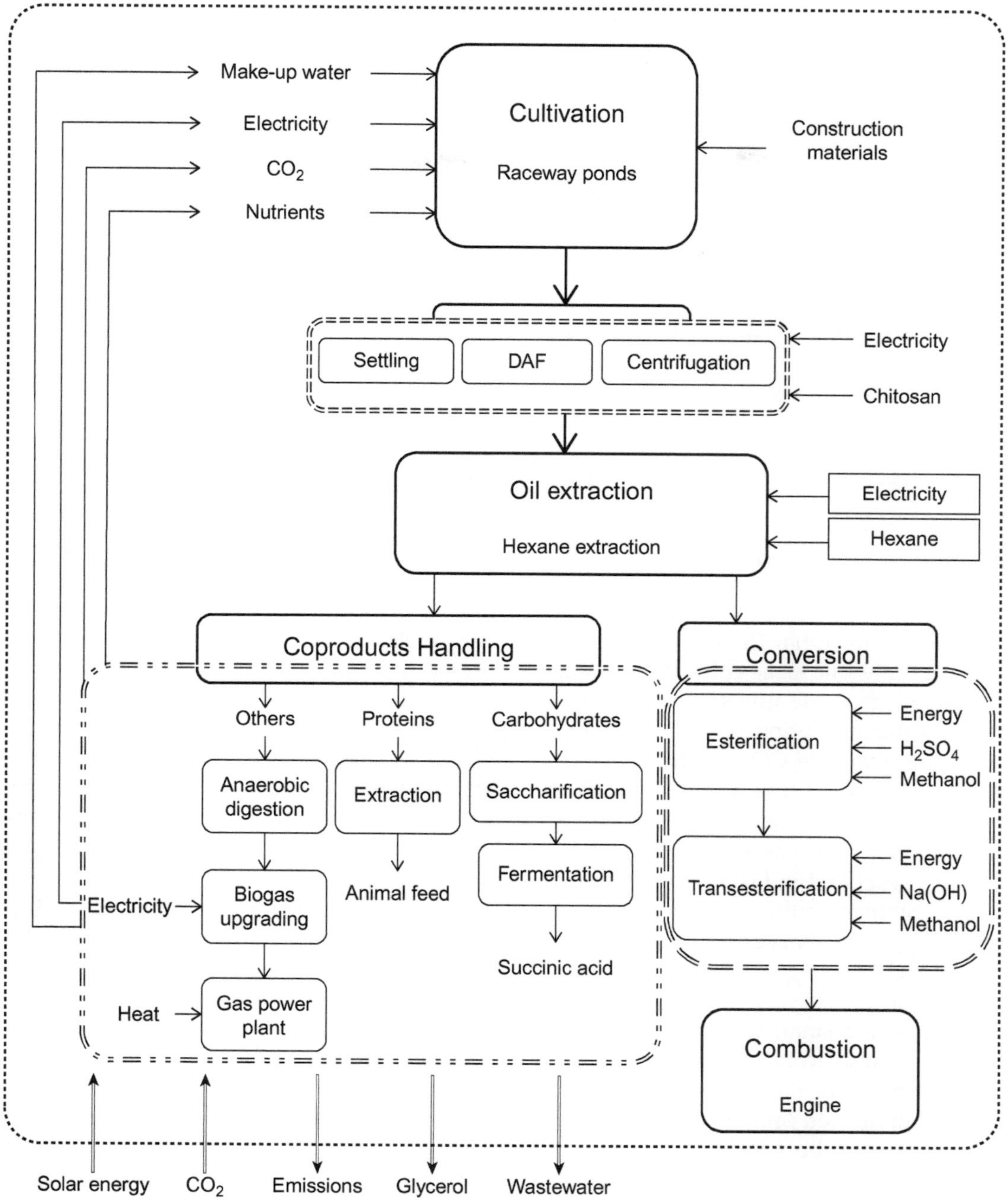

FIG. 7.2 System boundary of the algal biodiesel, animal feed, and succinic acid production [36].

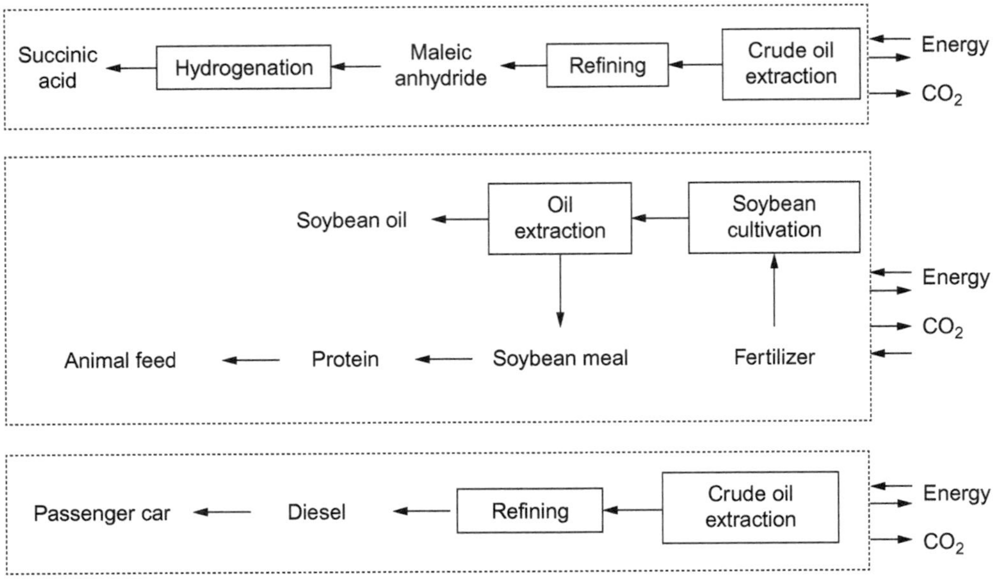

FIG. 7.3 Conventional diesel, soybean protein, and succinic acid production system [36].

fresh water in India. During cultivation, microalgae absorb CO_2 for photosynthesis and a feasible economic algae cultivation system requires a source of carbon which has higher concentrations than the atmospheric CO_2. Flue gas from power plants that burn fossil fuels could be a promising option to provide high concentration (up to 20%) of CO_2 [44]. Therefore, a readily available CO_2 stream that can be utilized for the raceway ponds and has been previously cooled down and stripped from NO_X and SO_X emissions before entering the system boundary has been assumed in this study. Additionally, recycled pumping and pumping from off-site were included in this study by considering similar pumping designs that are described in Davis et al. [45].

The biomass cultivated should be harvested from its growing medium for further processing. The harvesting stage is considered to be a challenge and is labeled as the "bottleneck" of algae-based fuels [46]. Many challenges such as the small size of the microalgae particles, low concentrations of algal broth [39], a mass density that is similar to water and a small surface charge that prevents compacting [47] occur in harvesting. As a result of these problems, various technical and conceptual technologies have been proposed for harvesting algal biomass. However, depending on the characteristics of microalgae and the type of desired final products, the appropriate harvesting pathways and the downstream process have to be selected [48]. This stage is crucial for biodiesel production because the transesterification processes require a biomass with very low water content. Microalgae particles are floating in the water medium suspension, so most of the algae harvesting methods can be expressed by fluid dynamics. Gravitational precipitation of particles would follow Stoke's law, which defines the relation between velocities of sedimentation, size of the particles and the fluid's density. Other methods are more efficient and much faster than natural settling. Coagulants can cause small microalgae particles to aggregate together and form considerably bigger particles of algae, hence increasing the "effective particle size" [39]. Centrifugation, on the other hand, increases the speed of sedimentation by adding centrifugal acceleration to the effective forces. Flotation is a gravity and physicochemical method that is being performed by a bubbling procedure of air or a gas through a gas-liquid suspension. Bubbles would ascend from the bottom of the suspension towards the top surface and the solid particles would attach on the surface of bubbles, based on their hydrophobicity. Bubbles carry the attached particles during levitation. Eventually, there would be an accumulation of the carried particles on the top surface of the liquid. These accumulated floating particles can be recovered by skimming or overflowing mechanisms [49]. The harvesting stage of this study has been modeled based on three consecutive stages: settling tanks coupled with the cultivation ponds, a dissolved air flotation unit, and centrifugation. Dewatering, homogenization, and pumping of liquids were also included in the harvesting section.

7.2.2 Oil Extraction and Transesterification

Oil extraction is defined as the process of separating triglyceride (TAG) lipids from the harvested and concentrated algal biomass and it could be done through a variety of mechanical or chemical manipulation techniques. Oil extraction using hexane solvent has been considered in this study using a stripper column model [27].

Biodiesel or fatty acid methyl ester (FAME) is formed by transesterification of acylglycerols (glycerides) namely monoacylglycerols (MAG), diacylglycerols (DAG), and triacylglycerols (TAG), as well as free fatty acids (FFA) and phospholipids (PL). Triglycerides are considered to be the most favorable type of lipids with the highest yield for production of biodiesel [50]. Fatty acids are constituents of lipids and, in general, a fatty acid molecule is comprised of a hydrocarbon chain attached to a carboxyl functional group. If the structure consists of at least one double bond, then it is an unsaturated fatty acid, and it can bond with hydrogen; otherwise it is saturated. Vegetable oils and animal fats mostly contain triglycerides which can be broken down by natural enzymes into other acylglycerols and FFA. Various non-acylglycerol types of lipids may be present in the extracted bio-oil such as polar lipids, FFAs, ketones, pigments, etc. These would have an adverse impact on the transesterification process and often there is a need for purification before the transesterification stage [51].

Current commercial transesterification of lipids is in the presence of alkaline (KOH/NaOH) catalysts. Methanol is generally used for large-scale production due to its availability and lower cost compared to other options. The main factors affecting transesterification are the TAG content, the amount of inhibitors and water content. The transesterification process is usually modeled based on industrial procedures that have already been used for conversion of first-generation bio-oils, such as rapeseed and soybean. The extracted oil from these biomasses is considered to have a low FFA content (often less than 1%) and high TAG lipids, respectively. Oil containing more than 2% FFA should undergo a pretreatment step (esterification) with methanol in the presence of a sulfuric catalyst to reduce the FFA content, followed by transesterification with methanol and sodium hydroxide [52,53]. For oil extracted from algae, the percentage of FFA and other inhibitors is usually much higher (up to 14%)

than the maximum allowable content [54,55]. Apparently, the inhibitors should be purified and removed from the oil before transesterification in order to prevent saponification and biodiesel hydrolysis. In order to avoid the undesired reactions of saponification and hydrolysis, a two-step transesterification has been suggested [56]. Thus more chemicals and energy are involved in the process. In this study an average value of 7% FFA content in algae oil was assumed. Materials and energy required for processing oil and biodiesel production with 7% FFA using a two-step process (acid esterification and alkali transesterification) modeled and reported in Zhang et al. [52] were used in this LCA study.

7.2.3 Algae Protein and Succinic Acid Production

Solvent extraction, enzymatic and pulsed electric field are the methods available for protein extraction. In this study, protein from the deoiled algae biomass was assumed to be separated by the solvent extraction method using ethanol (95%) and methanol (5%) [57]. This method is similar to that of the oil extraction method where the deoiled algae are mixed with solvent (ethanol and methanol) in a desired temperature and time. The extraction is carried out several times to recover the maximum amount of protein and later the solvent is recovered via evaporation. Therefore, it was assumed that the same energy from oil extraction would be required for protein extraction. The protein output from the extraction is considered to contain a similar quality compared to soybean protein used in animal feed.

The carbohydrate remaining in the biomass after oil and protein extraction was assumed to be converted to succinic acid through fermentation using succinic acid producing microorganisms. The processes involved in succinic acid production from carbohydrate are fermentation, ultra-filtration, crystallization, centrifugation, and drying. Initially the algae carbohydrate is fed into the fermentation tank along with other nutrients and the seed culture is added. After fermentation at the desired temperature and time, the broth is passed through a filtration to separate the solid and liquid fraction. Later the liquid fraction containing succinic acid is crystallized and centrifuged to separate the succinic acid, and dried to get the final product. As this succinic acid production technology is under research and no data on succinic acid production from algae carbohydrate are available, the life cycle data of biosuccinic production from starch using a similar technology reported in [58] was used in this study.

7.2.4 Diesel, Soy Protein and Succinic Acid (Conventional) Production

The scope of this study is to compare the system under study (algal biorefinery system) to conventional systems that are in use at present. Biodiesel, protein, and succinic acid production processes from algae are compared with fossil diesel, soy protein, and conventional succinic acid, respectively. The production process of fossil diesel involves crude oil extraction, transportation, and refining. For soy protein, the processes involved are soybean cultivation, transportation, oil extraction, protein meal separation, and drying. In the case of succinic acid, it is produced by catalytic hydrogenation of fossil-based maleic anhydride followed by crystallization and solid separation [59].

7.2.5 Coproducts Handling

There are various options for using the remaining algal biomass, such as combustion, biogas, and fertilizers. In this study, the algal biomass remaining after oil and protein extraction or oil, protein, and carbohydrate extraction was assumed to be used in an anaerobic digester to produce biogas that fuels the combined heat and power plant installed within the system to generate partial process heat and electricity. The glycerine produced during biodiesel production was assumed to be sold to the market, replacing conventional glycerine.

7.3 LIFE CYCLE INVENTORY

The environmental impacts of two algal biorefinery schemes were assessed in this study. In the first scheme biodiesel and algae protein (BDAP) are produced (Fig. 7.1) and the remaining biomass is used as fuel for energy generation within the plant. In the second scheme, biodiesel, algae protein, and succinic acid (BDAPSA) are produced (Fig. 7.2), and the remaining biomass is used as a fuel for energy generation. The impacts of both schemes are compared to their respective reference systems: fossil diesel, soy protein (FDSP-R) and fossil diesel, soy protein and succinic acid (FDSPSA-R) that is commercially available (Figs. 7.4 and 7.5). The lipid, protein, carbohydrate of the algae *C. vulgaris* biomass (dry mass basis) is considered

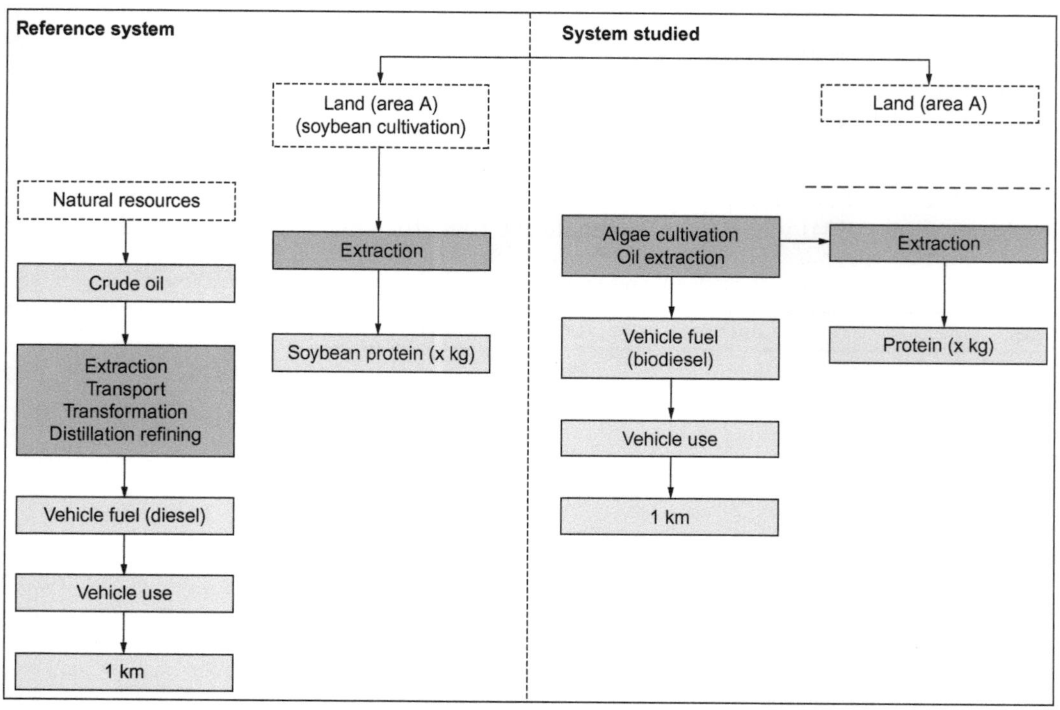

FIG. 7.4 System boundary of fuel and protein production from algal system and reference system.

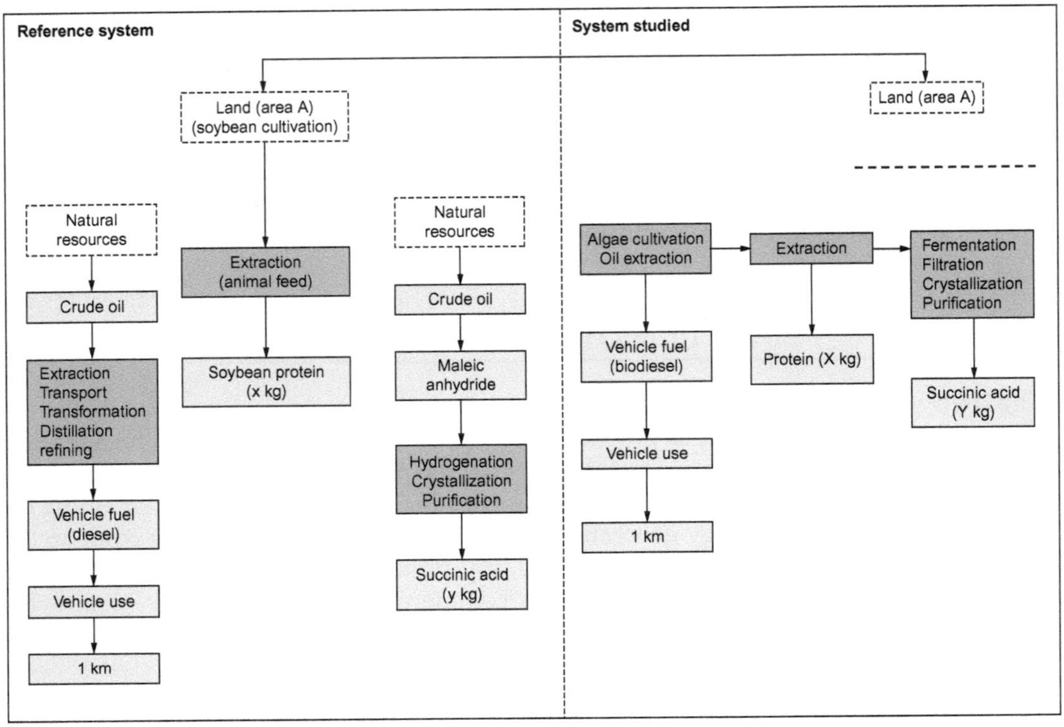

FIG. 7.5 System boundary of fuel, protein, and succinic acid production from algal system and reference system.

to be 22%, 50%, and 15%, respectively [60]. The data inventories of the two schemes are given in Tables 7.1 and 7.2 for a functional unit of 1 kg of biomass and the impacts are shown based on the service from the products (traveled distance for biodiesel and based on the weight for other products). For biodiesel, the well-to-wheel (WtW) methodology developed by Gnansounou et al. [34] was used to estimate the fuel efficiency and consequently the WtW

TABLE 7.1 Data Inventory of Biodiesel and Protein Production Case for 1 kg Algae Biomass

Stages	In/Out	Utilities/Materials	Unit	Base Case
Cultivation	Input	Water	kg	48
		Urea	g	19.08
		Diammonium phosphate	g	17.16
		Concrete	g	4.76E – 05
		Steel	g	1.62E – 07
		Plastic	g	1.10E – 05
		Cast iron	g	3.90E – 09
		Electricity	kWh	0.74
	Output	Algal broth	kg	49

TABLE 7.1 Data Inventory of Biodiesel and Protein Production Case for 1 kg Algae Biomass—cont'd

Stages	In/Out	Utilities/Materials	Unit	Base Case
Harvest	Input	Chitosan	g	9.95
		Electricity	kWh	0.75
		Algae broth	kg	49
	Output	Dry algae	g	1000
Oil extraction	Input	Hexane	g	0.64
		Electricity	kWh	0.02
		Heat	MJ	0.50
	Output	Algal oil	g	227.27
Protein extraction	Input	Deoiled algae biomass	g	772.73
		Ethanol	g	30.22
		Methanol	g	1.60
		Electricity	kWh	0.02
		Heat	MJ	0.50
	Output	Protein	g	425.01
		Algae biomass	g	347.49
Anaerobic digestion	Input	Electricity	kWh	0.07
		Heat	MJ	0.32
		Algae biomass	g	347.49
	Output	Electricity	kWh	0.39
		Heat	MJ	1.13
Biodiesel production	Input	Algal oil	g	227.27
		Methanol	g	27.03
		Sodium hydroxide	g	2.27
		Sulfuric acid	g	3.42
		Electricity	kWh	0.01
		Heat	MJ	1.21
		Water	kg	0.03
	Output	Biodiesel	g	216.45
		Glycerol	g	24.52
Transport	Biodiesel—from industry to outlet		km	100
	Protein—from industry to outlet		km	100

TABLE 7.2 Data Inventory of Algae Biodiesel, Protein, and Succinic Acid Production Case for 1 kg Algal Biomass

Stages	In/Out	Utilities/Materials	Unit	Base Case
Cultivation	Input	Water	kg	48
		Urea	g	19.08
		Diammonium phosphate	g	17.16
		Concrete	g	4.762E−05
		Steel	g	1.623E−07
		Plastic	g	1.1E−05
		Cast iron	g	3.896E−09
		Electricity	kWh	0.74
	Output	Algal broth	kg	49
Harvest	Input	Chitosan	g	9.95
		Electricity	kWh	0.75
		Algae broth	kg	49
	Output	Dry algae	g	1000
Oil extraction	Input	Hexane	g	0.64
		Electricity	kWh	0.02
		Heat	MJ	0.50
	Output	Algal oil	g	227.27
Protein extraction	Input	Deoiled algae biomass	g	772.73
		Ethanol	g	30.22
		Methanol	g	1.60
		Electricity	kWh	0.02
		Heat	MJ	0.50
	Output	Protein	g	425.01
		Algae biomass	g	347.49
Succinic acid production	Input	Algae biomass	g	347.49
	Output	Succinic acid	g	117.53
Anaerobic digestion	Input	Electricity	kWh	0.04
		Heat	MJ	0.19
		Algae biomass	g	202.73
	Output	Electricity	kWh	0.23
		Heat	MJ	0.65

TABLE 7.2 Data Inventory of Algae Biodiesel, Protein, and Succinic Acid Production Case for 1 kg Algal Biomass—cont'd

Stages	In/Out	Utilities/Materials	Unit	Base Case
Biodiesel production	Input	Algal oil	g	227.27
		Methanol	g	27.03
		Sodium hydroxide	g	2.27
		Sulfuric acid	g	3.42
		Electricity	kWh	0.01
		Heat	MJ	1.21
		Water	kg	0.03
	Output	Biodiesel	g	216.45
		Glycerol	g	24.52
Transport	Biodiesel—from industry to outlet		km	100
	Protein—from industry to outlet		km	100
	Succinic acid—from industry to outlet		km	100

GHG emissions of the biodiesel for traveled distance was calculated using Eqs. (7.1), (7.2), where biofuel, fuel blend and fossil fuel represent biodiesel, biodiesel blend and fossil diesel per km service.

$$\text{Fuel efficiency biofuel} = (\text{biofuel blend factor} \times \text{fuel efficiency fuel blend})/1 - [(1-\text{biofuel blend factor}) \times \text{fuel efficiency fuel blend}/\text{fuel efficiency fossil fuel}] \quad (7.1)$$

$$\text{GHG biofuel WtW} = \text{GHG biofuel WtT} \times \text{Fuel efficiency biofuel} \quad (7.2)$$

The geographic location of the algal biorefinery plant assumed in this study is India with a focus on the southern region (Tamil Nadu) and by considering utilization of local potentials in algae cultivation. This is due to the fact that the potential to grow algae is high in this region due to the coastal line, industry availability and also the fact that the region is well known for algae cultivation [38]. For this study, an algae plant with biodiesel production capacity of 100,000 tons per year and its coproducts were assumed [10].

The electricity generation scenario in southern India was used to model the electricity requirement. Coal, hydropower, natural gas, nuclear, diesel, biomass, and other renewables are the main sources of power generation and, according to the Indian Energy statistics, the energy mix for electricity generation (southern region) in 2011 was coal 44%, hydropower 21%, biomass and other renewables 22%, natural gas 9%, nuclear 2%, and diesel 2% [61]. An 18% average transmission and distribution loss in the southern region was also taken into account. The combined heat and power (CHP) plant that is utilized within the biorefinery generates heat and electrical with efficiency of 41% and 34%, respectively [25,62]. A theoretical maximum yield of 90%, 85%, and 70% were assumed for biodiesel, protein, and succinic acid production, respectively.

Data from the chosen system boundaries and inventory were modeled with SimaPro 7.3.3 using ReCiPe midpoint (E) methodology to calculate the CO_2 and fossil oil consumption. For the reference system data, fossil diesel and soybean meal production available in the LCA database were used for their environmental impacts. For conventional succinic acid production, the environmental impacts (CO_2 and fossil oil consumption) reported in [58] were used. In this reference the succinic acid was assumed to be produced via catalytic hydrogenation of maleic anhydride. However, due to the lack of data for these processes the environmental impact of maleic anhydride via partial oxidation of *n*-butane was used instead of succinic acid. To compare the life cycle impact among the products, value-based economic allocation was used considering the purchase price of the products (biodiesel: 65.5 Rs/kg, protein: 70 Rs/kg, and succinic acid: 100 Rs/kg).

7.4 SENSITIVITY ANALYSIS

The composition of the feedstock determines the amount of products that could be produced. Hence the composition of the feedstock plays a major role in the environmental impact of that system. Literature studies show a significant difference in the composition of algae *C. vulgaris* [6,63]. The effect of composition changes on environmental impacts was determined by sensitivity analysis by varying the composition at higher and lower levels, assuming the composition mentioned previously as the base case [lipids (25%), protein (50%), carbohydrates (15 %), and others (10%)]. The sensitivity analysis was performed for two scenarios: (case 1) lipids (30%), protein (40%), carbohydrates (20%), and others (10%) and (case 2) lipids (20%), protein (60%), carbohydrates (10%), and others (10%). The data inventories for the sensitivity analysis cases were generated similarly to the base case and the impact assessment was analyzed and compared with the reference systems for the 1 km functional unit.

7.5 IMPACT ASSESSMENT

The environmental impacts of fuels for WtW were estimated based on the fuel efficiency of the biodiesel blend and the inventory data of each case. The average fuel economy of diesel cars in India reported by the Society of Indian Automobile Manufacturers (SIAM) is 18.4 km/L or 0.0543 L/km. The fuel efficiency of B20 (20% biodiesel and 80% fossil-diesel in volume) at medium to full car load is less by 1–3.3% compared to that of 100% fossil diesel fuel economy [64]. Assuming an average of 2% lower efficiency, the fuel economy of B20 would be 0.0554 L/km. Using Eq. (7.1) and assuming 0.832 kg/L and 0.88 kg/L as the density of diesel and biodiesel, respectively, the fuel economy of biodiesel is 0.053 kg/km.

The life cycle impact assessment results (Fig. 7.6) show less carbon dioxide (25%) emission for biodiesel and protein production from algal system (base case—BDAP) compared to fossil diesel and soybean meal protein (reference case—FDSP-R). However, fossil oil consumption was higher (26%) for the case under study compared to the reference case. This could be

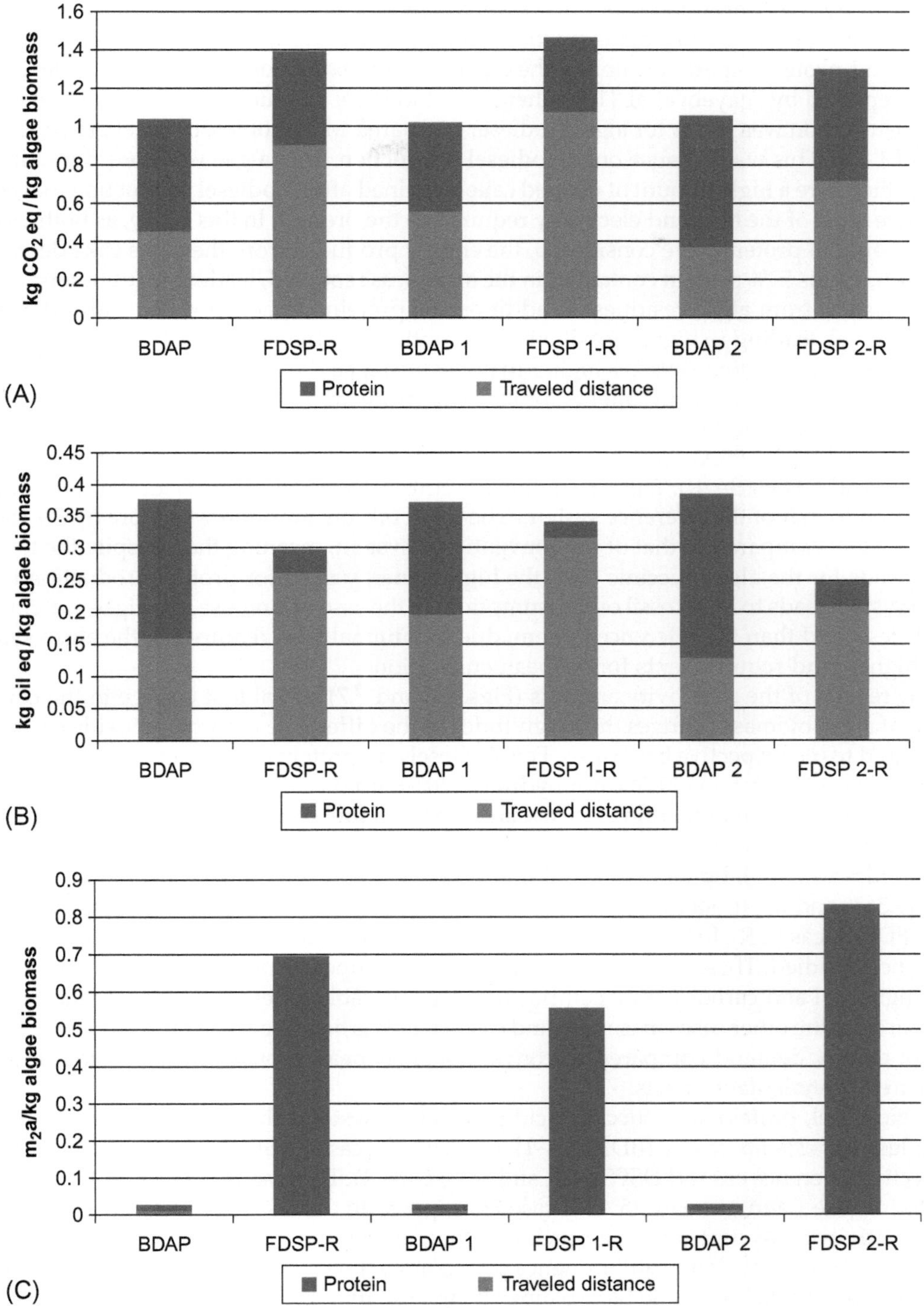

FIG. 7.6 LCA results for fuel and protein system: (A) climate change, (B) fossil oil consumption, and (C) land use.

explained by the energy intensive process involved in the protein extraction process compared to that of the soy protein extraction process which has a common extraction process for oil and protein separation unlike the algae protein extraction process. In an algae LCA study reported by Ajayebi et al. [10], where only biodiesel production was studied, the fossil oil consumption was lower for algal biodiesel compared to that of fossil diesel per functional unit of 1 km. This was because only biodiesel production from algae was considered in that study; therefore a high amount of deoiled cake remained after biodiesel production that could generate most of the heat and electricity required in the process. In this study, as both biodiesel and animal protein were considered, the energy produced from the algae cake decreased significantly, as 50% protein contained in the algae was removed, leading to less energy input to the system from algae residues. In addition the protein extraction process also demands more energy, leading to higher fossil oil consumption. On the other hand, the land required for the reference system is 96.1% higher than for the algae diesel and protein system. This is due to the higher land requirements for soybean production.

Fig. 7.7 shows the life cycle impact results of the algae system (biodiesel, protein and succinic acid—BDAPSA) and the reference system (conventional diesel, soybean meal and succinic acid—FDSPSA-R). The carbon dioxide emissions are less by 36% for the algae system compared to that of the reference system. The fossil oil consumption was more by 7% for the algae system compared to that of the conventional system, meaning that, despite the low energy needs for the algae succinic acid, the higher energy need for protein production in the algae system leads to high fossil oil consumption. In the case of land use, the algae system uses 96.3% less land than the reference system, due to minimal land required in the algae system and higher land requirements for soybean cultivation.

The results of the sensitivity analysis (Figs. 7.6 and 7.7) reveal that change in the composition of algae biomass changes the magnitudes of the difference in carbon dioxide emissions compared to the respective base cases. For biodiesel and protein production systems, in case 1 (BDAP—case 1) [protein (40%) carbohydrates (20%), lipids (30%), and others (10%)] the carbon dioxide emission reduction was 26% and the fossil fuel consumption was higher by 8.8% compared to the reference case (FDSP—case 1 R). For case 2 (BDAP—case 2) [protein (60%) carbohydrates (10%), lipids (20%), and others (10%)] the carbon dioxide emission reduction was 19.36% and the fossil fuel consumption was higher by 54.41% compared to the reference case (FDSP—case 2 R). In land use the changes were almost same for all the algae systems and references studied. These results show that the algae compositions (case 1) with lower protein and higher oil and carbohydrate composition are favorable to fewer environmental impacts compared to the other cases (base case and case 2). This is because algae protein extraction has a high energy demand compared to conventional soybean protein extraction that leads to higher environmental impacts.

In biodiesel, protein, and succinic acid production systems the carbon dioxide emissions were less by 42% for case 1 (BDAPSA-1) and 32% for case 2 (BDAPSA-2) compared to the respective reference cases (FDSPSA-2R and FDSPSA-2R). The fossil oil consumption was less by 3% for case 1 and more by 25% for case 2 compared to the respective reference case. From the sensitivity analysis it could be concluded that, among the systems, succinic acid production from algae has a significant influence on the impact reduction followed by biodiesel and protein production. This could be explained by the fact that it replaces the conventional fossil-based succinic acid that has high fossil oil use and energy demand.

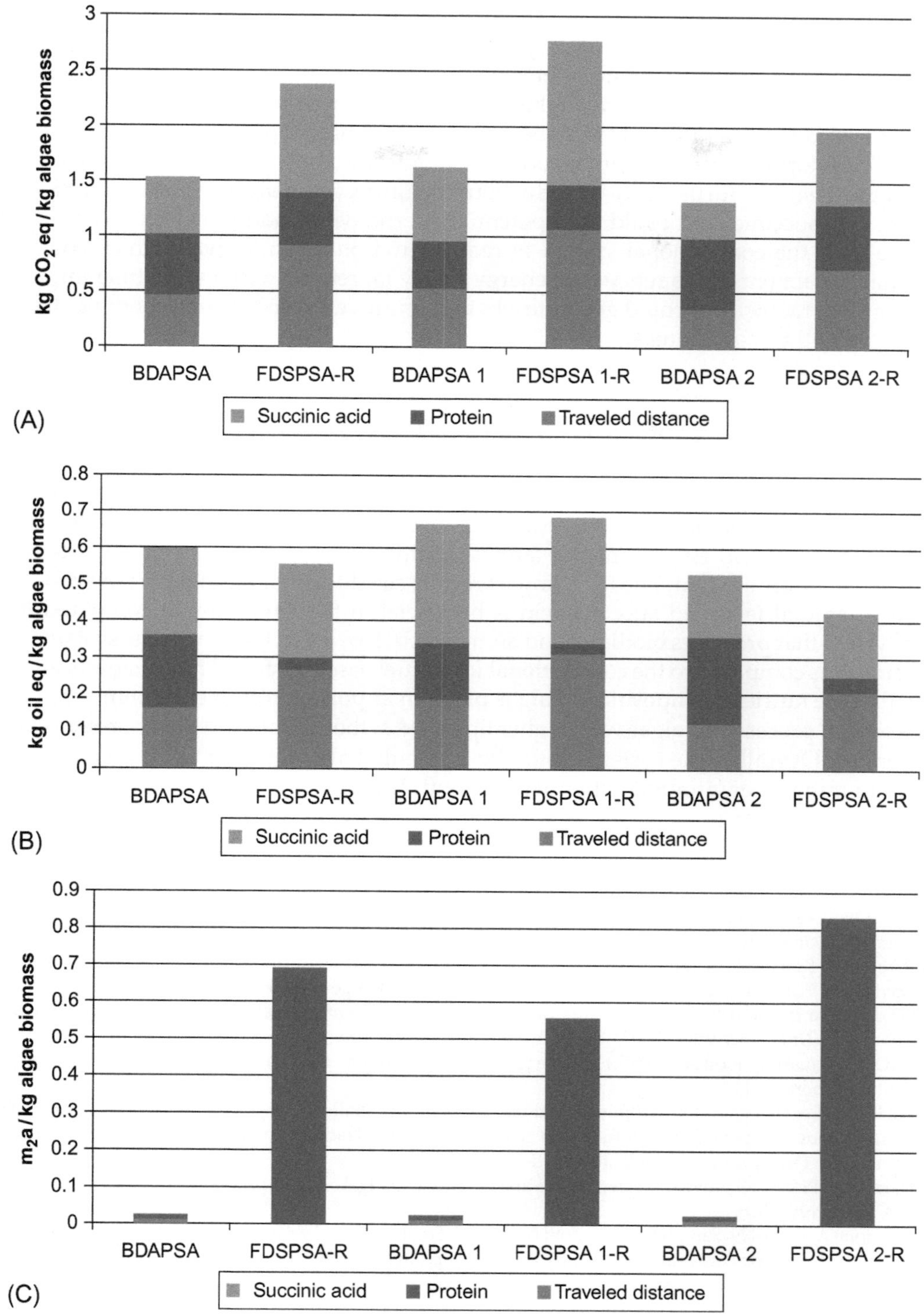

FIG. 7.7 LCA results for fuel, protein, and succinic acid system: (A) climate change, (B) fossil oil consumption, and (C) land use.

Overall, the algae system was better than the reference system in carbon dioxide emission and land use impacts and among algae systems, production of biodiesel, protein, and succinic acid (case 1) from algae was found to be favorable to the environment. With respect to algae composition, it is preferable to have higher lipid and carbohydrate than proteins. On the other hand, it is also important to note that the reference system to which the algae are compared are well developed ones and there is room for significant improvement for the algal system that would result in further impact reductions in future. Thus algae system with biodiesel, protein, and succinic acid could be a potential alternative or additional renewable system in India over the conventional system to reduce environmental impact, which could lead to partial fulfillment of the renewable energy policy target and at the same time implementation of such renewables could generate jobs and improve the social and economic status for the people in the coastal areas.

7.6 CONCLUSIONS AND PERSPECTIVES

Being a highly fossil-fuel dependent country, India has to explore the possibility of unconventional renewable systems for the production of biodiesel and biochemical to satisfy its energy, biochemical and environmental targets. In this connection, prospective LCAs of two algal systems were studied. The LCA impact results conclude that the algal system producing biodiesel, animal feed and succinic acid is beneficial to the environment compared to the algae system that produces biodiesel and animal feed. However, both the algae systems show fewer impacts compared to the conventional fossil-fuel based system. The sensitivity analyses show that the nutrient composition of algae plays an important role in environmental impact of the overall process and algae with higher lipid and carbohydrate and lower protein content is favorable. Overall, algae systems could be a potential alternative or additional renewable system to mitigate environmental impact in South India.

References

[1] US DOE. National algal biofuels technology roadmap. Biomass Program, U.S. Department of Energy, Office of Energy Efficiency and Renewable Energy; 2010.

[2] Ribeiro SK, Kobayashi S, Beuthe M, Gasca J, Greene D, Lee DS, et al. Transport and its infrastructure. In: Metz B, Davidson OR, Bosch PR, Dave R, Meyer LA, editors. Climate change 2007: mitigation. Contribution of Working Group III to the Fourth Assessment Report of the Intergovernmental Panel on Climate Change. Cambridge: Cambridge University Press; 2007.

[3] FAO. Safeguarding food security in volatile global markets. Rome: Food and Agriculture Organization of the United Nations; 2011.

[4] Elbehri A, Segerstedt A, Liu P. Biofuels and the sustainability challenge: a global assessment of sustainability issues, trends and policies for biofuels and related feedstocks. Trade and markets division, Rome: Food and Agriculture Organization of the United Nations; 2013.

[5] Darzins A, Pienkos P, Edye L. Current status and potential for algal biofuels production. Task 39 Report T39-T2, IEA Bioenergy; 2010.

[6] Guccione A, Biondi N, Sampietro G, Rodolfi L, Bassi N, Tredici MR. *Chlorella* for protein and biofuels: from strain selection to outdoor cultivation in a Green Wall Panel photobioreactor. Biotechnol Biofuels 2014;7:84.

[7] Zhu L. Biorefinery as a promising approach to promote microalgae industry: an innovative framework. Renew Sust Energ Rev 2015;41:1376–84.

[8] EIA. India is increasingly dependent on imported fossil fuels as demand continues to rise. Country analysis brief overview—India, Washington, DC: U.S. Energy Information Administration; 2014.

[9] Government of India Ministry of New and Renewable Energy. National policy on biofuels. New Delhi: Government of India Ministry of New and Renewable Energy; 2009.

[10] Ajayebi A, Gnansounou E, Raman JK. Comparative life cycle assessment of biodiesel from algae and jatropha: a case study of India. Bioresour Technol 2013;150:429–37.

[11] Ullah K, Ahmad M, Sharma VK, Lu P, Harvey A, Zafar M, et al. Algal biomass as a global source of transport fuels: overview and development perspectives. Prog Nat Sci: Mater Int 2014;24:329–39.

[12] Livestock Census. 19th Livestock census—all India report. New Delhi: Ministry of Agriculture Department of Animal Husbandry, Dairying and Fisheries; 2012.

[13] Iersel SV, Flammini A. Algae-based biofuels: applications and co-products. Series No. 44 Rome: FAO Environmental and Natural Resources Service; 2010.

[14] Pulz O, Gross W. Valuable products from biotechnology of microalgae. Appl Microbiol Biotechnol 2004;65:635–48.

[15] Spolaore P, Joannis-Cassan C, Duran E, Isambert A. Commercial applications of microalgae. J Biosci Bioeng 2006;101:87–96.

[16] Grand View Research. Bio succinic acid market analysis and segment forecasts to 2020. San Francisco, CA: Grand View Research; 2014.

[17] Werpy T, Petersen G. Top value added chemicals from biomass: volume I—results of screening for potential candidates from sugars and synthesis gas. Golden, CO: NREL; 2004.

[18] Cheng KK, Zhao XB, Zeng J, Zhang JA. Biotechnological production of succinic acid: current state and perspectives. Biofuels Bioprod Biorefin 2012;6:302–18.

[19] Campbell PK, Beer T, Batten D. Life cycle assessment of biodiesel production from microalgae in ponds. Bioresour Technol 2011;102:50–6.

[20] Sander K, Murthy G. Life cycle analysis of algae biodiesel. Int J Life Cycle Assess 2010;15:704–14.

[21] Sills DL, Paramita V, Franke MJ, Johnson MC, Akabas TM, Greene CH, et al. Quantitative uncertainty analysis of life cycle assessment for algal biofuel production. Environ Sci Technol 2012;47:687–94.

[22] Woertz IC, Benemann JR, Du N, Unnasch S, Mendola D, Mitchell BG, et al. Life cycle GHG emissions from microalgal biodiesel—a CA-GREET model. Environ Sci Technol 2014;48:6060–8.

[23] Batan L, Quinn J, Willson B, Bradley T. Net energy and greenhouse gas emission evaluation of biodiesel derived from microalgae. Environ Sci Technol 2010;44:7975–80.

[24] Clarens AF, Resurreccion EP, White MA, Colosi LM. Environmental life cycle comparison of algae to other bioenergy feedstocks. Environ Sci Technol 2010;44:1813–9.

[25] Frank ED, Han J, Palou-Rivera I, Elgowainy A, Wang MQ. Life-cycle analysis of algal lipid fuels with the GREET model. ANL/ESD/11-5, Argonne, IL: Argonne National Laboratory; 2011.

[26] Lardon L, Hélias A, Sialve B, Steyer JP, Bernard O. Life-cycle assessment of biodiesel production from microalgae. Environ Sci Technol 2009;43:6475–81.

[27] Stephenson AL, Kazamia E, Dennis JS, Howe CJ, Scott SA, Smith AG. Life-cycle assessment of potential algal biodiesel production in the United Kingdom: a comparison of raceways and air-lift tubular bioreactors. Energy Fuel 2010;24:4062–77.

[28] Gao X, Yu Y, Wu H. Life cycle energy and carbon footprints of microalgal biodiesel production in Western Australia: a comparison of byproducts utilization strategies. ACS Sustain Chem Eng 2013;1:1371–80.

[29] Khoo HH, Sharratt PN, Das P, Balasubramanian RK, Naraharisetti PK, Shaik S. Life cycle energy and CO2 analysis of microalgae-to-biodiesel: preliminary results and comparisons. Bioresour Technol 2011;102:5800–7.

[30] Lundquist TJ, Woertz IC, Quinn N, Benemann JR. A realistic technology and engineering assessment of algae biofuel production. Berkeley, CA: Energy Biosciences Institute; 2010.

[31] Mu D, Min M, Krohn B, Mullins KA, Ruan R, Hill J. Life cycle environmental impacts of wastewater-based algal biofuels. Environ Sci Technol 2014;48:11696–704.

[32] Passell H, Dhaliwal H, Reno M, Wu B, Ben Amotz A, Ivry E, et al. Algae biodiesel life cycle assessment using current commercial data. J Environ Manag 2013;129:103–11.

[33] Shirvani T, Yan X, Inderwildi OR, Edwards PP, King DA. Life cycle energy and greenhouse gas analysis for algae-derived biodiesel. Energy Environ Sci 2011;4:3773–8.
[34] Gnansounou E, Dauriat A, Villegas J, Panichelli L. Life cycle assessment of biofuels: energy and greenhouse gas balances. Bioresour Technol 2009;100:4919–30.
[35] Naik SN, Goud VV, Rout PK, Dalai AK. Production of first and second generation biofuels: a comprehensive review. Renew Sust Energ Rev 2010;14:578–97.
[36] Gnansounou E, Raman JK. Life cycle assessment of algae biodiesel and its co-products. Appl Energy 2016;161:300–8.
[37] Wiley P, Harris L, Reinsch S, Tozzi S, Embaye T, Clark K, et al. Microalgae cultivation using offshore membrane enclosures for growing algae (OMEGA). J Sustain Bioenergy Syst 2013;3:18–32.
[38] Chanakya H, Mahapatra DM, Ravi S, Chauhan V, Abitha R. Sustainability of large-scale algal biofuel production in India. J Indian Inst Sci 2012;92:63–98.
[39] Molina Grima E, Belarbi EH, Acién Fernández FG, Robles Medina A, Chisti Y. Recovery of microalgal biomass and metabolites: process options and economics. Biotechnol Adv 2003;20:491–515.
[40] Schenk P, Thomas-Hall S, Stephens E, Marx U, Mussgnug J, Posten C, et al. Second generation biofuels: high-efficiency microalgae for biodiesel production. Bioenerg Res 2008;1:20–43.
[41] Zaimes G, Khanna V. Microalgal biomass production pathways: evaluation of life cycle environmental impacts. Biotechnol Biofuels 2013;6:88.
[42] Chisti Y. Biodiesel from microalgae. Biotechnol Adv 2007;25:294–306.
[43] Wijffels RH, Barbosa MJ. An outlook on microalgal biofuels. Science 2010;329:796–9.
[44] Stepan DJ, Shockey RE, Moe TA, Dorn R. Carbon dioxide sequestering using microalgal systems. Technical Report, University of North Dakota; 2002.
[45] Davis R, Fishman D, Frank ED, Wigmosta MS. Renewable diesel from algal lipids: an integrated baseline for cost, emissions, and resource potential from a harmonized model. Technical Report, Golden, CO: Argonne National Laboratory and National Renewable Energy Laboratory; 2012.
[46] Uduman N, Qi Y, Danquah MK, Forde GM, Hoadley A. Dewatering of microalgal cultures: a major bottleneck to algae-based fuels. J Renew Sustain Energy 2010;2:012701.
[47] Teixeira MR, Rosa MJ. Comparing dissolved air flotation and conventional sedimentation to remove cyanobacterial cells of *Microcystis aeruginosa*: part I: the key operating conditions. Sep Purif Technol 2006;52:84–94.
[48] Olaizola M. Commercial development of microalgal biotechnology: from the test tube to the marketplace. Biomol Eng 2003;20:459–66.
[49] Matis KA, Gallios GP, Kydros KA. Separation of fines by flotation techniques. Sep Technol 1993;3:76–90.
[50] Halim R, Danquah MK, Webley PA. Extraction of oil from microalgae for biodiesel production: a review. Biotechnol Adv 2012;30:709–32.
[51] Moser B. Biodiesel production, properties, and feedstocks. In Vitro Cell Dev Biol Plant 2009;45:229–66.
[52] Zhang Y, Dubé MA, McLean DD, Kates M. Biodiesel production from waste cooking oil: 1. Process design and technological assessment. Bioresour Technol 2003;89:1–16.
[53] Zhang Y, Dubé MA, McLean DD, Kates M. Biodiesel production from waste cooking oil: 2. Economic assessment and sensitivity analysis. Bioresour Technol 2003;90:229–40.
[54] Berchmans HJ, Hirata S. Biodiesel production from crude *Jatropha curcas* L. seed oil with a high content of free fatty acids. Bioresour Technol 2008;99:1716–21.
[55] Kumar Tiwari A, Kumar A, Raheman H. Biodiesel production from jatropha oil (*Jatropha curcas*) with high free fatty acids: an optimized process. Biomass Bioenergy 2007;31:569–75.
[56] Chen L, Liu T, Zhang W, Chen X, Wang J. Biodiesel production from algae oil high in free fatty acids by two-step catalytic conversion. Bioresour Technol 2012;111:208–14.
[57] Kale A. Methods of protein extraction from substantially intact algal cells, US Patents 20120029170 A1; 2012.
[58] BREW. Medium and long-term opportunities and risks of the biotechnological production of bulk chemicals from renewable resources. Utrecht: Utrecht University; 2006.
[59] Cok B, Tsiropoulos I, Roes AL, Patel MK. Succinic acid production derived from carbohydrates: an energy and greenhouse gas assessment of a platform chemical toward a bio-based economy. Biofuels Bioprod Biorefin 2014;8:16–29.
[60] Becker EW. Micro-algae as a source of protein. Biotechnol Adv 2007;25:207–10.

[61] Energy Statistics. Central statistics office, national statistical organization, ministry of statistics and programme implementation. New Delhi: Government of India; 2012.
[62] Garofalo R. Algae and aquatic biomass for a sustainable production of 2nd generation biofuels. AquaFUELs Taxonomy, Biology and Biotechnology, Brussels: European Biodiesel Board; 2009.
[63] Chia MA, Lombardi AT, Melao Mda G. Growth and biochemical composition of *Chlorella vulgaris* in different growth media. An Acad Bras Cienc 2013;85:1427–38.
[64] Lapuerta M, Armas O, Rodríguez-Fernández J. Effect of biodiesel fuels on diesel engine emissions. Prog Energy Combust Sci 2008;34:198–223.

CHAPTER

8

Life Cycle Assessment and Land-Use Changes: Effectiveness and Limitations

*W. Ben Aoun**,†, *B. Gabrielle**

*Université Paris-Saclay, Thiverval-Grignon, France †Terres Inovia, Thiverval-Grignon, France

8.1 INTRODUCTION

The large-scale deployment of renewable energy, especially the use of biofuels in the transportation sector in many countries around the world, is expected to enhance energy security and to contribute to facing the challenge of climate change by mitigating anthropogenic greenhouse gas (GHG) emissions. It is also viewed as a way of supporting regional development and to create new opportunities for farmers, thereby improving their incomes [1,2]. Since the end of the last century, several policies were implemented to promote a rapid expansion of biofuel production [3] aiming at increasing the blending rate of renewable fuels with conventional ones.

However, from the end of the last decade, the development of first-generation biofuels such as ethanol and biodiesel from cereals, sugar plants, and oilseeds became largely controversial because of the unintended consequences of biofuel policies, notably on food prices and land use. Several pieces of research [4,5] highlighted the strong correlation between biofuel expansion, which creates an additional demand for agricultural crops, and the rise in food prices. Other studies [6–10] pointed out the increased pressure on agricultural lands, since the displacement of food crops by energy crops not only leads to direct land-use changes (dLUC) in the regions where biofuel feedstocks are grown but also to indirect land-use changes (iLUC) in other parts of the world, through the conversion of pastures, forests, and peat land to cropland.

http://dx.doi.org/10.1016/B978-0-444-63585-3.00008-5

These complex mechanisms are difficult to estimate and are usually associated with detrimental effects on the environment, such as increased emissions of GHG and biodiversity depletion from the conversion of natural ecosystems. Thus, iLUC effects are likely to severely degrade the environmental performance of biofuel chains [11–13].

The standardized method of life cycle assessment (LCA) is currently widely recommended for evaluating the environmental balance of biofuels, in particular for policy-making purposes (see Chapter 2). However, most published LCA studies do not take into account iLUC effects [14,15]. This is due to the inability of the classical (also called attributional) LCA framework to take into account such effects, since it ignores the market and economic implications of a given decision (e.g., to achieve a given blending target for biofuels at a national level). Economic equilibrium models and the so-called consequential approach of LCA have been promoted as more suitable alternatives to include these effects and produce a robust assessment of biofuels' environmental impacts [16–19]. Although there is a consensus on the fact that land-use change (LUC) effects need to be addressed and included in the environmental assessment of biofuels, the resulting indicators are quite heterogeneous and subject to a high uncertainty [19,20].

Policy makers, especially in Europe (EU) and the United States, are highly concerned with this controversial issue and are currently expecting more reliable scientific results to help frame their biofuels policies. For now, the production rate of first-generation biofuels has been slowed and some sustainability criteria, requiring abatement in GHG emissions in the short and medium term, have been outlined [21,22].

This chapter provides the reader with several updates on the issue of LUC and its consideration in LCA studies. Section 8.1 reviews the current definitions and types of LUCs, while Section 8.2 emphasizes the complexity of the underlying mechanisms. A review of the methods used to estimate LUCs due to biofuel development is presented in Section 8.3. Lastly, we identify the sources of variability and uncertainty of LCA studies published up to now.

8.2 A TYPOLOGY OF LUCs

In economic terms, the development of biofuels creates additional opportunities for economic agents. In fact, the increase in demand for a given biofuel feedstock will create a shortage of this product, increase its price and thereby provide an incentive for the farmer to increase its output. The latter may respond to this situation by intensifying crop management to improve yields, or alternatively transform uncultivated lands (e.g., fallow land, forests, grassland, or natural ecosystems) into arable land, and/or substitute food/feed crops by energy crops [23,24]. The expansion of land devoted to energy crops and the displacement of food crops trigger LUC mechanisms, of which we distinguish two types here: dLUC and iLUC (Fig. 8.1).

8.2.1 Direct Land-Use Change

Direct land-use change (dLUC) takes place when biofuel feedstock cultivation modifies the land use on the land where it is grown [19]. According to Gawel and Ludwig, this type of LUC

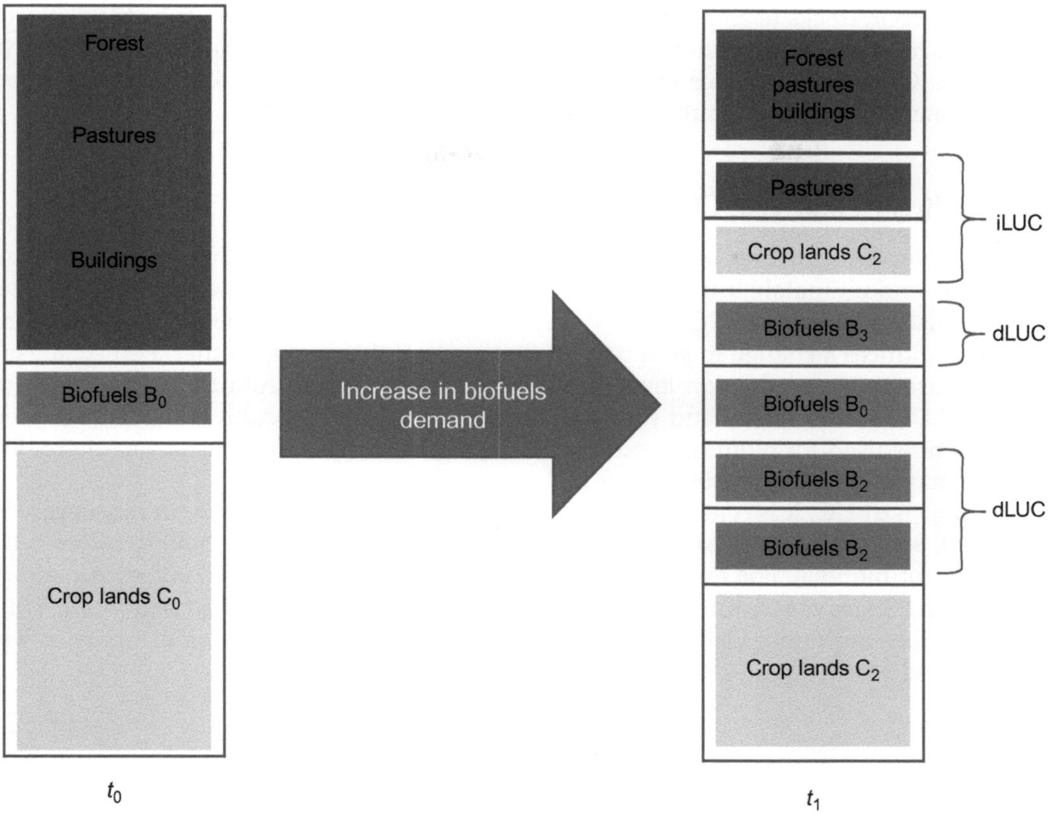

FIG. 8.1 Representation of changes in land use induced by the development of biofuels. dLUC and iLUC stand for direct and indirect land-use change, respectively. *Adapted from De Cara S, Gousssebaile A, Grateau R, Leverte F, Quemener J, Vermont B, et al. Revu critique des études évaluant l'effet des changements d'affectation des sols sur les bilans environnementaux des biocarburants. Etude réalisée pour le compte de l'ADEME par l'INRA. ADEME, Angers; 2012. p. 1–96.*

occurs when biomass cultivation displaces a different former land use (e.g., an arable crop grown after grassland) [25]. For Van Stappen et al., dLUC describes the introduction of a new cropping system in a site where this form of cultivation has not taken place before [26]. It may be estimated quantitatively from the changes in soil and vegetation carbon stocks. Fritsche et al. state that dLUC effects can be determined by comparing of the carbon balances of ecosystems between the new and previous land uses [24].

If the biofuel market only changes the valorization of a given crop (i.e., switching from food to energy use), local impacts are considered negligible. On the other hand, if energy crops displace other crops in a cropping system, the effects on the environment may be significant, as shown in Ref. [27].

Lastly, several studies [21,26,28,29] focused on GHG emissions due to dLUC. Their results show that GHG emissions due to dLUC can be positive or negative depending on the type of land use prior to the implementation of energy crops. If feedstock production occurs on land with high carbon stocks (e.g., pasture, peat land, and unmanaged forests), the dLUC effect is expected to have a large magnitude and strongly undermine the environmental performance of the crops. Conversely, when biofuel feedstocks are grown on degraded soil, dLUC can contribute to increasing the soil carbon content.

8.2.2 Indirect Land-Use Change

The development of first-generation biofuels inevitably increases the pressure on land worldwide, and ultimately brings into cultivation lands that otherwise would not have been put to this use [30]. Indirect LUC occurs when additional demand for bioenergy feedstock in a given area induces a change in land use in other parts of the world via market mechanisms which ensure stable production level of food/feed crops [19,26]. According to Gawel and Ludwig, iLUC occurs when land that was formerly used for the cultivation of food, feed, or fiber is now used for biomass production, shifting the original land use to an alternative area that may have a high carbon stock [25].

Contrary to dLUC, it is often impossible to quantify iLUC associated with bioenergy development, since it is a mechanism that can occur outside the country having fostered the production of biofuels. For example, Laurance showed that the increase of corn planting in the United States may contribute to more deforestation in the Amazon region [8]. Thus, iLUC can cause important GHG emissions, also with adverse effects on biodiversity as well as on soil and water quality.

8.3 COMPLEXITY OF LUC MECHANISMS

Theoretically, both direct and indirect LUC mechanisms appear quite simple. As shown in Section 8.2, any increase in biofuel production ultimately requires diverting cropland to the production of biofuel feedstocks, which inevitably causes dLUC and iLUC.

For example, an additional demand for a feedstock (crop A) used to produce biofuels is met through two main market responses: an increase in current yields and an expansion in the cultivation area of crop A to meet the demand for biofuel production. With the second option (land expansion), crop A historically grown on existing agricultural land can be diverted to biofuel production. This type of dLUC reduces the agricultural area of crop A used in food, which has to be produced on some other land, incurring an iLUC effect. Crop A may also be grown on nonagricultural land (e.g., fallow and grassland). This type of dLUC is generally not accompanied by iLUC. Expansion of crop A production can also be met by displacing other crops (crop B) on existing agricultural land. This can trigger iLUC in order to satisfy the demand in the displaced crops. One should mention that the production of biofuel from crop A is also accompanied by an increase in coproducts production such as meals (coproduct C) that can substitute for other products used in animal feeding from another crop (crop D). This substitution reduces the surface on which crop D is cultivated and therefore mitigates iLUC (Fig. 8.2) [22]. Here, we emphasize the importance of addressing LUC issues, especially for iLUC on a global scale to allow taking into account the overall consequences of biofuel production [14,26,31].

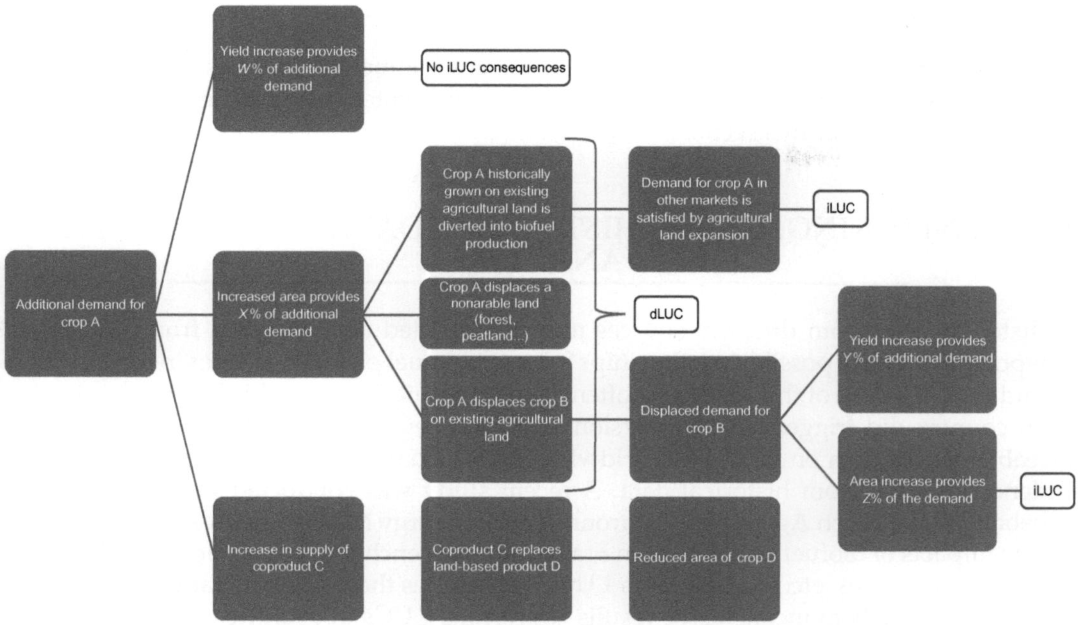

FIG. 8.2 Consequences of an increase in biofuels production. *Adapted from Bauen A, Chudziak C, Vad K, Watson PA. Causal descriptive approach to modelling the GHG emissions associated with the indirect land use impacts of biofuels: final report. E4tech, UK Department of Transport; 2010.*

The increase of pressure on land and the crop displacement that occurs in major exporting countries such as Europe and the United States change the market balance of products from these crops and thus affect their prices [19]. This has an effect on farmers' decisions regarding the allocation of land worldwide.

In other words, as long as crops are displaced, the effects of displacement trickle through the overall global agriculture system until it reaches a new equilibrium [30,31]. These new equilibria may promote substitution among several products (e.g., palm oil may substitute for oilseed rape oil used for biodiesel production). This makes the LUC mechanisms increasingly complex and their monitoring difficult, to the point that the estimation of their environmental impacts is impossible [32]. Furthermore, one should mention that LUCs are also driven by several other factors such as biophysical, demographic, and economic forces. Thus, attempting to attribute LUC to a single factor or the isolation of LUC only due to biofuel production reveals serious problems [29].

GHG emissions from dLUC may be assessed on the basis of the guidelines developed by the Intergovernmental Panel on Climate Change, which propose default emissions factors (Tier 1) but also recommend using country-specific validated data (Tier 2 or 3) wherever available [26]. Unfortunately, the complexity of the previously mentioned mechanisms leads to the fact that there is currently no consensus on one method for estimating GHG from iLUC [13,25], despite the general awareness that neglecting or over/underestimating of iLUC effects leads to wrong decisions and to an inefficient use of biofuels [33].

8.3.1 LUC Estimation

With the awareness of their importance in the environmental balances of biofuels, LUC effects are currently widely investigated, and are estimated using different approaches, which are reviewed in this section.

8.4 MONITORING: USE OF HISTORICAL DATA AND STATISTICAL ANALYSIS

Historical data from different sources may be collected and analyzed from a statistical viewpoint to identify possible relationships between biofuel production rates in a given country and LUC. The use of this method is often justified by the fact that if biofuel production in a given country did trigger land conversion elsewhere, evidence for LUC effects should be traceable in past data on land-use worldwide [32,34]. Some studies have attempted to find evidence for LUC from historical data. A recent study was conducted by In Numeri [35] on behalf of the French Agency for Environment and Energy Management (ADEME) to identify the impacts of biofuel production in France on the French and international markets (imports, exports, prices, etc.) as well as on LUC. It concludes that LUCs in France are relatively limited, but it leads to inconclusive results concerning LUCs in countries outside the European Union. Also on behalf of ADEME, Chakir and Vermont [36] analyzed the evolution of land use and dLUC generated by the development of energy crops and food crops in France during the last two decades, based on data from annual land-use surveys [37,38]. This study showed that until 2004, the increase of energy and food crop areas was limited to agricultural land while from 2006 on the expansion of these surfaces also impacted permanent grassland.

This approach was used by Kim and Dale [34] to detect evidence for iLUC that might be caused by biofuel production in the United States through a statistical analysis. This kind of analysis seems not to be sensitive enough to detect iLUC due to biofuel development. In contrast, Overmars et al. [32] used the same approach with a set of assumptions and concluded that emissions from iLUC related to biofuel production in Europe could shift the GHG balance from a net abatement to a net surplus of emissions relative to fossil fuels.

These retrospective and ex-post analyses are useful to illustrate the complexity of LUC mechanisms, but usually do not allow the isolation of LUC due to biofuels development from simple statistical analyses [14,19,32]. Assumptions (e.g., on where the iLUC is likely to occur) must be made in order to obtain some meaningful conclusions.

8.5 EXPERT-BASED OPINIONS

As shown in Section 8.4, statistical analyses of historical data are not sufficient to isolate and quantify the impact of biofuel production on LUC. Some understanding of the underlying mechanisms and the consultation of experts remain essential to be able to locate the LUC (notably iLUC) and predict their magnitude. This method, also called *causal descriptive*, is known for the transparency of assumptions, often based on intuitive cause-effect relations, and for simplification of market mechanisms [22,23,39].

Such approaches are often used in consequential LCA. For example, Reinhard and Zah [31] made some assumptions based on expert opinions to define a priori the crops displaced by biofuels feedstock as well as the origin of the products to be imported to offset the decline in oilseed rape oil diverted into biodiesel production in Switzerland. The estimation of LUC effects through this approach still remains coarse and fraught by a lack of accuracy because market mechanisms are not fully considered.

8.6 ECONOMIC EQUILIBRIUM MODELS

In practice, it is impossible to isolate the impact of biofuels development on LUC from historical data or expert-based opinions, as there are other activities that can lead to exchanges in land use. Moreover, these methods simplify market mechanisms so that the prediction of LUC (especially of iLUC) might not be accurate enough [23]. Actually, modeling seems to be the most successful method for measuring both direct and indirect LUC [40]. Quantitative assessments based on models have been the policy makers' preferred methodology, even if the latter tend to blame their lack of transparency compared to LCA. Today general consensus exists about using the economic approach to address iLUC [14]. This approach consists of using economic equilibrium models, which are complex optimization models based on the assumptions of perfect markets reaching equilibrium when demand equals supply in the studied economy [41]. The response of supply and demand to price changes is the basis of the estimate of the LUC. These models make it possible to pinpoint the consequences of an additional demand for biofuels on land use at global scale, provided they include a land-use module and some degree of spatial differentiation between world regions. Here, we distinguish two types of equilibrium models: partial and general equilibrium models.

8.6.1 Partial Equilibrium Models

Partial equilibrium models address a particular economic sector. Those which represent agriculture describe the different compartments of commodities supply (yields, areas allocated to different cultures, exports) and demand (human/animal demand for food/feed, nonfood demand, and imports). They estimate land demand for individual crops and allow them to compete for land through cross-price elasticity. They subsequently calculate prices that balance supply and demand in all markets represented and their evolution over time within a given time horizon [39]. The main partial equilibrium models currently existing are FAPRI FASOM, CAPRI, IMPACT, GLOBIOM, AGLINK-COSIMO, MATSIM-LUCA, Nexus Land-Use, Magpie, and MIST [2,42–44].

8.6.2 General Equilibrium Models

General equilibrium models address all economic sectors and are developed to describe international trade. Interactions between different markets are recognized endogenously in the model. It can be assumed that land is relatively easily transformed from one use to another through the definition of a constant elasticity of transformation. These models also include a representation of yield response to price and make the differentiation between the yields from new lands and those from land already cultivated. Thus, the models allow farmers to choose between increased business through the adjustment of production factors use (labor, fertilization, etc.) or the expansion on other land to meet the extra demand due to the development of biofuels.

The main general equilibrium models are GTAP, LEITAP, MAGENT, and MIRAGE [19]. Here, one should emphasize the necessity of incorporating geo-referenced information as inputs to such economic models, especially regarding land cover and land availability. It is also crucial to use biophysical models in combination with economic models in order to provide necessary information on yields and GHG emissions. For example, the GLOBIOM partial equilibrium model is coupled to the crop model EPIC [45], while the Magpie economic model is coupled to the land surface model LPjML [44,46].

8.7 ACCURACY OF BIOFUELS CHAINS' LCAs: IMPORTANCE OF ACCOUNTING FOR LUC EFFECTS

If the effects of LUCs due to the increase in feedstock demands are ignored, the majority of studies aiming to evaluate the environmental performance of first-generation biofuels concluded a positive balance allowing mitigating GHG emissions compared to the environmental performance of fossil fuels. However, this claim became controversial when the issue of LUC effects emerged, since taking into account such effects in the LCA of biofuels undermines the benefits of their substitution for fossil fuels.

Several studies [18,19,47,48] have sought to review and compare existing assessments of iLUC effects due to biofuels expansion. They highlighted the difficulty encountered in isolating the lands diverted to crop production in response to the increase in biofuels demand, regardless of the method used to estimate LUCs (as detailed in Section 8.6). These studies also emphasized the high variability of the current estimation of iLUC factors (i.e., annualized GHG emissions from LUC related to the production of 1 MJ of biofuel). This leads to a significant uncertainty about the accuracy of the current estimations.

The variability in iLUC effects estimation stems from differences at several levels. First, the iLUC factor differs from one crop to another. For example, De Cara et al. [19] showed that existing studies estimate a low iLUC factor associated with the expansion of wheat production compared to the factors related to soybean expansion. The same study evidenced significant effects of biofuel type and generation on iLUC factors. The literature review showed that bioethanol has a lower iLUC factor compared to biodiesel. The same conclusion was found when comparing first-generation biofuels to second-generation ones. Moreover, the location of biofuels supply and demand of biofuel also plays a determinant role on the number of iLUC factors.

The approach used to estimate LUC is also one of the major sources of results variability. The so-called heuristic (causal descriptive) approach, which is often used in consequential LCAs and based on the simplification of market mechanisms, leads to lower iLUC factors than those obtained through the use of economic equilibrium models [19,49]. Here, one should mention that for the same biofuel chain, iLUC estimation may widely vary according to the economic model used to approach it, as shown by Wicke et al. [47] and Broch et al. [48]. The differences not only stem from the type of model used (i.e., general equilibrium vs. partial equilibrium) and assumptions made to represent markets response to price signals and producers/consumers preferences, but also from the use of different databases on carbon stocks changes, fertilizer use, and gaseous emissions of reactive nitrogen [39].

Thus, the available estimations of LUC effects are uncertain and confirm the difficulty of isolating the effects of biofuels deployment on land use because of the complexity of the mechanisms involved and the manifold confounding factors. Pending more accurate estimates, such effects should still be taken into account, even in the form of overall ranges in order to capture these potential effects and provide a more complete picture of the various environmental impacts of biofuel chains.

8.8 CONCLUSIONS AND PERSPECTIVES

LUCs play a key role in LCAs, significantly affecting their results. Regulations on biofuels in the United States as well as in the European Union consider the effects of LUC on the GHG balance of biofuels, with strong implications in terms of eligibility to public subsidies. Accounting for LUC effects implies an estimation of the consequences of biofuel feedstock production on soil carbon dynamics. While this can be relatively easily estimated via default values for direct LUC, dealing with indirect LUC is trickier since these effects occur outside the country under study. Several methodologies are reviewed in this chapter, namely: statistical approach, expert opinion, and finally economic equilibrium models. It is stated that the latter shows more relevance and consistency compared to the others. Chapter 9 gives more details on these equilibrium models. The perspectives are to improve the specifications of the models, and the geographical as well as the temporal resolutions of the modeling.

References

[1] Bessou C, Ferchaud F, Gabrielle B, Mary B. Biofuels, greenhouse gases and climate change. A review. Agron Sustain Dev 2011;31:1–79.

[2] Blanco M, Adenäuer M, Shrestha S, Becker A. Methodology to assess EU biofuel policies: the CAPRI approach. Luxembourg: European Union; 2013. p.1–67.

[3] Sorda G, Banse M, Kemfert C. An overview of biofuel policies across the world. Energy Policy 2010;38:6977–88.

[4] Guindé L, Jacquet F, Millet G. Impacts du développement des biocarburants sur la production française de grandes cultures. Revue d'Etudes en agriculture et environnement. Rev Agric Environ Stud 2008;89:55–81.

[5] Elobeid A, Carriquiry M, Dumortier J, Rosas F, Mulik K, Fabiosa JF, et al. Biofuel expansion, fertilizer use, and GHG emissions: unintended consequences of mitigation policies. Econ Res Int 2013;708604.

[6] Searchinger T, Heimlich R, Houghton RA, Dong F, Elobeid A, Fabiosa J, et al. Use of U.S. croplands for biofuels increases greenhouse gases through emissions from land-use change. Science 2008;319:1238–40.

[7] Searchinger TD, Hamburg SP, Melillo J, Chameides W, Havlik P, Kammen DM, et al. Fixing a critical climate accounting error. Science 2009;326:527–8.

[8] Laurance WF. Switch to corn promotes Amazon deforestation. Science 2007;318:1721.

[9] Fargione J, Hill J, Tilman D, Polasky S, Hawthorne P, et al. Land clearing and the biofuel carbon debt. Science 2008;319:1235–8.

[10] Harvey M, Pilgrim S. The new competition for land: food, energy, and climate change. Food Policy 2011;36: S40–51.

[11] Drabik D, de Gorter H. Emissions from indirect land use change: do they matter with fuel market leakages? Rev Agric Appl Econ 2013;16:3–15.
[12] Plevin RJ, O'Hare Michael, Jones AD, Torn MS, Gibbs HK, et al. Greenhouse gas emissions from biofuels' indirect land use change are uncertain but may be much greater than previously estimated. Environ Sci Technol 2010;44:8015–21.
[13] Melillo JM, Reilly JM, Kicklighter DW, Gurgel AC, Cronin TW, Paltsev S, et al. Indirect emissions from biofuels: how important? Science 2009;326:1397–9.
[14] Di Lucia L, Ahlgren S, Ericsson K. The dilemma of indirect land-use changes in EU biofuel policy—an empirical study of policy-making in the context of scientific uncertainty. Environ Sci Pol 2012;16:9–19.
[15] Muench S, Guenther E. A systematic review of bioenergy life cycle assessments. Appl Energy 2013;112:257–73.
[16] Kløverpris J, Wenzel H, Nielsen PH. Life cycle inventory modelling of land use induced by crop consumption. Int J Life Cycle Assess 2008;13:13–21.
[17] Kløverpris JH. Consequential life cycle inventory modelling of land use induced by crop consumption: industrial. Ph.D. dissertation. Danmarks Tekniske Universitet, DTU Management Engineering; 2010.
[18] Sanchez ST, Woods J, Akhurst M, Brander M, O'Hare M, Dawson TP, et al. Accounting for indirect land-use change in the life cycle assessment of biofuel supply chains. J R Soc Interface 2012;9:1105–19.
[19] De Cara S, Goussebaïle A, Grateau R, Leverte F, Quemener J, Vermont B, et al. Revu critique des études évaluant l'effet des changements d'affectation des sols sur les bilans environnementaux des biocarburants. Etude réalisée pour le compte de l'ADEME par l'INRA, Angers: ADEME; 2012. p.1–96.
[20] Fritz S, See L, Valin H. Current issues and uncertainties in estimating global land availability for biofuel production. Biofuel 2013;4:343–5.
[21] European Commission. Directive 2009/28/CE du Parlement européen et du Conseil du 23 avril 2009 relative à la promotion de l'utilisation de l'énergie produite à partir de sources renouvelables et modifiant puis abrogeant les directives 2001/77/CE et 2003/30/CE; 2009. p.1–47.
[22] Bauen A, Chudziak C, Vad K, Watson PA. Causal descriptive approach to modelling the GHG emissions associated with the indirect land use impacts of biofuels: final report. E4tech, UK Department of Transport; 2010.
[23] Ben Aoun W, Gabrielle B, Gagnepain B. The importance of land use change in the environmental balance of biofuels. OCL 2013;20:D505.
[24] Fritsche UR, Sims REH, Monti A. Direct and indirect land-use competition issues for energy crops and their sustainable production—an overview. Biofuels Bioprod Biorefin 2010;4:692–704.
[25] Gawel E, Ludwig G. The iLUC dilemma: how to deal with indirect land use changes when governing energy crops? Land Use Policy 2011;28:846–56.
[26] Van Stappen F, Brose I, Schenkel Y. Direct and indirect land use changes issues in European sustainability initiatives: state-of-the-art, open issues and future developments. Biomass Bioenergy 2011;35:4824–34.
[27] Achten WMJ, Verchot LV. Implications of biodiesel-induced land-use changes for CO_2 emissions: case studies in tropical America, Africa, and Southeast Asia. Ecol Soc 2011;16:14.
[28] Louhichi K, Valin H. Impact of EU biofuel policies on the French arable sector: a micro-level analysis using global market and farm-based supply models. Rev Agric Environ Stud 2012;93:233–72.
[29] Gnansounou E, Dauriat A, Villegas J, Panichelli L. Life cycle assessment of biofuels: energy and greenhouse gas balances. Bioresour Technol 2009;100:4919–30.
[30] Delucchi M. A conceptual framework for estimating the climate impacts of land-use change due to energy crop programs. Biomass Bioenergy 2011;35:2337–60.
[31] Reinhard J, Zah R. Global environmental consequences of increased biodiesel consumption in Switzerland: consequential life cycle assessment. J Clean Prod 2009;17:S46–56.
[32] Overmars KP, Stehfest E, Ros JPM, Prins AG. Indirect land use change emissions related to EU biofuel consumption: an analysis based on historical data. Environ Sci Pol 2011;14:248–57. http://dx.doi.org/10.1016/j.envsci.2010.12.012.
[33] Zilberman D, Hochman G, Rajagopal D. Indirect land use change: a second-best solution to a first-class problem. AgBioforum 2010;13:382–90.
[34] Kim S, Dale BE. Indirect land use change for biofuels: testing predictions and improving analytical methodologies. Biomass Bioenergy 2011;35:3235–40.
[35] In Numeri. Analyse retrospective des interactions du développement des biocarburants en France avec l'évolution des marches français et mondiaux (productions agricoles, prosuits transformées et co-produits) et

les changement d'affectation des sols. Etude réalisée pour le compte de l'ADEME par In Numeri, Angers: ADEME; 2012.

[36] Chakir R, Vermont B. Etude complémentaire à l'analyse rétrospective des interactions du développement des biocarburants en France avec l'évolution des marchés français et mondiaux et les changements d'affectation des sols. Angers: ADEME; 2013. p.1–69.

[37] AGRESTE. Enquête TERUTI – série 1992 à 2004. Service de la statistique et de la prospective du ministère de l'agriculture, de l'alimentation, de la pêche, de la ruralité et de l'aménagement du territoire. Paris; 2004.

[38] AGRESTE. Enquête TERUTI-LUCAS – nouvelle série 2005 à 2010. Service de la statistique et de la prospective du ministère de l'agriculture, de l'alimentation, de la pêche, de la ruralité et de l'aménagement du territoire; 2010.

[39] Nassar AM, Harfuch L, Bachion LC, Moreira MR. Biofuels and land-use changes: searching for the top model. Interface Focus 2011;1:224–32.

[40] Edwards R, Mulligan D, Marelli L. Indirect land use change from increased biofuels demand—comparison of models and results for marginal biofuels production from different feedstocks. EC Joint Research Center; 2010. p.1–150.

[41] Höglund J, Hansen K, Gustavsson M, Ahlgren S, Börjesson P, Sundberg C, et al. Biofuels and land use in Sweden—an overview of land-use change effects. Report no 2013:7, Sweden: f3 The Swedish Knowledge Centre for Renewable Transportation Fuels; 2013.

[42] Blanco M, Burrell A, Gay H, Henseler M, Kavallari A, M'Barek R, et al. Impacts of the EU biofuel target on agricultural markets and land use: a comparative modelling assessment. Seville; 2010. p.1–230.

[43] Souty F, Brunelle T, Dumas P, Dorin B, Ciais P, Crassous R, et al. The nexus land-use model version 1.0, an approach articulating biophysical potentials and economic dynamics to model competition for land-use. Geosci Model Dev 2012;5:1297–322.

[44] Lotze-Campen H, von Lampe M, Kyle P, Fujimori S, Havlik P, Meijl H, et al. Impacts of increased bioenergy demand on global food markets: an AgMIP economic model intercomparison. Agric Econ 2014;45:103–16.

[45] Havlík P, Schneider UA, Schmid E, Böttcher H, Fritz S, Skalský R, et al. Global land-use implications of first and second generation biofuel targets. Energy Policy 2011;39:5690–702.

[46] Brunelle T, Dumas P. Can numerical models estimate indirect land use? Nota Di Lavoro 65. Fondazione Eni Enrico Mattei;. 2012.

[47] Wicke B, Verweij P, van Meijl H, Vuuren DP, Faaij APC. Indirect land use change: review of existing models and strategies for mitigation. Biofuel 2011;3:87–100.

[48] Broch A, Hoekman SK, Unnasch S. A review of variability in indirect land use change assessment and modeling in biofuel policy. Environ Sci Pol 2013;29:147–57.

[49] Malça J, Freire F. Life-cycle studies of biodiesel in Europe: a review addressing the variability of results and modeling issues. Renew Sust Energ Rev 2011;15:338–51.

CHAPTER

9

Modeling Land-Use Change Effects of Biofuel Policies: Coupling Economic Models and LCA

L. Panichelli, E. Gnansounou*†

*International Consultant (Freelance), Milano, Italy †Bioenergy and Energy Planning Research Group (BPE), Ecole Polytechnique Fédérale de Lausanne (EPFL), Lausanne, Switzerland

9.1 INTRODUCTION

9.1.1 Background

Land-use change (LUC) induced by feedstock provision for biofuels has become a major focus of attention for policy and science in recent years. From 2006 to 2008 agricultural commodity prices have risen above historical levels. While the causes of this price rise are multiple, including higher petroleum prices and their correlation with agricultural and energy markets, climatic changes affecting agricultural yields, market speculation on goods availability, and production and consumption expectations, most attention was turned to the role of biofuel policies [1].

First discussions focused on the impact of biofuel mandates on the price of primary agricultural crops used as feedstock for biofuels [2]. The uncertainty linked to the extent to which each farmer responds to higher crop prices leads to the use of economic models. A study reviewing previous research published in 2013 by consultancy Ecofys [3] estimates that the impact of EU biofuels demand from 2000 until 2010 has increased world grain prices by about 1–2% and oilseed prices by around 4%. It also estimates that without any cap on crop-based biofuels, EU policy could raise grain prices by 1%, and oilseed prices by 10% by 2020.

Economic models were used to assess the supply response to biofuel policies, including effects on international commodity prices, intensification, production and acreage change

Life-Cycle Assessment of Biorefineries
http://dx.doi.org/10.1016/B978-0-444-63585-3.00009-7

[4]. It was acknowledged that LUC from the provision of land for feedstock for biofuels may generate large quantities of GHG emissions and that including these emissions in the biofuel GHG emission balance may offset the emission reduction benefit of substituting fossil fuels by biofuels. This motivated several research and governmental institutions with experience in the development and use of economic models to provide economic model-based assessments of the impact of biofuels policies and their two major market mediated impacts: increased food prices and increased GHG emissions from LUC. Addressing these issues requires analyzing complex biophysical, economic and policy interactions in space and time linking global energy, agriculture, and land sectors. It involves understanding the mechanisms regulating global agricultural markets and LUC dynamics, identifying structural changes in markets and land supply and assessing the marginal effects of land provision for biofuels' feedstock in the long and short–medium terms. The issue of indirect LUC GHG emissions further motivated the need to link global economic models and life-cycle assessment (LCA) to calculate worldwide indirect LUC GHG emissions [5,6] and GHG emission savings from replacing fossil fuels by biofuels. Finally, the need to calculate changes in area and agricultural land supply induced by a marginal change in demand for crops for biofuels motivated the link of global economic models with the consequential LCA, the LCA approach to modeling the consequences of a decision [7].

Several reviews of economic agricultural models dealing with biofuels and LUC have been performed [8–11] that reveal the considerable improvement in addressing biofuels LUC GHG emissions, the large variability of results and the importance of data, models and assumptions in the uncertainty linked to current iLUC GHG emission values. While none of the approaches is perfect, the exploration and adaptation of existing models and the appearance of new models and coupled models shed some light on the possible effects of biofuel policies on LUC, agricultural commodity prices and GHG emissions. On the other hand, some controversial issues were raised regarding the validity of model results and how these results should be used in policy. Adaptations of existing models included more detailed treatment of the biofuel sector, link to other national and international markets (agriculture, energy, forestry, land), and more detailed land allocation and land supply structures. While several models have been used to address the LUC effects of biofuels, only a few were coupled with LCA.

This chapter focuses on the linkage between LCA and economic models to address LUC effects of biofuel policies. The next sections provide a description of the theoretical and methodological foundations to address LUC effects of biofuel policies and present an assessment of main economic models used for this purpose. General model structures are compared based on: type of model, sectors and regions involved and time accounting. A review of the main coupled economic-LCA approaches used to address biofuels LUC GHG emissions describes how the model works, presenting the default and typical results and how these results are being used in policy.

9.1.2 Market-Mediated LUC Induced by Biofuel Policies

Fig. 9.1 depicts the market mediated mechanisms and main influencing variables in the analysis of LUC effects of biofuel policies. Demand for biofuels is influenced by policy

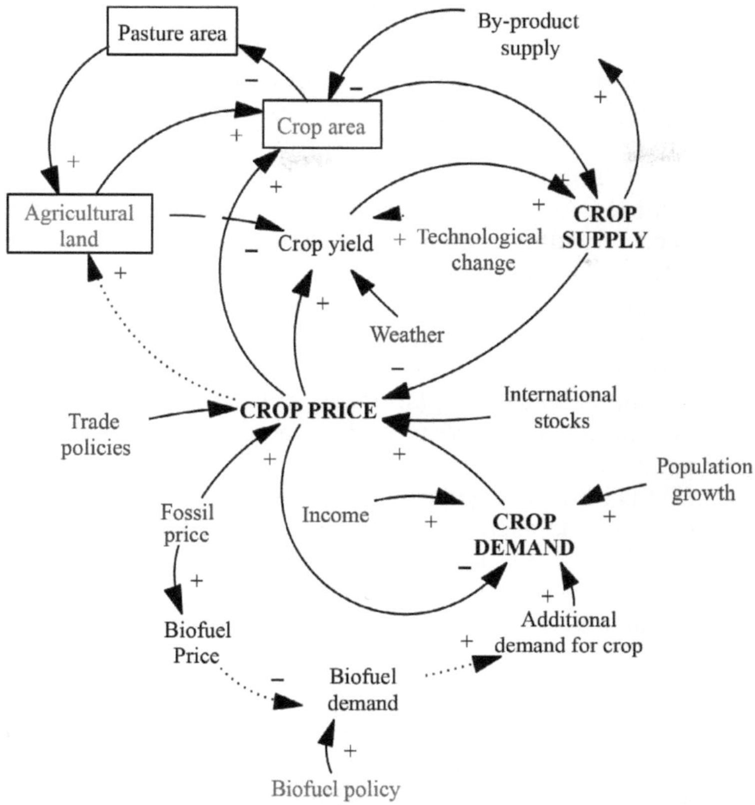

FIG. 9.1 Market-mediated LUC induced by biofuel policies.

mandates and the prices of the biofuel and fossil fuel substitute. Crop prices depend on the supply and demand equilibrium in the market but are also affected by trade policies, prices of inputs and substitutes, speculation on stocks and the price of fossil fuels. Based on market equilibrium theory and with all other things being equal, an increase in the demand for crops as feedstock for biofuels is prone to increase the price of the crop. Producers would increase output to adjust supply to the new equilibrium price. One of the main issues in addressing LUC impacts from biofuels is to determine how the additional demand for crops will be satisfied. Farmers can apply different strategies to increase agronomic and economic output, that can include the intensification of the production (increase agronomic yield per unit of area) or the expansion of the crop area by incorporating new agricultural land or by displacing other agricultural activities.

Agricultural yields depend on economic and noneconomic factors. Yields respond to crop prices and production costs that affect the management practice of the farmer (eg, technological level, cultivation methods, agro-chemicals use). Noneconomic factors include the effect of technological advances, weather conditions and climate change impacts.

Expanding production by area involves the conversion of land that was previously used for other purpose. There are basically two ways of supplying land to produce the required extra amount of crop for biofuels. This extra area of crop may come from the conversion of natural biomes (eg, forest, grassland, wetlands, peatlands) or from the displacement of other agricultural activities (eg, other food crops, cattle ranging). As agricultural markets are globally interconnected, the displacement of natural biomes or other agricultural land can occur in other countries. Additionally, all other things being equal, land is also required to relocate the displaced agricultural activities that again can take place in other regions. The extent and location of these land relocations are difficult to quantity and observe. Expansion of crop area into natural land is expected to occur in regions with available agricultural land, where the cost of land expansion into natural lands are lower and land policy is weak. Expansion of biofuel crops into other agricultural activities is expected to occur in countries where available agricultural land is scarce. Displacements are also expected to occur in countries that produce competing or substitute agricultural commodities (eg, soybean/oil seed producing countries) and producing countries that are traditionally linked through trade import-export agreements (EU-US-Brazil-Argentina-Indonesia).

How farmers would respond to changes in the international price of crops is of major importance as this leads to different types of LUC effects. LUC can occur in different locations and in different types of land uses, generating different types of impacts. Indeed, economic, social and environmental impacts related to increased crop prices are linked to different supply responses.

- Intensification processes reduce pressure on land expansion but, on the other hand, may lead to monoculture, use of genetically modified material and increased use of agrochemicals. Intensification practices may affect the local flora and fauna, the quality of soils, air and water resources that may increase health problems in local communities, the degradation of natural resources and biodiversity losses.
- Displacing other agricultural activities may result in the subsequent problem of relocating these activities. Shifting production may result in the short term in a reduced supply of other products that may translate into higher prices of other agricultural products. Food security issues may then arise if the displaced products are basic food products and disequilibrium occurs between supply and demand in the local market. If the displaced production is relocated in new areas or a new shift in production occurs, then indirect land use effects should be assessed. If the product is an agricultural commodity and depending on the market structure of the sector, LUC effects should be monitored on a worldwide scale.
- Expanding into new agricultural areas implies conversion of natural biomes, and as such, the loss of natural ecosystems (and their services including carbon stocks, biodiversity, and water and soil conservation).

While multiple direct and indirect LUC effects can occur, policy and research debate, however, has mainly focused on the effect of LUC on GHG emissions. To this end, Righelatto [12] was the first to give evidence of the magnitude of direct LUC GHG emissions when substituting natural land by crops for biofuels. He argued that if the main objective of biofuel policy was to mitigate climate change, policy-makers might better focus on increasing the efficiency of fossil fuel use, conserving the existing forests and savannahs, and restoring natural forest and grassland habitats to nonfood cropland. Fargione et al. [13] additionally claimed that clearing lands to produce biofuel feedstock created a carbon deficit. For instance, sugarcane ethanol produced on natural Cerrado lands would take about 17 years to repay its carbon debt, while corn ethanol produced on US central grasslands would result in a repayment time of about 93 years. The worst-case scenario is converting Indonesian/Malaysian tropical peat land rainforest to palm oil biodiesel production, which would require about 420 years to repay. Finally, Searchinger et al. [6] was the first to show the magnitude of iLUC GHG emissions for US corn ethanol. The study showed that iLUC GHG emissions for US corn ethanol would represent 104 gCO_2eq/MJ alone if the emissions from land clearing were amortized over a 30-year period. Since then several model developments and applications were directed to address direct and indirect LUC GHG emissions of biofuel policies. Accounting for LUC GHG emissions in the biofuels life cycle required linking economic, spatial and LCA models.

9.1.3 The Economic—Spatial—Consequential LCA Approach to LUC GHG Emissions

The international linkage of agriculture and energy commodity markets, the uncertainty in the supply response of agricultural producers to increased demand for biofuels feedstock, and the difficulty of isolating the effects of biofuels policies from other causes of prices rises and LUC motivated the use of global economic models to address the LUC effects of biofuels.

The isolation of the effect of the biofuel policy is implemented through a *ceteris-paribus* assessment. The economic model is run two times, typically with and without a biofuel mandate. The only change is the additional demand for the crop (all other things being equal). Model runs are then compared to calculate the change in prices, production, land supply and land-use induced by the additional demand for the crop for biofuel. A base-line scenario is constructed upon historical economic data that is typically compared with a scenario with biofuels. Through runs comparison the effect of the biofuel policy can be isolated, provided that all other things remain equal.

The intended purpose of the method is to understand the level of pressure that biofuels may be exerting over agricultural land and commodity prices. The truth, however, is that it is impossible to know what will happen if the policy is not implemented. So, modelers attempt to better describe the structure of the system while creating scenarios about the possible future with and without a biofuel policy. Dealing with such a complex system gives room to different representations of reality while dealing with the uncertainty about the future and the lack of data. As a result models with different structures and data have been used to assess LUC effects of biofuel policies. Annex 9.1 provides an overview of selected applications of models to assess LUC effects of biofuel policies.

Most sophisticated modeling frameworks required coupling global economic, spatial explicit and biophysical models. Fig. 9.2 depicts the main model structure, inputs and outputs for the coupled model to assess LUC GHG emissions from biofuel policies.

Kloverpris [5] first proposed the theoretical foundations for coupling economic modeling with LCA as a method to account for marginal increments in demand for crops. The approach was illustrated by an example of increase of wheat demand in selected countries, and was adopted as a methodology for assessing biofuels LUC-GHG emissions [14]. The need to

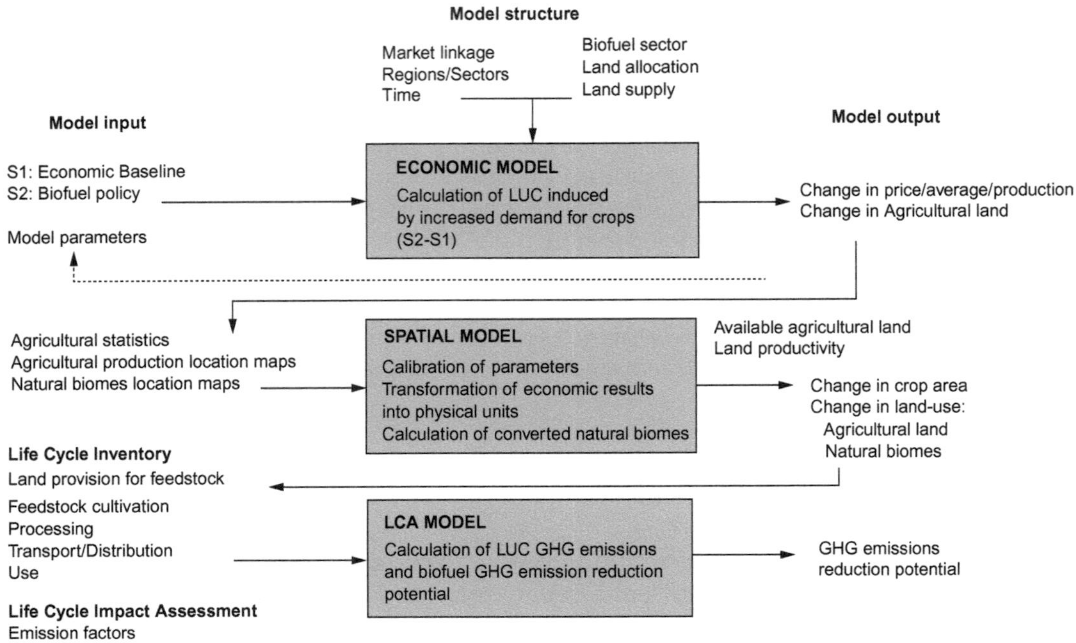

FIG. 9.2 Coupled economic-spatial-LCA modeling approach.

isolate the effect of the biofuel policy and evaluate effects of the extra demand for crops for biofuels required a marginal approach. The conventional approach to estimate GHG emission balance of biofuels relied on the attributional LCA (ALCA), which evaluates the environmental impact of a product through the quantification of input and output flows of the production system based on average data and fixing the system boundaries to those flows physically connected to the product under study. ALCA of biofuels does not account for price variations, changes in demand or technological improvements. The consequential LCA (CLCA), on the other hand, is based on a system-wide approach where system boundaries are expanded to consider the consequence of a decision. Sanchez et al. [15] proved a comprehensive analysis of how ALCA and CLCA can be combined to assess iLUC GHG emissions from biofuels, exploring critical issues such as time treatment and system boundaries. Different time and scale issues needed to be considered. Time is involved in the model calculation method, the definition of the policy framework, the lagged supply of land and the amortization/discounting of GHG emissions. The link among international agricultural markets creates the possibility that LUC may occur elsewhere. Consequently, a global approach was required to trace international LUC.

Provided that a fundamental aspect of LUC is the location of the land conversion, geographic information was needed to extract important biophysical variables affecting agricultural production and land conversion decisions. The need to account for location factors then stimulated the link of economic and spatial explicit models. The coupling mechanism works by linking economic model output to input data of the life cycle inventory of the biofuel. The economic spatial LCA coupling is performed by means of geographical information that allows converting monetary value into physical units. Through the use of geographic information systems (GISs) and agricultural statistics, the output of economic modeling is translated into actual changes in crop area and agricultural land. The GIS serves also to trace the types of converted natural biomes due to the expansion of agricultural land. Assessments at different spatial and time scales were possible, providing a highly complex representation of agricultural and land markets, and their link to biofuels policies and GHG emissions. By far the two most explored scenarios account for the effect of EU and US biofuel mandates, including a variety of first- and second-generation feedstocks.

The next section provides examples of the main economic models used to assess the market-mediated effects of biofuel policies.

9.2 MAIN ECONOMIC MODELS

Different economic models have been developed, adapted and use for modeling LUC effects of biofuel policies. Table 9.1 gives an overview.

The models in Table 9.1 all rely on the same equilibrium theory, where the global economy (or a sector of the economy) is modeled as a system of supply and demand equations that find equilibrium by adjusting price. When demand for the crop is increased, the crop price increases, reducing demand and increasing supply, until a new equilibrium price is obtained in each market sector and region. Production shifts may occur until all production is relocated through intensification (yield increment) or expansion processes (area increment), at which point supply and demand equilibrates in all sectors and regions of the global economy. The resulting new equilibrium is compared to a baseline, in order to compute the effect of additional demand for crops for biofuels.

Theory on market equilibrium is combined with real economic data in computable equilibrium models. In the case of the LUC effect of biofuel policies, two types of computable economic models are generally used, namely partial and general equilibrium (GE) models. Sector-specific effects are treated in partial equilibrium (PE) models, which focus on a specific market. The main PE models applied to address biofuels LUC are global agricultural sector models and, to a lesser extent, energy models have also been used. A broader perspective of

TABLE 9.1 Economic Models Used to Assess LUC Effects of Biofuel Policies

Economic Approach	Selected models (used to assess LUC effects of biofuel policies)	Key model characteristics (in selected studies to quantify biofuels LUC)
General equilibrium	GTAP, LEITAP, EPPA, DART, FARM	Global, static or recursive dynamic, non-spatial, focused in all economic sectors, aggregated actors and regions, policy oriented
Partial equilibrium	AgLink-COSIMO, ESIM, FAPRI (MIRAGE), CAPRI, GCAM, IMPACT	Global (regional) or national, recursive dynamic, non-spatial, focused in agricultural markets, aggregated actors and regions, policy oriented
Optimisation	FASOM (GLOBIOM/EUBIOM), LUCEA, P&G, POLYSIS	Global (regional) or national, recursive dynamic, spatial/non-spatial, linear programming, focused in forestry/agricultural sectors, aggregated actors and regions

market-mediated impacts of biofuels is addressed in GE models. GE models represent the global market economy and provide a better description of market linkages to assess economy-wide feedback effects. In the case of biofuels, GE models allow an accounting for biofuel interactions with the agricultural (biofuel feedstocks) and energy markets (fuel substitutes and inputs).

Economic models are calibrated to a reference year, using real economic data defining the baseline that is used for comparison (eg, the latest GTAP database is based on the 2004 economy). Quality and availability of data to calibrate models is one of the main reasons why results differ. The choice of the reference year is subjected to the availability of economic data (eg, the previous GTAP database was based on the 2001 economy). Generally, institutions running these models are constantly updating databases when new consistent information is generated and becomes available.

Economic models need to calibrate utility and production functions. Model parameters, so-called elasticities, reflect the effect of the relative change between two economic variables. The common approach to model the possibility of substitution between goods and production factors is by using the constant elasticity of substitution (CES) function. The degree of substitution between inputs can range from complementary to exclusive products, represented by the Leontief and Coub-Doublas production functions, respectively. The higher the elasticity, the more the other variable will be affected. The constant property of the CES function means the rate of change between the two variables is always constant. Global models need to calibrate function for each sector and region of the global economy, which requires large amounts of consistent data, which is not always available. In the case of GE models, these relations should be known and calibrated for several interrelated markets. Then assuming constant relative changes is a common assumption in order to be able to handle the data, assure calibration and interpretation of results.

The most important elasticities affecting global LUC induced by biofuels demand include:

- Biofuel cross-price elasticity: Measures the rate of substitution between the biofuel and the fossil fuel alternative. (ie, the effect of the fossil fuel price on the demand for biofuel).
- Crop price-demand elasticity: Crop price response to changes in crop demand.
- Armington Elasticity: Substitution between local and foreign goods (key parameter to understand international LUC).
- Price-crop acreage elasticity: Measures the rate of substitution between economic land uses, and generally is modeled as a nested constant elasticity of transformation (CET) function. This means a CET with different levels of land substitutability.
- Agricultural land supply elasticity: Measures the possibility to expand available agricultural land based on the change in the aggregated land rent (see Van Meijl [16] approach to agricultural land supply).
- Price-yield elasticity: Measures the effect of changes in crop prices in the agronomic yield of the crop (ie, intensification effect). It is commonly assumed exogenous and is a major topic of discussion (see for example Malins et al. [17].)

Accounting for coproducts in biofuels production may also lead to significantly different estimates of LUC. According to Taheripour et al. [18] accounting for coproducts results in lower LUC, as the extra area of land required to produce the biofuel is, at least partially, compensated for by the provision of feedstock for the livestock industry. This results in a decrease in the demand for land for crops used as feedstock for livestock production. The study shows that models with and without coproducts reveal different portraits of the economic impacts of the US and EU biofuel mandates for the world economy in 2015. The model version with coproducts shows smaller changes in the production of cereal grains and larger changes for oilseed products in the US and EU, and the reverse for Brazil. Models that omit coproducts are found to overstate cropland conversion from US and EU mandates by about 27%.

The other reason why models differ is model structure. The type of linkage of the agricultural sector with other markets (energy, forestry, inputs, land), the sectors and regions involved, the level of aggregation and the method to account for time all influence model outputs. The EU Joint Research Center (JRC) [4] provided a comprehensive review of models used for assessing changes in prices and required area for the production of energy crops. Their review included GE (LEITAP, GTAP) and PE models (AGLINK-COSIMO, CAPRI, IMPACT, CARD) and provides a description of the main characteristics of each economic model, the inclusion of the bioenergy sector and the application of the models to analyze LUC effects of bioenergy production. According to the review, most agricultural PE models do not include second-generation biofuel crops. Second-generation biofuels tend to require lower energy inputs per unit of area, increasing the efficiency of biomass production [19]. Moreover, these crops have higher agricultural yields, or are obtained from wastes and residues that do not enter into direct competition for agricultural land. Then it is expected that moving forward to second-generation biofuels will result in lower LUC.

Most PE models like AGLINK-COSIMO, FAPRI and IMPACT are agricultural economic models. The link to the energy, forestry and land sectors sometimes is missing. Several models have coupled other models or adapt economic models to include features from other sectors. Energy models (eg, POLES, PRIMES) and forestry economic models (eg, EUBIOM, GLOBIOM) may help to extend existing agricultural models to account for a better representation of the linkage of biofuels to other markets and feedstock types. In this sense, GE models give a more integrated picture, accounting better for the market interaction. In GTAP for instance, biodiesel demand depends on the price of biodiesel relative to gasoline and biodiesel supply depends on the price of the crop relative to the price of land.

According to the JRC review, one of the major problems of the reviewed economic models (except for GTAP) is the model linearity. Roughly, all models project crop area changes that are proportional to the change in demand for a particular biofuel. While this is attributed to the lack of economic data to calibrate nonlinear functions, one should expect the supply of area for crops to be nonlinear. That is because one expects land productivity to decrease as the agricultural frontier expands and then, more land is required for increasing levels of demand for crops. Moreover, agricultural yields are expected to face diminishing returns so that less improvements in yield are obtained with increased use of agricultural inputs.

The JRC report also argues that the critical issue why models differ in terms of LUC relies on the rate between the increase in demand for crop and the increase in crop yield. Low yield response to increase crop demand (price), results in higher LUC and vice versa. IMPACT model applications resulted in low LUC because the model assumes high price-induced yield responses. GTAP assumes quite modest contributions from increased yields. However, modest crop price-yield responses are compensated for by lower yield assumption in new land conversions. Then in the end, the overall effect of price on yield in the GTAP model does not differ much from IMPACT. On the other hand CARD and AGLINK-COSIMO models assume relatively low price-yield responses, resulting in higher LUC.

A critical issue in price-yield and price-acreage responses is the possibility of agricultural area to expand into new areas. Most models assume agricultural land to be fixed, so that LUC occurs only at the inside of agricultural land. This means that competition arises among different economic land uses of the agricultural land, but no extra area is transformed to agricultural land use. Agricultural land uses are generally classified in economic models as cropland, pasture and managed forests. However, each model assumes different land-use classifications and different approaches to deal with the competition among agricultural economic land uses and the supply of new land to agriculture. In models reviewed by Edwards et al. [20], all economic models except for GTAP model land supply as a function of the particular crop price increased. In GTAP the supply of land to agriculture depends on the crop yield at the margin.

9.3 MAIN COUPLING APPROACHES

The issue of indirect LUC motivated the development and adaptation of models and the development of integrated and coupling approaches. Accounting for LUC GHG emissions in biofuels life cycle emissions required coupling LCA models with other modeling approaches, the development of simpler but integrated models, or the expansion of existing models. Coupled models are more complex than integrated ones. Because outputs remain independent from each of the models that have been coupled, the models can be adapted to address problems at different scales. Moreover, complex descriptions of the economic and biophysical drivers of biofuel production and their impact on LUC-GHG emissions can be created by linking models that treat each issue specifically. Table 9.2 provides examples of the main coupled models used to study LUC GHG emissions from biofuels.

Assessments of GHG emissions from biofuels in the United States have primarily coupled the GREET LCA model with global and national economic models to account for LUC GHG emissions of US biofuel mandates. Searchinger et al. [6] coupled the FAPRI global PE model [25] with the GREET LCA model to estimate iLUC GHG emissions induced by US corn ethanol production required to achieve the US ethanol mandate. The FAPRI economic model estimates two types of iLUC: the amount of land required to relocate displaced corn production for other purposes and soybean displaced from rotation in the same land. Shares of converted natural land types are assigned based on the proportion of lands that has been transformed into cropland in the past. The amount of land converted of each land type is used as input for GREET. GREET provides emission factors for changes in land carbon stocks to calculate LUC GHG emission of US corn ethanol. Moreover, the California Air Resources Board (CARB) linked GTAP and GREET models to assess LUC GHG emissions for biofuels to be supplied under the Low Carbon Fuel Standard (LCFS) Program [22]. Similarly to the Searchinger et al. approach, the GE model estimates LUC induced by the extra demand for feedstock for biofuels required to achieve the CARB policy mandate.

The US Environmental Protection Association (EPA) implemented a more sophisticated modeling framework, coupling PE and GE models, optimization, and biophysical models [21]. The impact of US biofuel mandates on global and national LUC is estimated by linking

TABLE 9.2 Model Coupling Approaches: Biofuels LUC GHG Emissions

Author	Models	Description	Feedstock/Origin	Biofuel Policy Scenario	Scale
Searchinger [6]	FAPRI	Partial equilibrium (Agriculture)	Corn/US	US Ethanol (2022)	Global
	GREET	Life-Cycle Assessment			US+
EPA [21]	GTAP	General Equilibrium (Economic)	Corn/US Soybean/US Sugarcane/BR Switchgrass/US Waste Grease/US Corn residue/US	US Ethanol US Biodiesel (2022)	Global
	FAPRI	Partial Equilibrium (Agriculture)			Global
	FASOM	Spatial Optimization (Agriculture)			US
	GREET	Life-Cycle Assessment			US+
CARB [22]	GTAP	General Equilibrium	Tallow/US Corn oil/US Soybean/US Sugarcane/BR Corn/US	US Ethanol US Biodiesel (2020)	Global
	GREET	Life-Cycle Assessment			US+
Wageningen [23]	LEITAP	General Equilibrium	Cereals, Vegetable oils, Sugar/EU+Imports (BR)	EU Biodiesel EU Ethanol (2030)	Global
	IMAGE	Biophysical			Global
	CLUE	Spatial			EU
Popp [24]	LPJmL MAgPIE ReMIND	Spatial (Biophysical) Land-use Optimization (Economic-Agriculture) Economic (Energy)-Biophysical (Climate Change)	Cellulosic (Miscanthus/poplar/eucalyptus)	Bioenergy demand for climate change mitigation scenario (2100)	Global

GTAP with FAPRI, a GE and PE model, respectively. LUC at the country level is estimated with the FASOM, a spatial-explicit optimization model of the US agricultural sector. The historical share of converted land uses and emission factors are provided by Winrock International [26] based on global georeferenced data of land conversions (2001–2004) and carbon stock changes between land-use types. Similarly, this data is used as input into the GREET model to estimate biofuel GHG emission balance.

In the case of the EU, MAGNET (previous LEITAP), IMAGE and Dyna-CLUE models were coupled to assess the spatial patterns of biofuel crops distribution [27,28]. The dynamic version of the CLUE model is a spatially explicit model for the analysis of spatial patterns of LUC [29,30]. CLUE mainly uses regression analysis to estimate land-use transitions based on a set of location variables that are assumed to guide land-management decisions on land allocation. The coupled economic biophysical and spatially explicit model (MAGNET IMAGE and CLUE, respectively) constructs a spatially explicit, multiscale, quantitative description of LUCs through the determination and quantification of location factors of land uses based on the actual land-use structure. The approach determines the location of crop expansion and consequently the direct LUC induced by biofuel feedstock production.

A coupled approach has also been proposed by Popp et al. [24] to address global bioenergy LUC GHG emissions. The approach links a global vegetation and hydrology model (LPJmL), a global land-use optimization model (MAgPIE), and the global energy–economy–climate model (ReMIND). LPJmL is a spatially explicit model that simulates biophysical, biogeochemical, and hydrological processes in the global land surface. MAgPIE is a mathematical programming model covering the most important agricultural crop and livestock production types in 10 economic regions worldwide. The nonlinear programming model minimizes the total cost of production for a given amount of regional food and bioenergy demand. Finally, ReMIND is an integrated modeling framework that links an energy system model with a macroeconomic intertemporal growth model and a climate-change model that simulates the effect of GHG emissions on global mean temperature.

Endogenous estimation of LUC-GHG emissions has also been improved in economic models [31]. Ahammad and Mi [32] extended the GTEM model to allow modeling LUC and associated GHG emissions. The main advantage of integrated models is that they give a more consistent overall representation of the system and allow accounting for internal feedbacks. Moreover, because the integrated model is developed in the same modeling environment, consistency is increased and adapting and transforming outputs beyond the original intent of the model is avoided. Integrated models are more robust in terms of linking different independent modules, but this consistency is gained at the expense of simpler representations of the system components. Some simplifications are appropriate in the quest to understand complex systems. Relying on simple structures that mimic the overall behavior of the system can be an advantage to avoid model coupling, but provides a simplistic approach for quantitatively analyzing complex interactions. In the case of LUC GHG emissions from biofuels, models' coupling has been the preferred approach.

9.4 TYPICAL IMPLEMENTATION AND RESULTS

A typical assessment of biofuel LUC GHG emissions includes the following steps:

- Definition of biofuel scenario
- Model run with/without biofuel scenario
- Computation of LUCs induced by the change in feedstock demand
- Definition of converted natural lands
- Application of carbon emission factors
- Amortization of emissions

Once the biofuel scenario is developed, economic models are used to compute LUCs induced by the change in feedstock demand. A typical implementation involves a "shock" of the baseline feedstock demand with an additional demand for biofuels feedstock. The way of introducing this shock depends on the model structure. Model runs, with and without biofuel demand, are then compared to compute the change in the area of agricultural products and the change in agricultural land.

The difference in model structure, data and parameters and the application of models to assess different policy scenarios made results incomparable. By selecting the main set of economic models being used in this context, the EU JRC launched a modeling initiative to assess indirect LUC induced by increased biofuels demand, based on the same biofuel scenarios. Edwards et al. [20] summarized JRC research efforts to harmonize biofuel scenarios and compare the economic models used. Scenarios that were developed included marginal extra quantities of around 1 million tons of biofuel applied to the model: (1) ethanol demand in the US and EU, (2) biodiesel and palm oil demand in the EU. The total estimated global iLUC induced by biofuel scenarios is estimated based on the:

- required feedstock: Agreed biofuel scenarios (tons),
- fraction of feedstock saved by coproducts (%),
- fraction of feedstock "saved" by reduction in food use (%),
- feedstock yield: Baseline production/Baseline area (tons/ha),
- area "saved" by yield increase: Baseline area × average of fractional yield increase (per region per crop) weighted by baseline area (ha/toe),
- area saved by total net yield effects (ha/toe).

JRC first indicates that model results are basically linear with the size of the "shock" in each biofuel scenario (required feedstock). Then similar biofuel scenarios should result in similar LUC estimates. LUC results differ based on the assumption of different scenarios, the use of different economic data and years for baselines, model structures and calibration of parameters in each model.

The JRC study reported model results in terms of LUC (kha/Mtoe) provided that all the models used generate outputs at that level. All biofuel scenarios result in significant LUC. In general, LEITAP projects the highest LUC while IFPRI-MIRAGE (DG-TRADE mixed scenario of sugarcane ethanol and rapeseed biodiesel) and IFPRI-IMPACT models project the lowest LUC (Table 9.3).

TABLE 9.3 Total Estimated Global iLUC in JRC Modeling Approach [20]

Scenarios	Total Estimated Global iLUC (kHa/Mtoe)	Location of LUC	Comment
EU ethanol (wheat)	223–743	Largest share: outside the EU	Exception: FAPRI/LEITAP (forces all production to come from within the EU).
EU biodiesel (rapeseed)	242–1928	Largest share: outside the EU	LEITAP: highest value Scenario for EU biodiesel in Germany
US ethanol (corn)	107–863	AGLINK-COSIMO/GTAP largest share: outside the US LEITAP: 90% within the US	
IND/MY Palm oil	103–425	LEITAP: all of the LUC would occur in Indonesia GTAP: largest share would occur outside the Malaysia/Indonesia region	Only modeled by LEITAP and GTAP
BR sugar-cane	134	All of the ILUC occurring in Brazil	Only modeled by AGLINK-COSIMO

For the same scenario, each model projects different shares of LUC sources. The IFPRI-IMPACT model reported in the JRC study has no correction for coproducts and projected greater yield improvements than the other models. Then for the US/EU ethanol scenario the coproduct effects and the greater yield result in an underestimation of LUC. FAPRI EU-wheat ethanol results involve an increase in meat production with increased area for cattle ranching that result in significant LUC. Then, while the models were run for the same scenario, these differences explain the wide range of LUC estimates.

The projection of the location of LUC also varies between models. This is important because, depending on where LUC occurs, different land types will be substituted. In the EU scenarios LUC occurs mainly outside the EU, except for LEITAP that forces land to come within the EU. In the US ethanol scenario, AGLINK-COSIMO and GTAP projects the largest share of LUC to come from outside the US, while LEITAP estimates that 90% of the LUC will be within the US. The same difference in LUC location (ie, national, international) occurs for the palm oil scenario. Then the model structure with respect to trade assumptions (import/export) is crucial to estimate the location of the LUC. According to LEITAP and GTAP simulations, the possibility of importing biofuels and feedstock from other regions (eg, Indonesia) reduces the impact of producing these fuels or feedstock in the EU.

The estimation of LUC GHG emissions further involves the link of LUC to GHG emissions. For a fixed amortization time, results then differ based on the type of converted land use and the carbon stock change in soils and biomass between converted land uses. Not all economic models provide endogenous estimations of LUC GHG emissions or have been linked to LCA models. The JRC study, for instance, assumes an average carbon stock change of 40 tonC/ha to roughly estimate the range of GHG emissions that correspond to the total estimated LUC reported by each modeling exercise. This carbon stock change value is the average between the lowest and the highest values reported in the literature: 10 tonC/ha for abandoned EU cropland according to GreenAgSim and 95 tonC/ha for conversion to cropland in the EU and North America, according to IPCC default values (also used in Searchinger et al. [6]). iLUC carbon intensities range from around 15 gCO_2eq/MJ to close to 350 gCO_2eq/MJ and results largely differ between models and scenarios. In general, LEITAP produces the highest iLUC GHG emission estimates for all scenarios while IMPACT generated the lowest iLUC GHG emission estimates.

Since the first iLUC GHG emissions estimation by Searchinger for US corn ethanol (104 gCO_2/MJ), several studies performed economic model-based estimates of LUC-related GHG emissions that allowed the range of LUC-GHG emissions induced by biofuel policies to be narrowed down. For instance, improvements in the GTAP model allowed reducing the iLUC factor of US corn ethanol first to 32 gCO_2/MJ (used in the CARB approach) and then to 15 gCO_2/MJ [33], and, while using the MIRAGE model, Laborde [34] found a value of 7 gCO_2/MJ for LUC-related GHG emissions.

The uncertainty linked to iLUC GHG estimations has been explored by Plevin et al. [35]. Using a simplified model and key assumptions, the authors showed that the 95% confidence interval for carbon stock, model behavior and the amortization period would yield significantly different iLUC factors ranging from 21 to 142 gCO_2eq/MJ per year. Furthermore, they found an upper and a lower bound of 340 gCO_2eq/MJ and 10 gCO_2eq/MJ, respectively.

Using the MIRAGE-BioF model, Al Riffai et al. [36] found a 20-year iLUC factor of 18–20 gCO_2eq/MJ for European Union biofuel policy, depending on trade assumptions. Using the same model, Havlík et al. [37] focused on global biofuel expansion and found 30-year iLUC coefficients for first-generation feedstocks of 53–62 gCO_2eq/MJ, depending on scenarios. Melillo et al. [38] assessed cellulosic biofuel crops finding a 30-year iLUC factor of 180–190 gCO_2eq/MJ, more than double the displaced gasoline emissions, depending on the characteristics of land-use expansion. Finally, Laborde [34] also reported LUC-related GHG emissions for biofuels under the European biofuels policy, for a variety of feedstocks. While the study found average LUC emissions for the European mandate to be 38 gCO_2eq/MJ (by 2020, for the "no change in trade regime" scenario) bioethanol LUC emissions range from 7 to 14 gCO_2eq/MJ, for sugar beet and wheat-based ethanol, respectively, and biodiesel LUC emissions range from 52 to 56 gCO_2eq/MJ for sunflower and soybean-based biodiesel, respectively.

These results clearly illustrate the variability of outcomes with respect to feedstocks, amortization period, and crop-growing region. Wicke et al. [11] acknowledged that key uncertainties in iLUC modeling relate to (1) the underlying datasets, (2) the amount, location and type of projected LUC, (3) by-/coproducts allocation, (4) future production and trade patterns of bioenergy, (5) technological change over time, and (6) the dynamic nature of LUC.

The variability of iLUC GHG emissions is also explored in recent reviews and comparisons of iLUC GHG emission results. Broch [10] investigated the effects of modeling approaches and emission factor databases on iLUC variability in relevant studies, with particular emphasis on the US policy. De Cara et al. [39] performed a meta-analysis of 71 references leading to 239 values of dLUC GHG emissions and 561 values of d + iLUC GHG emissions. The study provides a median value of the d + iLUC factor for different feedstocks (soybean: 79 gCO_2eq/MJ, palm oil: 55 gCO_2eq/MJ, rapeseed and corn: 54 gCO_2eq/MJ, sugarcane: 30 gCO_2eq/MJ, wheat: 17 gCO_2eq/MJ and sugar beet: 8 gCO_2eq/MJ), biofuels (first-generation biodiesel: 61 gCO_2eq/MJ, first-generation ethanol: 31 gCO_2eq/MJ, second-generation ethanol (18 gCO_2eq/MJ), supply regions (global coverage: 46 gCO_2eq/MJ, Latin America: 74 gCO_2eq/MJ, North America: 56 gCO_2eq/MJ, Asia: 53 gCO_2eq/MJ, Europe: 28 gCO_2eq/MJ) and origin of the biofuel demand (North America: 72 gCO_2eq/MJ, Europe: 43 gCO_2eq/MJ). The study acknowledges that when including the d + iLUC factor in the total GHG emissions life cycle, the total emissions are greater than that of the reference fossil fuel (83.8 gCO_2eq/MJ) for more than half of the collected evaluations. Moreover, for 80% of the collected evaluations, total net emissions are above the EU-RED 35% reduction threshold. Finally, Ahlgren and Di Lucia [8] also provide an assessment of iLUC GHGH emissions from biodiesel and bioethanol feedstocks with particular emphasis on the possibility of these biofuels complying with the EU GHG saving requirements. Maize stover and sugar beet were identified as feedstocks with low iLUC GHG emissions in all studies, while the largest variation in results was seen for wheat ethanol and soybean biodiesel. The inclusion of iLUC emissions in policy and regulation is critical provided that if current d + iLUC factor values were introduced into EU policy, no (first-generation) biodiesel fuel would be able to comply with the EU GHG saving requirements. The authors then proposed that alternative approaches to the iLUC factor, such as iLUC mitigation options, should be implemented in the EU.

9.5 IMPLEMENTATION IN BIOFUELS POLICY AND REGULATION

The importance of LUC induced by biofuel policies, the difficulty in producing model-based estimations and the diversity of results obtained have forced policy makers to be cautious on the inclusion of iLUC GHG emissions in biofuel policy [40]. The acknowledgment of the potential LUC effect of biofuel policies caused governments in biofuel-consuming countries (eg, United States) and regions (eg, European Union) to start revising their biofuel market penetration targets [41]. On the other hand, biofuel producing and exporting countries (eg, Brazil, Argentina, Indonesia), who were worried about biofuel sustainability and business, started generating more detailed and reliable data and methods to reduce the uncertainty linked to the estimate while contesting global approaches to address LUC GHG emissions [42].

GHG emission thresholds for biofuels have emerged as a way to address the urgent issue of LUC induced by biofuels policies. Other policies to meet climate goals, such as the implementation of a carbon price for land use, can be effective to control GHG emissions. However, it seems that such a policy is unlikely to be adopted globally within any realistic time frame.

In the United States, the Renewable Fuel Standard (RFS2) regulation implemented by the US EPA and California's LCFS implemented by the CARB require that life-cycle GHG emissions of biofuels must include not only direct but also indirect LUC GHG emissions.

Whether and how to include iLUC emissions in Europe's Renewable Energy Directive (RED) and the UK's Renewable Transport Fuel Obligation (RTFO) are under review and development. The process has raised questions about whether, and if so, how regulations should account for or influence actions and actors outside policies' immediate jurisdictions. Finkbeiner [43], for instance, acknowledges that none of the relevant international standards of LCA require the inclusion of iLUC GHG emissions. Instead, science and policy should focus on proactive real-world mitigation of iLUC rather than reactive and theoretical iLUC factors. To this end, approaches are already under development to propose ILUC mitigation measures [44].

9.5.1 The US Case

In the United States, the Energy Policy Act of 2005 created a Renewable Fuel Standard (RFS1) that required 2.78% of gasoline consumed in the US in 2006 to be renewable fuel. The EPA finalized this requirement for RFS1 in April of 2007. Congress expanded US renewable fuel usage with the Energy Independence and Security Act (EISA) of 2007. The Act included a provision for a new Renewable Fuel Standard (RFS2) [45], which in 2010 increased the required volume of renewable fuel to 36 billion gallons by 2022 (or about 7% of expected annual gasoline and diesel consumption above a business-as-usual scenario).

The California's' LCFS is a mix of command and control regulation and emissions trading. It uses market-based mechanisms that allow providers to choose how they will reduce emissions while responding to consumer demand. Four categories of renewable fuel are specified in the RFS2. In order to be classified under one of these categories, a fuel must meet a threshold percentage reduction in life-cycle GHG emissions relative to a 2005 petroleum baseline. The GHG emission reduction thresholds are as follows: 20% for renewable fuel (produced from renewable biomass), 50% for advances biofuels (excluding corn starch based ethanol) and biomass-based diesel, and 60% for cellulosic biofuels

The Regulatory Impact Analysis (RIA) [21] assessed the impacts of an increase in production, distribution, and use of the renewable fuels sufficient to meet the RFS2 volumes. Among other impacts, the RIA assesses the life cycle GHG emissions of the biofuels being proposed for the rule. The RFS2 rule is expected to reduce GHG emissions by 138 million metric tons in 2022; equivalent to taking about 27 million vehicles off the road.

The EPA approach link several models and data sources to assess GHG life-cycle emissions of fuels qualifying under the RFS2. The FASOM US economic spatial explicit model is used to estimate domestic LUC while FAPRI global PE model is used to estimate indirect LUC. GREET biophysical model is combined with IPCC, DAYCENT and MOVES emission factors to calculate LUC GHG emissions.

The EPA modeling approach is used to include within the life-cycle GHG emissions estimate an iLUC factor from LUC resulting from increased biofuel demand. The iLUC factor is considered as an indicator of a feedstock's contribution to the overall policy's effect that evaluates whether a specific biofuel pathway life- cycle GHG emissions meet the reduction in GHG emissions established in the RFS2 rule. Estimates of iLUC GHG emissions are associated with specific biofuel feedstocks and conversion processes. Biochemical and thermochemical processes to obtain biofuels from switchgrass and corn residues result in the higher GHG reductions, when including iLUC emission factors for these feedstocks (-10 gCO_2eq/MJ for corn residues and 12 gCO_2eq/MJ for switch grass respectively). Corn ethanol and soybean biodiesel result in the lower GHG reductions, especially when including iLUC GHG emissions. Soybean biodiesel results in the higher iLUC GHG emissions (41 gCO_2eq/MJ). The iLUC factor of ethanol varies largely depending on the type and origin of the feedstock. US corn ethanol, US switchgrass ethanol and Brazilian sugar cane ethanol result in an iLUC factor of 30, 12, and 4 gCO_2eq/MJ, respectively.

California implemented through the CARB the LCFS as part of California Global Warming Solutions Act of 2006. The LCFS allows supplying low carbon fuels as pure biofuels or mixed with fossil fuels (gasoline or diesel) in order to meet the policy target of reducing the state-average GHG intensity of California's transportation fuels by 10% by 2020.

Based on the linkage of GTAP and GREET models, CARB performs several analyses of life-cycle GHG emissions of specific fuels able to replace gasoline or diesel in the US. Several different processing pathways (type of plant, location of production) are specified, with direct emissions varying significantly between each other. Fuel carbon intensities (CI) in the CARB rule show upper and lower CI values for each type of fuel.

An iLUC factor is added to the life cycle GHG-intensity of the biofuel, as follows: 30 gCO_2eq/MJ for US corn ethanol, 46 gCO_2eq/MJ for Brazilian sugarcane ethanol, and 62 gCO_2eq/MJ for US soybean biodiesel. Britz and Hertel [46] additionally used the CARB modeling framework to estimate iLUC GHG emissions from rapeseed oil use in the EU, finding an iLUC factor of 47 gCO_2eq/MJ.

CARB's estimate of the iLUC factor for corn ethanol is identical to EPA's estimate (30 gCO_2eq/MJ), but is higher for soybean biodiesel than EPA's value (62 and 40 gCO_2eq/MJ over 30 years for EPA and CARB, respectively). However, the CARB's estimate for soybean iLUC could vary from 20 to 70 gCO_2eq/MJ, depending on the calibration of historical LUC data and land carbon emission factors.

Then, while EPA's estimation relies on relatively detailed set of data for carbon stock changes and cropland expansion dynamics, it revealed the high sensitivity of results to key modeling assumptions [47]. A shortcoming of the EPA analysis is that LUC impacts of alternative fuels are evaluated one feedstock at one time [48]. However, the RFS2 mandate will likely lead to simultaneous shifts in production of multiple types of biofuels produced from a variety of feedstocks.

9.5.2 The EU Case

In 2003, the European Union introduced a target in its Biofuels Directive involving reaching a 5.75% share of biofuels in the EU's transport sector by 2010. In 2009, as part of the 2020 climate and energy strategy, the Directive 2009/28/EC, so called RED introduced a 10% renewable energy target (comprised essentially by biofuels) in the transport sector, to be reached by 2020.

The EU-RED also set out sustainability criteria to ensure that the use of biofuels in transport and bioliquids for electricity and heating is done in a way that guarantees real carbon savings and protects biodiversity. The criteria include requirements on the minimum greenhouse gas (GHG) emission savings that biofuels and bioliquids need to achieve compared to fossil fuels. Economic operators can show they comply with the sustainability criteria through national systems or so-called voluntary schemes recognized by the European Commission. Only biofuels and bioliquids that comply with the criteria can receive government support or count towards national renewable energy targets. To be considered sustainable, biofuels must achieve GHG savings of at least 35% in comparison to fossil fuels. This savings requirement rises to 50% in 2017. In 2018 onwards, it rises again to 60% but only for new production plants.

The EU-RED also provides a standard methodology for the calculation of GHG emission saving: All life cycle emissions from the production and use of biofuel are considered, including annualized emissions from carbon stock changes caused by LUC. Annualized LUC GHG emissions shall be calculated by dividing total LUC emissions equally over 20 years, while the reference land use shall be the land use in January 2008 or 20 years before the raw material was obtained, whichever was the later. The EU-RED also accounts for a bonus of 29 gCO_2eq/MJ of biofuel or bioliquid if biomass is obtained from restored degraded land and adds additional criteria for land-use:

- Biofuels cannot be grown in areas converted from land with previously high carbon stock such as wetlands or forests.
- Biofuels cannot be produced from raw materials obtained from land with high biodiversity such as primary forests or highly bio-diverse grasslands.

The EU-RED provides default and typical life cycle GHG emission values (without including LUC) for selected biofuel types. Biofuel types include ethanol (sugar beet, sugar cane, wheat and corn), biodiesel (rape seed, sunflower, soybean, palm oil, waste vegetable oil), hydro-treated vegetable oil (rape seed, sunflower, palm oil), pure vegetable oil (rape seed) and biogas (municipal organic waste). Default and typical values are also disaggregated for cultivation, processing, transport and distribution emissions, while leaving the economic operators the task of estimating their own direct LUC GHG emissions. To this respect, the JRC developed standard methodologies for the estimation of global GHG emissions from LUC [49] and the calculation of carbon stock changes in biofuels sustainability schemes [50], that were officially adopted by the EC in June 2010 [51].

In 2010, the EC also acknowledged that iLUC can increase the CO_2 emissions of biofuels and recommended addressing this issue under a precautionary approach. Indeed, iLUC GHG emissions are not directly estimated nor included in the EU-RED. The RIA of the EU-RED [41] provided a detailed assessment of whether iLUC GHG emissions should be addressed, and if so, in which way. The EC explored options aimed at minimizing iLUC impacts, as follows:

A. Take no action for the time being, while continuing to monitor,
B. Increase the minimum GHG saving threshold for biofuels,
C. Introduce additional sustainability requirements on certain categories of biofuels,
D. Attribute a quantity of GHG emissions to biofuels reflecting the estimated iLUC,
E. Limit the contribution from conventional biofuels to the RED current levels.

The RIA showed that a balanced approach based on option E, accompanied by complementary elements of options B and D and additional incentives for advanced biofuels, would be the best way to minimize estimated iLUC emissions. In order to move forward, the EC presented in 2012 a legislative proposal to address some of these concerns while preserving existing investments. It proposed capping conventional biofuels and promoting advanced biofuels. Advanced biofuels however are not yet produced on a large scale in the EU, the technologies are not fully mature and investment is lacking. Following the US approach, the European Parliament called for a conventional biofuels cap, a sub-target for advanced biofuels and the consideration of iLUC factors, while stressing the need for a post-2020 policy.

Given the uncertainty linked to model-based estimations of iLUC GHG emissions from biofuels, the EU adopted a precautionary approach instead of including iLUC GHG emissions into the regulatory framework in terms of quantitative cap. The February 2015 Biofuel Policy Briefing of the EU [52] gives a picture of iLUC estimates in the EU, reporting average, higher and lower estimates from a set of selected studies [53]. The EU acknowledges that differences between the studies available are much larger than the differences between feedstocks in a given study. Average estimates of net GHG emission from iLUC range from −42% for sugarcane ethanol to +14% for sunflower biodiesel. In spite of limitations in terms of scope, methodology, consistency and data, the Policy Briefing considers that studies allow the determination of a reasonable minimum range of iLUC GHG emissions.

In September 2015, the EU adopted Directive 2015/1513, so-called iLUC Directive, to reduce the risk of iLUC from biofuels. The iLUC Directive enter into force in October 2015 and includes a number of additional reporting obligations regarding the scientific research and the best latest available scientific evidence. The iLUC Directive also obliges the EC to assess the possibility of setting out criteria for the identification and certification of low ILUC-risk biofuels and bioliquids that are produced in accordance with the EU sustainability criteria.

9.6 CONCLUSIONS AND PERSPECTIVES

Assessing LUC effects of biofuel policies has become one of the major issues of debate concerning the possibility of such policies to reduce GHG emissions. Outstanding efforts have been put in developing and adapting models to deal with the international and complex linkages of biofuel policies and LUC. However, some major challenges still remain:

- The development and adoption of a standard methodology to calculate GHG emissions from LUC,
- The development of more dynamics modeling approaches that account for simultaneous changes over time to capture the dynamic behavior of LUC,
- The disaggregation of data to account for regional specificities, especially in the land provision and cultivation stages, that largely vary depending on the location of the feedstock for biofuels,
- The development of clear coupling and integration methods to address global LUC GHG emissions and the coupling with modeling approaches other than equilibrium economic models,
- The availability of data and the uncertainty link to different economic data sources that largely affect the baseline for comparison and the calibration of parameters in economic models,
- The development of an approach to monitor LUC effects of biofuel policies and integrate them into policy and regulation.

The issue of LUC induced by biofuels policies raises up the importance of the sustainability of biofuel pathways produced with different feedstocks in different locations and their potential to reduce GHG emissions. In the policy perspective, a general consensus needs to be reached with respect to the way in which GHG thresholds for biofuels are included into policy and regulation and a need to consider alternative policies to reduce GHG emissions (especially those from LUC).

Moreover, the reviewed modeling exercises evidence the difficulty to isolate the marginal effect of biofuel policies on global LUC and to compare LUC GHG emissions that rely on different model structures and assumptions. The isolation of the effect of biofuels is an approach to estimate iLUC factors to be included in policy. However, it is still discussable if one should charge biofuels for changes occurring outside the possible boundaries of national regulation and whether biofuels should be treated as an isolated agricultural-based product.

Addressing the sustainability of agricultural-based biofuels in the framework of an holistic approach that account for other drivers of LUC including the use of land for different purposes may result in a more consistent approach to minimize negative impacts from LUC. In other words, isolating the effect of biofuels only make calculations more complex, as a method needs to be specified to assign GHG emissions to the biofuels. Policy and science need to address the sustainability of the whole agricultural and land sector though integrated agricultural and land-use planning. The sustainability of these sectors exceeds the limited scope of addressing the sustainability of specific biofuel pathways.

ANNEX 9.1 SELECTED MODEL APPLICATIONS TO ASSESS THE LUC EFFECTS OF BIOFUEL POLICIES

Model Acronym	Model Name	Selected Applications
GENERAL EQUILIBRIUM		
GTAP (https://www.gtap.agecon.purdue.edu/models/) [54]	Global Trade Analysis Project	GTAP-E: Energy sector [55], GTAP-BIO: Biofuel sector [56], Disaggregation GTAP-BIO [57], GTAP-Dyn: [58]
LEITAP (http://www.lei.wur.nl/UK/newsagenda/Dossiers/Biobased_economy.htm) [59]	Agricultural Economics Research Institute Trade Analysis Project	Bio-LUC [60], EU biofuel policies on world agricultural production and land use [23,61–63]
PARTIAL EQUILIBRIUM		
AGLINK-COSIMO [64]	Worldwide Agribusiness Linkage Program + COmmodity SImulation MOdel	OECD Biofuel policies impact on markets [2,65]
ESIM [66]	European Simulation Model	Impact on agricultural markets: EU mandate in 2020 [67]
FAPRI-CARD (http://www.fapri.iastate.edu/models/) [68]	Food and Agriculture Policy Research Institute- Center for Agricultural and Rural Development	US corn ethanol on US Grain, Oilseed, and Livestock Markets [69], US corn ethanol iLUC GHG [6]
CAPRI (http://www.ec4macs.eu/home/capri-news.html) [70]	Common Agricultural Policy Regional Impact	EU mandate: [46]
MIRAGE [71]	Modeling International Relationships in Applied General Equilibrium	Bio-LUC-GHG EU Mandate [72]
IMPACT (http://www.ifpri.org/themes/impact/impactresearch.asp) [73]	International Model for Policy Analysis of Agricultural Commodities and Trade	Biofuels and food prices [74]
OPTIMISATION		
FASOM [75]	Forest and Agricultural Sector Optimisation Model (US)	US Agricultural and Forestry Impacts of US Energy Independence and Security Act [76], Domestic impacts in EPA modeling approach [21], Bio-LUC-GHG emissions estimation: FASOM-GHG [77]
GLOBIOM (http://www.iiasa.ac.at/Research/FOR/globiom.html) [78]	Global Biomass Optimisation Model	GHG emissions reduction from bioenergy—Social welfare [79], EU Biofuel target-global LUC [37]
EUFASOM [80]	Forest and Agricultural Sector Optimisation Model (EU)	Bio-LUC [81]

Continued

Model Acronym	Model Name	Selected Applications
POLYSIS (http://www.agpolicy.org/polysys.htm) [82]	Policy Analysis System	Second generation biofuels-US biofuel targets-LUC-GHG-Effects in US [83]
REAP[84]	Regional Environment and Agriculture Programming	First generation biofuels-US biofuel targets-LUC-GHG-Effects in US [83]
LIFE-CYCLE ASSESSMENT		
GREET (http://www.transportation.anl.gov/software/GREET/)	Greenhouse Gases, Regulated Emissions, and Energy Use in Transportation	GHG emission balance of US biofuels [6,21]
ecoinvent® (http://www.ecoinvent.ch/)	Ecoinvent	GHG emission balance of (mainly EU) biofuels [85]

References

[1] Gnansounou E, Panichelli L, Villegas JD. Sustainable liquid biofuels development for transport: frequently asked questions and review of initiatives. Technical Report, Lausanne: Bioenergy and Energy Planning Research Group, EPFL, 2008. p. 51.

[2] OECD. Agricultural market impacts of future growth in the production of biofuels. Working Party on Agricultural Policies and Markets. Paris, France: OECD, Directorate for food, agriculture and fisheries, committee for agriculture; 2006.

[3] Hamelinck C. Biofuels and food security: risks and opportunities. Utrecht, NL: Ecofys; 2013. p. 25, BIENL13469.

[4] Witzke P, Banse M, Gömann H, Heckelei T, Breuer T, Mann S, et al. Modelling of energy-crops in agricultural sector models—a review of existing methodologies. In: Pérez Domínguez I, Müller M, editors. JRC scientific and technical reports. Luxembourg: European Commission, Joint Research Centre, Institute for Prospective Technological Studies; 2008. p. 59.

[5] Kløverpris J, Wenzel H, Nielsen P. Life cycle inventory modelling of land use induced by crop consumption: Part 1: Conceptual analysis and methodological proposal. Int J Life Cycle Assess 2008;13:13–21.

[6] Searchinger T, Heimlich R, Houghton RA, Dong F, Elobeid A, Fabiosa J, et al. Use of U.S. croplands for biofuels increases greenhouse gases through emissions from land-use change. Science 2008;319:1238–40.

[7] Martin E, Chester M, Vergara S. Attributional and consequential life-cycle assessment in biofuels: a review of recent literature in the context of system boundaries. Curr Sust Renew Energy Rep 2015;2:82–9.

[8] Ahlgren S, Di Lucia L. Indirect land use changes of biofuel production—a review of modelling efforts and policy developments in the European Union. Biotechnol Biofuels 2014;7:35.

[9] Panichelli L, Gnansounou E. Impact of agricultural-based biofuel production on greenhouse gas emissions from land-use change: key modelling choices. Renew Sust Energ Rev 2015;42:344–60.

[10] Broch A, Hoekman SK, Unnasch S. A review of variability in indirect land use change assessment and modeling in biofuel policy. Environ Sci Pol 2013;29:147–57.

[11] Wicke B, Verweij P, van Meijl H, van Vuuren DP, Faaij AP. Indirect land use change: review of existing models and strategies for mitigation. Biofuels 2012;3:87–100.

[12] Righelato R, Spracklen DV. Carbon mitigation by biofuels or by saving and restoring forests? Science 2007;317:902.

[13] Fargione J, Hill J, Tilman D, Polasky S, Hawthorne P. Land clearing and the biofuel carbon debt. Science 2008;319:1235–8.

[14] Kløverpris JH, Baltzer K, Nielsen P. Life cycle inventory modelling of land use induced by crop consumption: part 2: example of wheat consumption in Brazil, China, Denmark and the USA. Int J Life Cycle Assess 2010;15:90–103.

[15] Sanchez ST, Woods J, Akhurst M, Brander M, O'Hare M, Dawson TP, et al. Accounting for indirect land-use change in the life cycle assessment of biofuel supply chains. J R Soc Interface 2012;9:1105–19.

[16] van Meijl H, van Rheenen T, Tabeau A, Eickhout B. The impact of different policy environments on agricultural land use in Europe. Agric Ecosyst Environ 2006;114:21–38.
[17] Malins C, Searle S, Baral A. A Guide for the Perplexed to the Indirect Effects of Biofuels Production. Brussels: International Council on Clean Transportation; 2014. p. 166.
[18] Taheripour F, Hertel TW, Tyner WE, Beckman JF, Birur DK. Biofuels and their by-products: global economic and environmental implications. Biomass Bioenergy 2010;34:278–89.
[19] Smeets EMW, Faaij APC, Lewandowski IM, Turkenburg WC. A bottom-up assessment and review of global bio-energy potentials to 2050. Prog Energy Combust Sci 2007;33:56–106.
[20] Edwards R, Mulligan D, Marelli L. Comparison of models and results for marginal biofuels production from different feedstocks. Indirect Land Use Change from increased biofuels demand. Brussels, 2010.
[21] EPA. Renewable fuel standard program: (RFS2) regulatory impact analysis. Washington, DC: Assessment and Standards Division, Office of Transportation and Air Quality, U.S. Environmental Protection Agency; 2010. p. 1107.
[22] CARB. Low carbon fuel standard (LCFS). California: Air Resource Board, California Envrionmental Protection Agency; 2009. p. 63.
[23] Banse M, van Meijl H, Tabeau A, Woltjer G, Hellmann F, Verburg PH. Impact of EU biofuel policies on world agricultural production and land use. Biomass Bioenergy 2011;35:2385–90.
[24] Popp A, Dietrich JP, Lotze-Campen H, Klein D, Bauer N, Krause M, et al. The economic potential of bioenergy for climate change mitigation with special attention given to implications for the land system. Environ Res Lett 2011;6:034017.
[25] FAPRI. Documentation of the FAPRI modeling system. In FAPRI-UMC, (Ed.), Report #12-04 Park DeVille Drive, Suite E, Columbia: Food and Agricultural Policy Research Institute, 2004. p. 238.
[26] Winrock. The impact of expanding biofuel production on GHG emissions: accessing and interpreting existing data. Arlington VA: Winrock International; 2009. p. 72.
[27] Hellmann F, Verburg PH. Impact assessment of the European biofuel directive on land use and biodiversity. J Environ Manag 2010;91:1389–96.
[28] Hellmann F, Verburg PH. Spatially explicit modelling of biofuel crops in Europe. Biomass Bioenergy 2011;35:2411–24.
[29] Veldkamp A, Lambin EF. Predicting land-use change. Agric Ecosyst Environ 2001;85:1–6.
[30] Veldkamp A, Fresco LO. CLUE: a conceptual model to study the conversion of land use and its effects. Ecol Model 1996;85:253–70.
[31] Lee HL, Hertel TW, Sohngen B, Ramankutty N. Towards an integrated land use data base for assessing the potential for greenhouse gas mitigation. GTAP Technical Paper, 2005.
[32] Ahammad H, Mi R. Land use change modeling in GTEM: accounting for forest sinks. In: Australian Bureau of Agricultural and Resource Economics , editor. EMF 22: climate change control scenarios. California: Stanford University; 2005.
[33] Tyner W, Taheripour F, Zhuang Q, Birur D, Baldos U. Land use changes and consequent CO2 emissions due to US corn ethanol production: a comprehensive analysis. West Lafayette, IN, USA: Center for Global Trade Analysis—Purdue University; 2010. p. 90.
[34] Laborde D. Assessing the land use change consequences of european biofuel policies. In: ATLASS Consortium—International Food Policy Institute (IFPRI); 2011. p. 111.
[35] Plevin RJ, Kammen DM. Indirect land use and greenhouse gas impacts of biofuels. Encyclopedia of biodiversity (second edition). Waltham: Academic Press; 2013. pp. 293-297.
[36] Al-Riffai P, Dimaranan B, Laborde D. Global trade and environmental impact study of the EU biofuels mandate. Washington, DC: International Food Policy Research Institute; 2010. p. 125.
[37] Havlík P, Schneider UA, Schmid E, Böttcher H, Fritz S, Skalský R, et al. Global land-use implications of first and second generation biofuel targets. Energy Policy 2011;39:5690–702.
[38] Melillo JM, Reilly JM, Kicklighter DW, Gurgel AC, Cronin TW, Paltsev S, et al. Indirect emissions from biofuels: how important? Science 2009;326:1397–9.
[39] De Cara S, Goussebaïle A, Grateau R, Levert F, Quemener J, Vermont B. Revue critique des etudes evaulant l'effect des changements d'affectation des sols sur les bilans environmentaux des biocarburants. In: Service Agriculture et Forêt DPeED, ADEME (Angers) (Eds.), Grignon: Etude réalisée pour le compte de l'ADEME par l'INRA (contrat 10-60-C0039), 2012. p. 96.

[40] Yeh S, Witcover J. Indirect land-use change from biofuels: recent developments in modeling and policy landscapes. International Food and Agricultural Trade Policy Council, 2010. p. 6.
[41] EC. Impact assessment accompanying the document proposal for a directive of the European Parliament and of the Council amending Directive 98/70/EC relating to the quality of petrol and diesel fuels and amending Directive 2009/28/EC on the promotion of the use of energy from renewable sources. Brussels: European Commision, 2012. p. 129.
[42] Carballo SM, Marco NF, Anschau RA, Hilbert J. Argentine potentiality to develop sustainable bioenergy projects. Methodology to determine driving forces of land use changes using gis tools. In: ASME , editor. Proceedings of ES2009-03-02 Energy Sustainability 2009, San Francisco, California, USA; 2009. p. 13.
[43] Finkbeiner M. Indirect land use change? Help beyond the hype? Biomass Bioenergy 2014;62:218–21.
[44] Brinkman MLJ, Wicke B, Gerssen-Gondelach SJ, van der Laan C, Faaij A. Methodology for assessing and quantifying ILUC prevention options. The Netherlands: Copernicus Institute of Sustainable Development—Utrecht University; 2015. p. 28.
[45] EPA. Regulation of fuels and fuel additives: changes to renewable fuel standard program: final rule. Renewable Fuel Standard Program (RFS2). Washington, DC: Federal Register; 2010. pp. 14669–15320.
[46] Britz W, Hertel TW. Impacts of EU biofuels directives on global markets and EU environmental quality: An integrated PE, global CGE analysis. Agric Ecosyst Environ 2011;142:102–9.
[47] Mosnier A, Havlík P, Valin H, Baker JS, Murray BC, Feng S, et al. The net global effects of alternative U.S. biofuel mandates: fossil fuel displacement, indirect land use change, and the role of agricultural productivity growth. Report NI R 12-01. Durham, NC: Nicholas Institute for Environmental Policy Solutions, Duke University, 2012. p. 31.
[48] Baker J, Murray B, McCarl BA, Rose SK, Schneck J. Greenhouse gas emissions and nitrogen use in U.S. agriculture: historic trends, future projections, and biofuel policy impacts. Report NI R 11-08. Durham, NC: Nicholas Institute for Environmental Policy Solutions, 2011. p. 51.
[49] Hiederer R, Ramos F, Capitani C, Koeble R, Blujdea V, Gomez O, et al. Biofuels: a new methodology to estimate GHG emissions from global land use change: a methodology involving spatial allocation of agricultural land demand and estimation of CO2 and N2O emissions. Joint Research Center (JRC)—European Commission, 2010. p. 172.
[50] Carré R, Hiederer R, Blujdea V, Koeble R. Background Guide for the calculation of land carbon stocks in the biofuels sustainability scheme. Drawing on the 2006 IPCC guidelines for national greenhouse gas inventories. Luxembourg: Office for Official Publications of the European Communities; 2010. p. 109.
[51] EC. Guidelines for the calculation of land carbon stocks for the purpose of Annex V to Directive 2009/28/EC. In European Commission (Ed.). Brussels, 2010. p. 23.
[52] Bourguignon D. EU biofuels policy: dealing with indirect land use change. Briefing February 2015. Bruxeles. Bruxeles: European Parliamentary Research Service; 2015. p. 10.
[53] Croezen HJ, Bergsma GC, Otten MBJ, van Valkengoed MPJ. Biofuels: indirect land use change and climate impact. Delft: CE Delft; 2010. p. 62.
[54] Hertel TW, Tsigas ME. Chapter 2: "structure of GTAP" In: Hertel TW, editor. Global trade analysis: modeling and applications. Cambridge: Cambridge University Press; 1997.
[55] Burniaux J-M, Truong TP. GTAP-E: an energy-environmental version of the GTAP model. GTAP Technical Paper. West Lafayette: Center for Global Trade Analysis, Department of Agricultural Economics, Purdue University, 2002. p. 69.
[56] Taheripour F, Hertel TW, Tyner WE, Beckman JF, Birur DK. Biofuels and their by-products: global economic and environmental implications. GTAP Working Paper. West Lafayette IN: Center for Global Trade Analysis, Department of Agricultural Economics, Purdue University, 2008. p. 28.
[57] Birur DK, Hertel TW, Tyner WE. Impact of biofuel production on world agricultural markets: a computable general equilibrium analysis. GTAP Working Paper. West Lafayette IN: Center for Global Trade Analysis, Department of Agricultural Economics, Purdue University, 2008. p. 63.
[58] Ianchovichina E, McDougall R. Theoretical structure of dynamic GTAP. GTAP Technical Paper No 17. West Lafayette: Center for Global Trade Analysis, Department of Agricultural Economics, Purdue University, 2000. p. 71.
[59] Woltjer G, Banse M, van Meijl H, Tabeau A. Alternative approaches to extend GTAP to biofuel crops. In: 10th annual GTAP conference, Purdue University, Indiana, USA; 2007. GTAP.

[60] Woltjer G, Banse M, van Meijl H, Tabeau A. Biofuels and long-term land use dynamics. In: GTAP conference, Helsinki, Finland; 2008. GTAP.
[61] Banse M, van Meijl H, Tabeau A, Woltjer G. Impact of EU biofuel policies on world agricultural and food markets. The Hague: Agricultural Economics Research Institute (LEI); 2007. p. 29.
[62] Banse M, van Meijl H, Tabeau A, Woltjer G. Will EU biofuel policies affect global agricultural markets? Eur Rev Agric Econ 2008;35:117–41.
[63] Banse M, van Meijl H, Woltjer G. The impact of first and second generation biofuels on global agricultural production, trade and land use. In: 11th annual conference on global economic analysis, Helsinki, Finland; 2008. GTAP.
[64] Conforti P, Londero P. AGLINK: the OECD partial equilibrium model. Roma: National Institute of Agricultural Economics INEA; 2001. p. 14.
[65] OECD. Biofuel support policies: an economic assessment. Paris, France: OECD; 2008. p. 146.
[66] Banse M, Grethe H, Nolte S. European simulation model (ESIM) in GAMS: model documentation. Göttingen and Berlin: DG AGRI, European Commission; 2005.
[67] EC. The impact of a minimum 10% obligation for biofuel use in the EU-27 in 2020 on agricultural markets. Economic analysis, perspectives and evaluations. Brussels: European Commission Directorate—General for Agriculture and Rural Development; 2007. p. 10.
[68] Meyers WH, Helmar MD, Devadoss S. FAPRI trade model for the soybean sector: specification, estimation, and validation. Ames, Iowa, US: Center for Agriculture and Rural Development (CARD), Iowa State University; 1991. p. 76.
[69] Tokgoz S, Elobeid A, Fabiosa J, Hayes DJ, Babcock BA, Yu T, et al. Emerging biofuels: outlook of effects on U.S. grain, oilseed, and livestock markets. Washington, DC: FAPRI; 2007. p. 57.
[70] Britz W, Witzke HP. CAPRI model documentation 2008: Version 2. Available from: http://wwwcaprimodelorg/docs/capri_documentationpdf2008.
[71] Bchir H, Decreux Y, Guerin J-L, Jean S. MIRAGE: a computable general equilibrium model for trade policy analysis. CEPII Working Paper No 2002-17. Paris: French research center in international economics (CEPII), 2002. p. 59.
[72] Al-Riffai P, Dimaranan B, Laborde D. Global trade and environmental impact study of the EU biofuels mandate, In: ATLASS Consortium, Washington, DC; 2010. IFPRI.
[73] Rosegrant MW, the IMPACT Development Team. International model for policy analysis of agricultural commodities and trade (IMPACT): model description. Washington, DC: International Food Policy Research Institute; 2012. p. 55.
[74] Rosegrant MW. Biofuels and grain prices: impacts and policy responses. testimony for the U.S. Senate Committee on Homeland Security and Governmental Affairs. Washington, DC: International Food Policy Research Institute; 2008.
[75] Adams DM, Alig RJ, Callaway JM, McCarl BA, Winnett SM. The forest and agricultural sector optimization model (FASOM): model structure and policy applications. Washington, DC: United States Department of Agriculture, Forest Service, Pacific Northwest Research Station; 1996. p. 68.
[76] Beach RH, McCarl BA. U.S. agricultural and forestry impacts of the energy independence and security act: FASOM results and model description. Washington, DC: Research Triangle Institute (RTI) International; 2010. p. 178.
[77] Beach RH, Adams D, Alig R, Baker J, Latta GS, McCarl BA, et al. Model documentation for the forest and agricultural sector optimization model with greenhouse gases (FASOMGHG). In US EPA—Environmental Protection Agency—Climate Change Division (Ed.). Washington, DC: RTI International, 2010. p. 188.
[78] Havlík P, Mosnier A, Fuss S, Valin H, Obersteiner M, Schneider U, et al. GLOBIOM. In (IIASA) IIfASA (Ed.). BF WS. Champaign-Urbana, 2010. p. 38.
[79] Havlík P, Schneider UA, Schmid E, Böttcher H, Kindermann G, Leduc S, et al. GHG mitigation through bioenergy production versus carbon sinks enhancement: a quantitative analysis. Climate change: global risks, challenges & decisions. Copenhagen: University of Copenhagen; 2009.
[80] Schneider UA, Balkovic J, De Cara S, Franklin O, Fritz S, Havlik P, et al. The European forest and agricultural sector optimization model—EUFASOM, In: 16th annual Conference of the European Association of Environmental and Resource Economists (EAERE). Gothenburg, Sweden; 2008. EAERE.
[81] Schleupner C, Schneider UA. Effects of bioenergy policies and targets on European wetland restoration options. Environ Sci Pol 2010;13:721–32.

[82] APAC. The POLYSYS modeling framework: a documentation. Knoxville, Tennessee: University of Tennesse—Institute of Agriculture—Agricultural Policy Analysis Center (APAC), 2012. p. 169. Available from http://www.agpolicy.org/tools/doccom.pdf.

[83] BRDB. Increasing feedstock production for biofuels: economic drivers, environmental implications, and the role of research. In Biomass Research and Developement iniciative (BR&Di) (Ed.). The Biomass Research and Development Board (BRDB), 2008. p. 197. Available from: http://www.usbiomassboard.gov/pdfs/increasing_feedstock_revised.pdf.

[84] Johansson R, Peters M, House R. Regional environment and agriculture programming model (REAP). Technical Bulletin No (TB-1916) USDA, 2007. p. 118.

[85] Jungbluth N, Faist Emmenegger M, Dinkel F, Stettler O, Doka G, Chudacoff M, et al. Life cycle inventories of bioenergy. Data v20, Uster: Swiss Center for Life Cycle Inventories, 2007. p. 727.

CHAPTER

10

Towards an Integrated Sustainability Assessment of Biorefineries

D. Tuazon, E. Gnansounou*[†]

[*]Ecole Polytechnique Fédérale de Lausanne (EPFL), Lausanne, Switzerland
[†]Bioenergy and Energy Planning Research Group (BPE), Ecole Polytechnique Fédérale de Lausanne (EPFL), Lausanne, Switzerland

10.1 INTRODUCTION

Biorefinery and biofuel performance has been commonly evaluated on economic and environmental bases, as has been evidenced through several studies and through the use of economic valuation instruments—such as net present value or other methods of temporal adjusted valuation—and, for environmental performance, life cycle assessment (LCA), almost exclusively. The use of such evaluations has been important in making the case for further study, operational design and implementation for biorefineries and biofuel production. Furthermore, although economic and environmental assessment may be present in the same study, the results are often presented exclusively of one another with little to no attempt to consider coupling effects. However, in order to go beyond current assessment methods and address maturing and emerging imperatives within multiple sociopolitical scales, it is necessary to investigate the holistic sustainability assessment of biorefineries (and, hitherto, biofuels).

Aspects of sustainability and the assessment thereof in biorefineries have already been considered in some isolated instances. For example, conformance or rating standards for biofuels (national, regional, or global) have been conceived through a combination of environmental and social considerations. Regulatory compliance is often regarded as a necessary imperative and not explicitly included in many sustainability analyses, but rather treated as constraints. The roundtable for sustainable biomaterials (RSB) has created a set of sustainability standards [1] which set out both necessary and aspirational guidelines that are applicable to bioenergy products and associated byproducts. An examination of the literature, however, shows that such efforts are often treated as overarching boundaries or constraints, rather than

http://dx.doi.org/10.1016/B978-0-444-63585-3.00010-3

actually being the basis of assessment or even integrated consideration in the design or implementation of biorefineries.

In this chapter the development towards the integrated sustainability assessment of biorefineries is explained, moving beyond the current common methods of evaluation, including LCA. The limitations of LCA set the context and case for integrated sustainability assessment, which is then expanded through a discussion of key environmental, economic, and social issues associated with biofuels and biorefineries. Given the multidisciplinary nature of sustainability, it is important to consider multicriteria and multiactor assessments and how they can contribute to the integrated sustainability assessment. Finally, the synthesis of different perspectives around sustainability assessment is used to make the case for a prospective framework for the integrated sustainability assessment of biorefineries.

10.2 SUSTAINABILITY DEFINITION

Although sustainability has gained considerable understanding and importance at both conceptual and practical levels within the last three decades at least, this maturity in understanding is not accompanied by a common definition. Indeed, sustainability has evolved and continues to be understood and interpreted in different ways, despite there being some discrete "baseline" concepts or themes from which several other frameworks or tools have derived their meaning.

The definition of sustainability is discussed at length in the literature such as Tuazon et al. [2], and it is important to discern between the different levels of defining sustainability, from fairly broad defining principles, to more organized frameworks and constituent, specific metrics and indicators. Different definitional levels are required for different aims, but keeping in mind the sustainability definitional hierarchy (depicted in Fig. 10.1) is essential for approach consistency, as well as linking holism and specificity.

The first formalized definition of sustainability is typically quoted from the Brundtland Report [3] and refers to the ability to maintain perpetuating intergenerational equity without significant or with no loss of opportunity. Although this is an important concept in sustainability, the operationalization of sustainability is usually through capital-based sustainability frameworks. The predominant frameworks used in industrial applications, including biofuels, include the triple bottom line [4] or The Five Capitals framework [5].

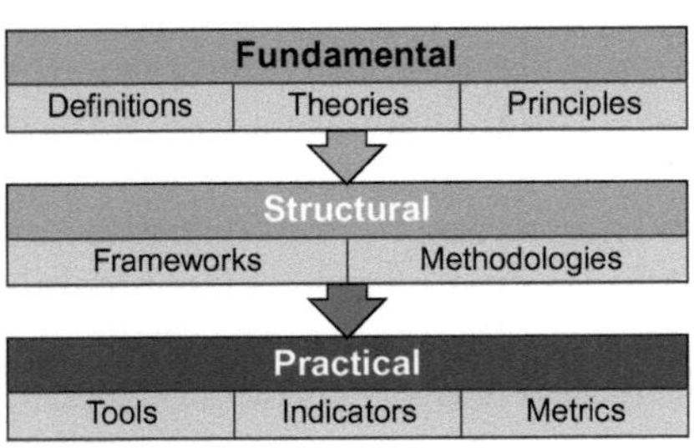

FIG. 10.1 Hierarchy of sustainability definitions, concepts, structures, and applications [2].

The triple bottom line is often regarded as one of the most accessible frameworks of sustainability for engineering-based applications [6,7] and is often used as the basis for headlining sustainability achievements reported by industry partners as well as regulators. The three capitals—economic, environmental, and social—are most commonly perceived as nonhierarchical, equally perceived sets of equity of which their overall balance is essential for achieving sustainability, as seen in Fig. 10.2. In practice, however, it is not uncommon to see a nominal precedence of capitals (or weak sustainability), particularly as the ability to compare between equities is not widely agreed upon both in practice or in the literature, and indeed it has been concluded that simple attempts at normalization to common currencies can be counterproductive to effective, sustainability-based decision making [8].

The discussion in this chapter will use the triple bottom line as the basis for defining and explaining sustainability, as it is the most commonly adopted definition in the predominant literature and policies regarding biofuels, and indeed this mirrors the approach taken by several regulatory bodies and companies. In considering sustainability using the triple bottom line approach, the sustainability of biofuels has been discussed at length in the literature, for example by Liew et al. [9], who provides an extensive overview of the current literature around the sustainability of biofuels according to process type, and Azapagic [10] has articulated a number of sustainability issues with respect to integrated biorefineries, which are summarized in Table 10.1 and discussed further in the following subsections. It is important to note that despite the attempt to segment the discussion of sustainability into three constituent capital domains, many of the issues being discussed in this chapter can veritably be considered to be within more than one capital, or at the very least have closely coupled or indirect effects covered in another capital. In addition, some factors, such as infrastructure and technology, can be very difficult to classify in a single capital, as the effects in a particular capital can change depending on the types of impact or benefit, the actors concerned and the broad context. Exemplifying this kind of difficulty is the graphical interpretation of the sustainability principles of the Roundtable for Sustainable Biofuels, presented in Fig. 10.3. In this interpretation, food security is classed as a social and economic issue as there are factors in food security that affect local or marginalized communities, but in addition has coupled effects in the price of food, which has wider societal

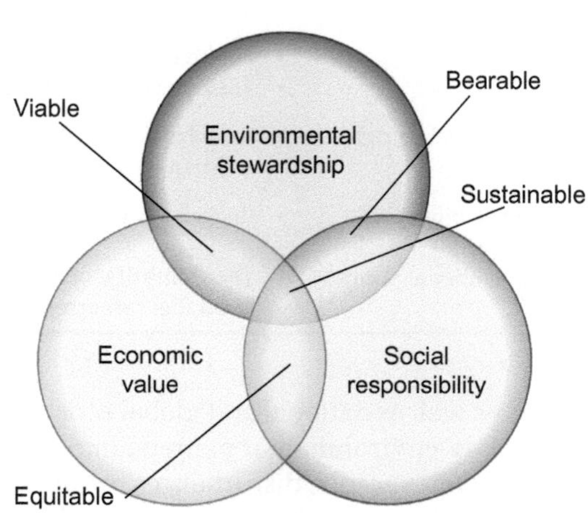

FIG. 10.2 Representation of the triple bottom line sustainability framework [2].

TABLE 10.1 Sustainability Considerations for Integrated Biorefineries According to Azapagic [10]

Headline Issue	Remarks
Environmental considerations	
Greenhouse gas emissions	• Consideration of life-cycle GHG emissions of biofuels, including production impacts resulting from grid electricity use or farming practices
Land use change	• Indirect GHG emissions due to changes in carbon stocks • Important for various potential locations, for example, forested land
Biodiversity	• May result in an improvement or degradation of biodiversity owing to the original state of biofuel crop land, for example, farming land, monoculture plantations, etc.
Water use	• Water is mainly accounted for in the irrigation of crops used for biofuel • Water use within biorefineries is relatively low
Other environmental impacts	• Acidification • Eutrophication • Human and ecotoxicity • Most miscellaneous environmental impacts are in the biofuel crop cultivation processes
Economic considerations	
Feedstock costs	• Costs of transporting crops between source and biorefinery can have significant effects
Capital costs	• High uncertainty due to lack of implementation of integrated biorefineries
Biofuel costs	• Largely sensitive based on factors such as: ○ Oil prices ○ Land competition ○ Dynamics of equivalent food market ○ Enzyme costs
Social considerations	
Jobs and regional development	• Can provide both jobs and opportunities for rural development (infrastructure and local business) throughout the project life cycle • May have impacts on local population that are dependent on agricultural residues
Health issues	• Fugitive emissions from handling of biomass • Toxicity within agricultural practices (e.g., pesticide application)
Human and labor rights	• Includes issues related to employment of marginalized groups (e.g., women) and children
Land availability and food prices	• Preference of crops for fuel over food can have a large effect on food prices
Intergenerational issues	• The ability of biofuels to assist in improving sustainability and reduce the reliance on fossil fuel resources for further generations

effects as well as national and global economic implications. While the quality of air and water are mainly environmental matters, impacts of either may also manifest themselves as social issues if, for example, disturbances by biofuel operations on natural resources also have an effect on local communities and their well-being.

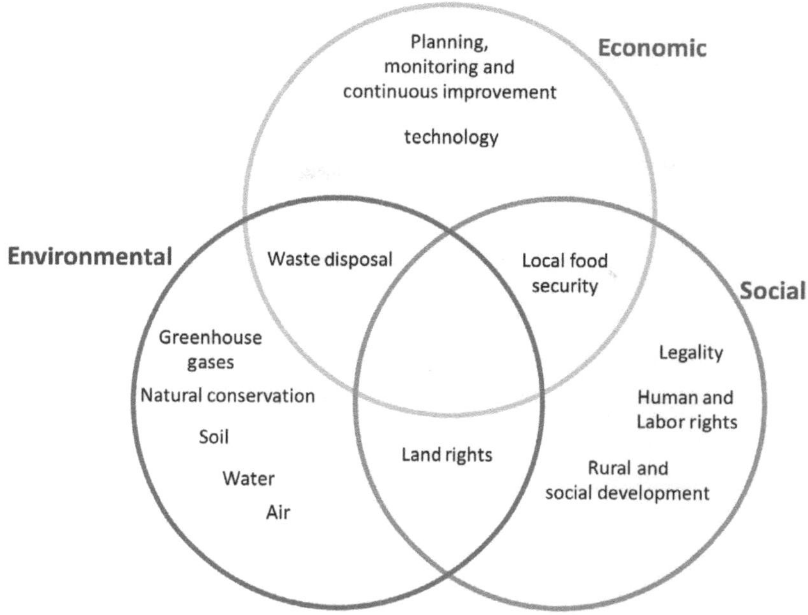

FIG. 10.3 Interpretation of the Sustainability Principles of the Roundtable for Sustainable Biofuels [1], classified according to the triple bottom line.

10.3 LIMITATIONS OF LCA

Since the early years where biofuel was promoted as an energy panacea, LCAs have been crucial in evaluating the environmental performance of biofuels, addressing both environmental criticisms of biofuels as well as showing the differences compared to fossil fuels. The robustness, reliability, and established process standards for conducting LCAs (via the ISO 14040 series) will continue to be important for the evaluation of environmental performance of biofuels, particularly given the large bevy of upcoming biorefining technologies of which measurable environmental impact data and real-world experience is absent. LCAs, however, have key limitations, especially when considering their use for the purposes of holistic sustainability assessment, or even complete environmental assessment; in this respect, the LCA forms a necessary but insufficient process.

10.3.1 Lack of Locational Context

Ultimately, sustainability assessment should be used to holistically justify the operation of biorefineries within a given society and environment. However, a key limitation of LCAs is that the analyses typically exist without specific context to a location, the society that exists there or even the environmental system. The use of global impact potential indicators, including the typical selection or subset thereof shown in Table 10.2, is useful for attempting to aggregate and contextualize the LCIs of biorefineries, but reporting on potentials does not typically consider the direct impacts on a local environment, nor does it consider issues

TABLE 10.2 List of Selected Global Impact Categories Typically Reportable in Biorefinery LCA Studies

	Global Impact Potential Category	Characteristic Substance
ADP	Abiotic depletion	Sb
AP	Acidification	SO_2 or H^+
EP	Eutrophication	PO_4
FAETP	Freshwater ecotoxicity	1,4-DCB (dichlorobenzene)
GWP	Global warming	CO_2
HTP	Human toxicity (*separable by uptake pathway*)	1,4-DCB
MAETP	Marine aquatic ecotoxicity	1,4-DCB
ODP	Ozone layer depletion	R-11 (freon-11)
POCP	Photochemical oxidant creation	C_2H_4
TETP	Terrestrial ecotoxicity	1,4-DCB

Data from Falano T, Jeswani HK, Azapagic A. Assessing the environmental sustainability of ethanol from integrated biorefineries. Biotechnol J 2014;9:753–65 and Curran MA. Life cycle assessment: principles and practice (manual no. EPA/600/R-06/60). Cincinnati, OH: Scientific Applications International Corporation; 2006.

related to absolute impacts or carrying capacity. An example of this shortfall can be seen in the work by Dias et al. [11], which includes an environmental assessment via LCA reported on a number of global impact potentials. The actual discussion of the effects of different configurations is limited, and there is no discussion of the effects of higher or lower potential values (or even key life cycle indicators) on ecosystems or other natural resources. The use of potentials to report key environmental impacts for the purposes of relative comparisons is risky, especially if the results are from an attributional LCA rather than a consequential one; subsequently, direct comparative results arising from impact potentials should not be used as the foundation of propositioning legislative change [12].

In many cases, especially where the biorefining technology has not yet been implemented or even been tested on a pilot scale, approaching the assessment without consideration of location, society, or environment may be satisfactory. However, it must then be stressed that such results are broadly indicative, especially in cases where potentially substantial environmental benefits are supposedly demonstrated. When considering a biorefinery within a given context, for example, for advancing to design or implementation, less importance should be placed on global impact indicators and more on the component life cycle inventories (carbon emissions, fugitive elements, etc.) and the effects they have on the local environment. A failure to consider impacts within a given environment can lead to overlooking acute or irreversible environmental damage, such as altering of ecosystems (including environments with human habitation), physical disturbances of land, or degradation of water bodies for human consumption or other activities [13,14].

10.3.2 Accounting for Economic and Social Factors

LCAs are not well-suited to account for social or economic impacts. In itself, this is not an unexpected conclusion, as the LCA has been clearly promoted as a tool to evaluate environmental performance. However, given the established rigor of process inherent in LCA, coupled with the worthy approach of considering the life cycle, there is a temptation to progress to integrated sustainability assessment by adapting the LCA to include social and economic factors as indicators, perhaps with minor modifications in the assessment process. This proposal, however, has several drawbacks, and it has been suggested that attempting to use LCA as the foundation for integrated sustainability assessment would result in a process which is too rigid [15].

LCA is based strongly in quantitative methods, which are not very well suited to social indicators, which are often difficult (or not useful) to quantify. Quantification of some social and economic indicators may actually reduce the significance of their impacts when associated with the real world, or may produce results which are useless or misleading, in particular when results are reported per unit of production and subsequently scaled up [13].

There have been attempts to assess social aspects in an LCA fashion, and many such tools have been proposed. However, there is little agreement on assessing the materiality of indicators (which may be biased through the ability to quantify indicators), ensuring that relevant stakeholders are accounted for, and adequate linking and accurate determination of impact pathways as well as causal relationships [16]. In addition, social LCAs may not agree with the standard approaches regarding functional units, system boundaries, time frames, and the generality of environmental data; far from actually attempting to integrate social factors into LCA, many are just separate tools (e.g., the social LCA work by Brent and Labuschagne [17]) which also achieve little towards sustainability integration.

10.3.3 Allocation

Allocation is nearly unavoidable for many biorefinery scenarios, as many such operations are hypothesized to produce multiple useful products apart from biofuels, including biochemicals or excess generated electricity. While there are guidelines for allocation in the ISO standards as well as studies of allocation methods [18], ultimately the choice and manner in which allocation is done are arbitrary, static and may not even reflect business approaches [19].

One of the most common approaches to allocation is by economic value; despite the standards recommending economic allocation as a last resort, the simplicity and flexibility of the approach has ensured its frequent use in the literature [20]. For systems where large quantities of products with very different economic values are produced, economic allocation may be a satisfactory method of distributing environmental burdens [20]. The common application of economic allocation should not be understood as it being the most "preferred" method in LCA practices; indeed, the best method of allocation should be considered on a case-by-case basis [20]. Given that the environmental performance can be markedly different depending on allocation method, a robust discussion of the method of allocation as well as sensitivity analyses are suggested each time [21]. Allocation failures, even unintentional, can hide or exaggerate particular environmental burdens, where these biases may be passed downstream or compromise reporting accountability [22,23].

Ultimately, the inconsistency of approach in biorefinery LCAs with respect to allocation makes it difficult to achieve a common basis of assessment and understanding of impacts. Economic LCA also suffers from the inability to take into account changes in commodity prices. The risk of excluding or otherwise unwittingly underreporting environmental burdens via allocation is also a concern, especially when those impacts are related to a given context or environment.

10.3.4 Risk of Rigidity

One of the greatest risks of relying on LCA heavily for integrated sustainability assessment is that due to the degree of established methodology in LCA and high quantification, holistic sustainability assessment may be rendered too rigid if it is created within the same methodological fashion; indeed, there are real risks already in relying heavily on LCAs for the purposes of environmental planning and assurance [24].

Apart from several social indicators, environmental indicators such as biodiversity, effects on ecosystems (and ecosystem services), health and status of water sources (e.g., groundwater aquifers) as well as amenity-based indicators are not easily quantified. Common practice finds that, unless these indicators are of absolute orientation (e.g., a discrete binary decision), often such indicators of uncertain measurement are ignored in analyses, and even disregarded from overall decision-making considerations. There have been attempts to quantify some of these indicators, which has been traditionally difficult to do, though in some cases these indicators are composed of simpler metrics which are not necessarily combinable. For example, biodiversity can be quantified through the count of taxa (both absolute numbers and different varieties) and the habitat which is occupied; and food security can be described using a combination of food prices (including statistical characterization), fraction of land required for food subsistence and amounts of staple food crop sales [25]. The problem with composite indicators is that they do not immediately harmonize well with existing singular indicators, and attempting to compose them into a single indicator introduces another layer of uncertainty. In addition, many indicators so created will need to be carefully considered when allocation is applied (if even appropriate).

LCAs are often based on assumptions which limit the dynamics of the system under study both spatially and temporally [26]. The kinds of problems this creates in understanding and interpreting the impacts are magnified when the results of an LCA are reported based on a per-unit basis and subsequently scaled out of context. The static limitations of LCA make it difficult to draw useful conclusions suitable for policy setting in order to support a desired ecosystem or even social services [26], although this is often what is propositioned and sometimes used in practice. There have been some attempts to overcome the temporally static limitations of LCAs by considering a dynamic life-cycle inventory over a given time period [27], but such attempts have had almost no uptake in most studies and LCA software packages.

Overall, it should not be denied that LCA is a useful and necessary tool for environmental assessment, but it is necessary to acknowledge its shortcomings, and even more important to understand that a simple extension of its approach is insufficient for holistic sustainability assessment. Despite the often sociopolitical demand for hard numbers in order to conclusively justify or avoid particular scenarios or consequences, attempting to quantify as many indicators as possible to be integrated into a framework rooted in the LCA method may result in an assessment which is too rigid for real-world application and disconnected from its uncertainties [24], which can also lead to significant barriers to implementation at best, or disastrous impacts at worst, when moving from general design of a biorefinery to implementation at a location. LCAs will continue to be a tool used alongside others to supply the overall sustainability assessment framework [13], but moreover, integrated sustainability assessment should go beyond LCA and allow not only the characterization of negative impacts to avoid, but also those of positive impacts to enhance a society or environment, contributions to sustainability goals (societal or corporate) and potential opportunities [15]. Such potential opportunities can include industrial symbiosis (creating regional synergies or controlling cumulative impacts), and provision or protection of ecosystem services (e.g., clean drinking water, avoiding downstream flooding, maintaining or improving soil fertility, etc. [26]).

10.4 OTHER ENVIRONMENTAL ISSUES

LCA is predominantly used to evaluate the environmental burdens of a scenario; however, other fundamental environmental issues arise when considering the impacts of biorefineries. Different environmental indicators that extend beyond LCA but are essential for the success of biorefineries are normally connected to the health of ecosystem resources and services. Such indicators may also be required for satisfying environmental regulations and reporting obligations. A list of such typical environmental indicators, as proposed by McBride et al. [28], is given in Table 10.3, where most indicators are related to natural resources, their reserves and their quality. Several of these are a function of the environment in which a biorefinery is situated, although the operation of the biorefinery and other associated services would have impacts on these indicators. Many biorefinery configurations which are at the conceptual level of analysis do not often discuss any of the factors in the table, except for greenhouse gas emissions, although the standard LCA analysis can point to possible sources of environmental burden which may be easily connected to these indicators. For example, in the LCA investigation of cellulosic ethanol production from empty fruit bunches [29], it was identified through the impact analysis that a large contribution of the environmental impacts resulted from fertilizers; consequential analysis of fertilizer use will then also necessitate an examination of nutrient balances as well as the effects on soil and water courses. The list of indicators in Table 10.3 is not exhaustive; however, Efroymson et al. [30] emphasize the importance of the local context (location, situation, life-cycle stage) on selecting the most salient environmental indicators or key issues for consideration. Efroymson et al. [30] also make a point that choices of environmental indicators can reveal biases which must be taken into account, and indeed the uncertainty in being able to quantify or estimate particular environmental impacts largely points to their often being ignored in many analyses, partly compromising transparency but also making it difficult to adequately negotiate the transition between concept and implementation. Much of the promotion of the environmental issues captured by the indicators in Table 10.3 (as well as others) can be attributed to separate speculative studies as well as experiences from first-generation biofuel production.

TABLE 10.3 Specific Indicators for Assessing Environmental Sustainability of Bioenergy Scenarios [28]

Indicator	Units	Potential Related Environmental Effects
Soil quality		
Total organic carbon (TOC)	Mg/ha	Climate change, N mineralization, humidification, water-holding capacity, infiltration, cation exchange capacity
Total nitrogen (N)	Mg/ha	Eutrophication, N availability
Extractable phosphorus (P)	Mg/ha	Eutrophication, P availability
Bulk density	g/cm^3	Water-holding capacity, infiltration, crop nutrient availability
Water quality and quantity (system balance)		
Nitrate concentration (balance and export)	mg/L; kg/ha/year	Eutrophication, hypoxia, potability
Total phosphorus concentration (balance and export)	mg/L; kg/ha/year	Eutrophication, hypoxia
Suspended sediment concentration (balance and export)	mg/L; kg/ha/year	Benthic habitat degradation, siltation
Herbicide concentration (balance and export)	mg/L; kg/ha/year	Habitat degradation, toxicity, potability
Peak storm flow of water courses	L/s	Erosion, sediment loading, infiltration
Minimum base flow of water courses	L/s	Habitat degradation, dissolved oxygen
Consumptive water use	m^3/day or m^3/ha/day	Water availability (for other uses)
Greenhouse gas emissions		
CO_2 equivalent emissions (CO_2 and N_2O)	kg C_{eq}/GJ	Climate change, plant growth
Biodiversity		
Presence or count of particular taxa	Presence/absence or count	Biodiversity, endangerment of species
Habitat area of particular taxa	ha	Biodiversity, endangerment of species
Air quality		
Tropospheric ozone	ppb	Human health, plant health
Carbon monoxide	ppm	Human health
Total particulate matter ($PM_{2.5}$ and PM_{10})	$\mu g/m^3$	Visibility, human health
Productivity		
Aboveground net primary productivity (ANPP)/yield	g C/m^2/year	Climate change, soil fertility, nutrient, and carbon cycling

10.4.1 Land Use Change

Despite the renewable nature of biofuels compared to standard fossil fuels, there has been (and continues to be) concern over the use of land for cultivating biofuel feedstock. Apart from social issues related to disempowerment of landowners as well as food versus fuel issues, there is great concern for maintaining the integrity of any land used for biofuel feedstock cultivation and critical ecosystem services in proximity. Irresponsible cultivation practices may lead to irreversible damage to the land, compromising further use and possibly having detrimental downstream effects. Land acquired for feedstock cultivation may have been originally prime agricultural land or forested; in the former case, this has social issues based on food versus fuel as well as displacement of the original land use, and in the latter there are significant issues related to deforestation and loss of carbon sinks.

Indirect land use change (ILUC), where land other than that already designated for biofuel crops is changed as such, is more difficult to measure, yet it may have a significant effect on the environmental success of a biofuel operation. ILUC quantification via carbon accounting was mainly spearheaded in the biofuels industry from papers arising from studies by Searchinger et al. [31] and Fargione et al. [32]. At the time these studies were published, they were labeled as alarmist and antibiofuel; however, they both drew attention to the need to account for ILUC, methods to achieve this, and integration of ILUC considerations in legislation. In addition, ILUC not only considers the potential carbon sink degradation from a biofuel operation, but the acute loss of those sinks, for example, extensive deforestation.

The measurement of ILUC is not uniform, though it can be estimated through a judicious combination of attributive and comparative LCAs [33]. Overall, there are conflicts between ILUC modeling, differences within various such models, and static regulatory ILUC factors which may form the basis of certification schemes [34–36]. There are also disagreements as to the main regulatory or industry instruments or methods that should be adopted to account for ILUC [37], but in spite of the uncertainty in and difficulties of ILUC integration into regulation (e.g., as discussed by Liska and Perrin [38]), it is argued that ILUC should not be a discriminatory barrier to the further implementation of biofuels [36]. Fritsche et al. [39] recommend that land use change related to carbon management should be heavily influenced by government land management policies. In contrast to this, another belief is that land use change carbon effects should be subject to less direct regulation and corporate incentives (or encouragement of corporate social responsibility) should be offered to companies in order to control overall land-use change impacts [40].

Brazil seems to be largely in control of any risks related to ILUC, with plenty of available land to support both an increasing biofuel market, as well as traditional and alternative agricultural activities, such as sugarcane for food production and cattle grazing [41]. However, this does not mean there is no environmental-related pressure on land use, as is evidenced by destructive deforestation in the Amazon (in a past case, for soybean agriculture) [42]. This is a clear example of having to be judicious about land use, and how an increasing biofuel market may add to this pressure for suitable land to be repurposed for it, which is further compounded by economic or social demands. As a contrast, there has been extensive literature and criticism leveled at the extensive deforestation in Indonesia that was done in order to promote oil palm–based biofuels, which resulted in the destruction of vast carbon sinks, as well as using potentially rich peat soils that may be suited for better uses [43–45].

The pressures associated with accounting for land-use change, as well as the implicit imperatives to be mindful of land utility overall, has given rise to studies which may encourage biofuel cropping on marginal land with little agricultural prospect, especially with degenerated soils [46,47]. Planning models have also been developed to plan the distribution of land for the purposes of cultivating bioethanol feedstocks, which offer a level of system connectivity and holism over more traditional isolated studies which may rely on static factors [48].

10.4.2 Biodiversity

Biodiversity can manifest as an issue in several ways. Some biofuel feedstock plantations are monocultures, and it is questionable whether these species were native to the area where the plantation was established. When the land was cleared to make way for these plantations or croplands, the existing land use may have promoted or contained a greater range of species as well as ecological services which may have been destroyed, degraded, or displaced in setting up the feedstock operation. Pristine, extensive or long-standing forests are of particular interest in order to preserve biodiversity, yet some of these may be under threat from expanding biofuel operations or indirect effects (e.g., other agricultural use).

Additional difficulties are encountered when attempting to characterize the extent of biodiversity. Taxa counts are a common method, but simply counting the number of taxa does not adequately characterize the range of species, nor the services they offer. A different method of characterization is thus to judiciously inventory ecosystem services (particularly those critical to the local ecosystem) and endangered species; neither of these are directly quantifiable. The difficulties associated with biodiversity measurement often complicate both its provability and acceptance in environmental impact statements, as well as continuous impact reporting, preventative action, and social acceptability.

The importance of biodiversity is often only enforced by matter of policy [49], although the presence of specific ecosystem services or critical species often dominates the biodiversity issue and thus is managed almost solely on the basis of this critical subset. One way to avoid the loss of biodiversity through potentially propagating invasive species or monocultures is to take into account the ecological survey of an area, then propose a biofuel cropping operation (including crop selection and agricultural practices) that would best be suited to support biodiversity, which doesn't appear to be common practice at the moment [50]. Still another proposal to both improve biodiversity and provide opportunities for bioenergy is to consider operations from the biomass of invasive species [51].

10.4.3 Water

Water used in biofuels is typically divided between the cropping and refinery operations, both of which may not be geographically proximate to each other. In addition, there are implications for both water use quantity and quality of both inflows (draw from external sources) and outflows (discharge to water courses).

It is generally agreed that the largest water usage, both in terms of external water demand as well as losses, is attributable to the cropping operation. For most crops, water is required for irrigation, but at the same time a lot of this water is lost through evaporation and thus permanently leaves the system. The seriousness of this loss may be exacerbated, for example, in areas where water starvation is a key risk, and may also be considerable when the cropping operation was specifically set up for biofuels, rather than adapted from existing and long-standing crop farming. Crop farming may also demand the use of fertilizers, which has water quality (nitrogen and phosphorus balance) issues due to surface run off and ground penetration. All of these factors may be further stressed in a given area if biofuel cropping is allowed to proliferate without consideration of the total amount of stress on the local water systems.

Biorefineries use, on average, less water overall compared to the accompanying cropping operation [52], but the use of this water is far more concentrated, which needs to be taken into account against the operating environment and society in which the refinery exists, in order to avoid overstress on water resources and water use conflicts.

10.4.4 Absolute and Permanent Environmental Impact

As biofuel performance can often be reported in terms of intensity rather than the absolute impact, this has unintended consequences if expansion of the biofuel operation is intended, or multiple operations are approved within the same area. The main issue is that, while the impact per unit of biofuel produced may seem efficient based on design calculations and previous designs, the total impact on the environment—air, soil, water, etc.—needs to be taken into account, and even more so if the operation is within a network of other extensive natural resource users. Some of the potential issues that may arise through these kinds of cumulative impacts are:

- *Air quality*. In particular, in sugarcane farming, a common technique in the harvesting process is the burning of the cane field to remove dry leaves and pests. Field burning has a high contribution to the overall environmental impact of biofuel production, but there are also social and health impacts to be considered. The actual impacts of field burning have been previously measured, but it may be the case that these estimates have not been fully accounted for, and could be much higher than initially believed [53].
- *Water stress*. Depending on the availability of water, for example, from groundwater, lakes, or rivers, water draw may be in direct competition with other large water users or the local community. For groundwater, it will be important to monitor well drawdown. Cropping operations are known to impart significant concerns due to the quality issues of surface runoff, which may be exacerbated in local water courses due to multiple farming operations. In areas with known incidents of drought or highly unpredictable rainfall patterns, it is important to consider the biofuel operation and its overall water impact within the entire system in order to avoid collective starvation or possible damage due to excessive water (e.g., inundation, eutrophication).
- *Land and soil effects*. Cropping and biorefining operations need to be examined for their effects on the local land and soil, usually in relation to the baseline quality of those natural resources and especially if they had a high initial agricultural potential (e.g., peat soils).

Where high climatic variance is expected, effects on the land due to excessive runoff or erosion are possible.

- *External GHG (greenhouse gas) emissions*. In particular, transportation of materials for biofuel manufacture, as well as the biofuel itself, is often achieved through means which consume traditional fossil fuels; a similar argument may be made for powering agricultural equipment in the cropping operation. It is important that the emissions from these steps are monitored as they reduce the overall environmental credentials of biofuels. This is sometimes difficult as these steps—for example, transportation—often comprise an action between major stages in the supply chain, which is most at risk of being inadequately accounted for.

10.5 ECONOMIC ISSUES

Although a key motivation of biofuels was to promote a more environmentally friendly alternative to fossil fuels, it is necessary that biofuels also compare favorably on economic terms. There are several dimensions which affect the relative fundamental economic viability of biofuels. At face value, biofuels and their operations must be economically feasible in and of themselves; however, that feasibility is often strongly affected by:

- The cost of feedstock, including, if necessary, the costs of intermediate steps required to bring biofuel crop to the refinery.
- The cost of the biofuel on the market, which is highly variable, depending strongly on fossil fuel price, local market conditions and the relative supply of both fossil fuels and biofuels.
- The intended end use of the biofuel and the characteristics of that market.

As a secondary consideration, the economic viability of biofuels must also take into account the costs of operating a biorefinery within a given sociopolitical environment, including taxes and local community investment. In particular, as well as being an energy panacea to reduce dependency on fossil fuels, biofuels may be widely viewed as a means of sustainably contributing to developing economies [54].

The comparison against fossil fuels is a very strong yardstick for alternative energy sources, including biofuels, due to the incumbent and prevalent nature of fossil fuels in the global economy. The accessibility and relative ease in supplying fossil fuels in the near future means that the economics of biofuels need to be exceptionally competitive to be acceptable, in addition to having adequate support for biofuel project capital investment. Pieprzyk et al. [55] suggest that marginal oil produced at high cost would be an ideal candidate for substitution by biofuels; as the theoretical peak oil point approaches, the viability of biofuel should generally improve. Subsidies of both fossil and biofuels also have large effects on their relative prices and hence the viability of biofuels; indeed, biofuel subsidies have in some cases improved the viability in order to promote an environmentally charged agenda rather than necessarily the economic sufficiency of biofuels. The danger in such a practice is that subsidies can create an unfortunate side effect of a false economy and hence a false sense of economic self-sufficiency of biofuels, especially when the uptake drivers are heavily biased towards economic factors without due consideration by regulators for the environmental or social implications.

Despite the interest in biofuels for replacing fossil fuels and many studies demonstrating reasonable to profitable prospects for biorefineries, especially upcoming second-generation (cellulosic) bioethanol, resistance to unfettered construction of new-generation biorefineries is largely tied to economic risk, which is a composite of environmental and social sustainability issues relating to biofuels, as well as uncertainties in market demand and supply. In particular, the cost implications around biofuels should be analyzed in a more inclusive manner which considers multiple stakeholder needs, the complete value chain from raw material to end use, implications of competition at each stage and improvement or feedback loops [56]. The uptake of biofuels by common consumers, for example, municipal automobile transport, is also currently dominated by the well-established gasoline-based engines (except for Brazil). A number of ingrained perceptions of engines utilizing alternative fuels have not helped in voluntary adoption of biofuel-based transport, and Salvo and Huse [57] suggest that an en masse switch from gasoline vehicles may only be achieved through considerable price discounts, a suggestion which will impact sharply on the already relatively marginal profitability of biofuels, as well as not necessarily addressing all current concerns of technologies which are not well-established. In addition, as 1G ethanol begins to be passed over in favor of 2G or further generation ethanol production, it is important to adjust any economic incentives accordingly (e.g., reduce those for 1G and improve them for 2G) in order to avoid blocking out the possibility of technology uptake of further generations of ethanol processing [58].

10.6 SOCIAL ISSUES

Since LCAs do not often include any indication of social impacts that a biorefinery—concept or implementation design—may have, the discussion, evaluation, and consideration of social sustainability factors are often ignored in many research projects. There is literature available which discusses the social and political impacts of biorefinery operations and biofuels, but these are usually general, dedicated research projects which are removed from the process design itself. Often such studies are independent, a follow-on consequence of the techno-economic study or design, or only performed to discharge legislative requirements. In the case of meeting legislative requirements, quite often the social issues are ensuring human and other associated rights of all stakeholders of a biorefinery operation. One way to broadly separate the different social issues is according to stakeholder group, for example as seen in Fig. 10.4. The separation by stakeholder type assists in addressing those issues through understanding the most critical aspects to be negotiated with each stakeholder, and finally how those issues connect throughout the supply chain. This is also important as different stakeholder groups may only be educated or aware of issues which are critical to that group.

Another common way that social impacts are sometimes "integrated" into studies which otherwise only focus on economic and environmental evaluation is through project or scenario constraints. For example, regulatory compliance may be regarded as a social

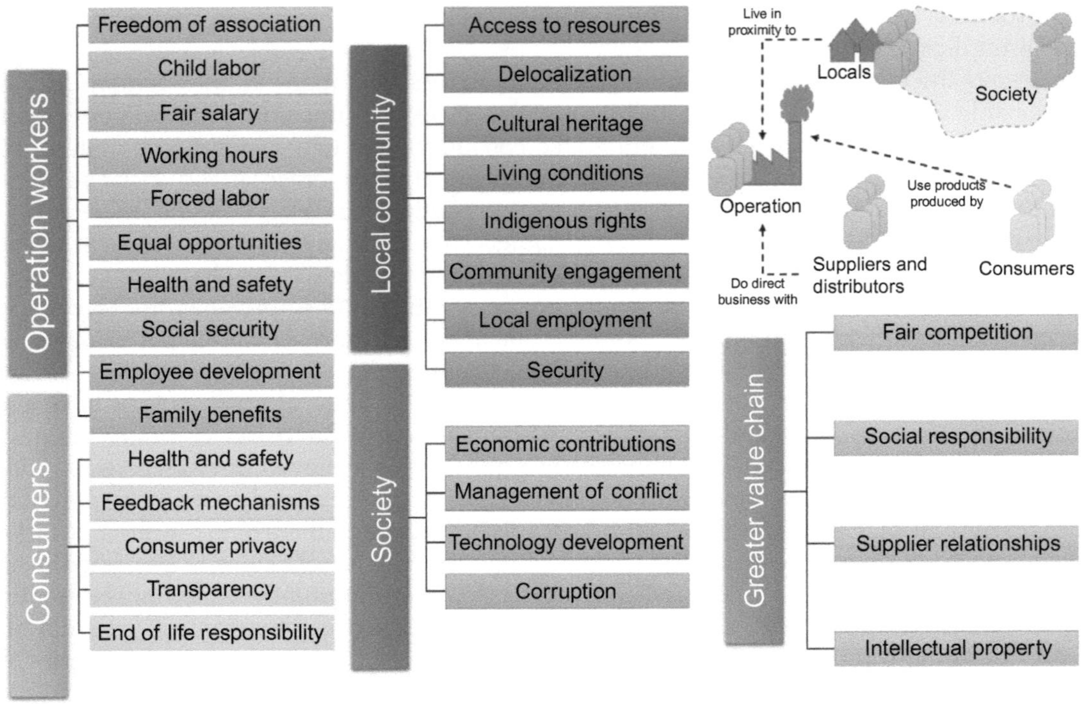

FIG. 10.4 Social issues pertinent to different categories of stakeholders in industrial systems, expressed through indicators [16].

indicator; the assumption in a given study is that an operation design may not result in breaking the law, for example, exceedance of emission guidelines, etc. In this way, the acceptable operating parameters of biorefinery scenarios may be defined through such constraints. This integration technique is not always possible (or optimal) for all social indicators, and in this case the constraint approach somewhat stifles the possibility of going beyond compliance, particularly if such an action may improve social well-being and community reputation.

In all of the common, holistic sustainability paradigms, social issues form an essential part of the overall framework with environmental and economic ones. However, in most analytical engineering-based studies, social issues are rarely investigated or are only given a cursory mention. The major reason for this is that social issues and factors (or the extents of them) are difficult to measure quantitatively. Several social issues can only be described qualitatively based on, at best, nominally comparable quality scales; in many cases, the most important distinction in determining the quality may be based on critical governance factors, such as legal compliance, that is, it is not necessarily important to establish the maturity or quality of an issue, but more important whether or not an issue is compliant with local laws, regulations, or social norms. Where such a binary distinction is realized for a given set of social issues, this may lead to:

- A simplification of the measurement of such issues, since a binary assessment of "compliant" or "not compliant" (or similar qualifiers) may be suggested; or
- Exclusion of social considerations based on the assumption of necessity; that is, if certain social issues show noncompliance with local laws in a given state or scenario, then further assessment (indeed, in any other part of sustainability) would be redundant.

In short, to most effectively understand and account for social issues, a reductionist approach to assessment will prove, for the most part, very ineffective, and assessment must take into account uncertainties and qualitative subtleties that are strongly associated with different groups of stakeholders.

10.6.1 Jobs and Regional Development

The introduction of major projects into small communities, including biofuel production, has the potential to bring jobs and local economic investment into host communities; indeed, such socioeconomic benefits and developments are expected to a large degree in order to compensate for disturbance as well as promote equity (or avoid exploitation).

The main issue associated with jobs in communities is adequate education to allow locals to benefit from jobs at biofuel operations. For this reason, in order to account for a lack of suitable locals and to expedite the process, expatriate workers may be introduced to the community that will work at the operations. However, the introduction of expatriates may be a source of conflict if they remain socially isolated from the local community. There are also disagreements as to the number of jobs actually generated by biofuel operations (whether it involves cropping, refining, or both, as well as miscellaneous utility roles), and whether these created jobs are sustainable in the long term (or are merely transient, as in the case explained by Acheampong and Campion [59]). The number of jobs potentially created by an operation is frequently modeled [60] and often used as a social lynchpin—or, indeed, the sole social promotor—to advance the biofuel production agenda in an area.

Regional development is best carried out in an integrated fashion with consultations between communities, local governments, and the corporations, perhaps by drawing inspiration from global level sustainability policies and guiding principles [61]. In this way, it is possible for biofuel production to occur in mutual agreement with communities, with as minimal disturbance as possible, but also actively introducing value to communities, such as sharing of resources (e.g., power infrastructure, water treatment infrastructure), or improving career and economic prospects of local communities (e.g., through education, apprenticeships, improved spending in local businesses or using local services).

10.6.2 Social Disempowerment

Many of the social issues which befall communities have often been used as a case for whole-chain corporate social responsibility activism. The disempowerment of communities in the overall establishment of projects, including biofuel production, is the topic of extensive focus especially in developing or underdeveloped regions, and especially those lacking adequate governmental support. This is somewhat similar to "fair trade" initiatives, and there biofuel standards—such as BonSucro—which attempt to ensure (or highly encourage efforts toward) socially agreeable production of biofuels which respects communities/societies (as well as the environment) so affected. It is not clear, however, whether standards are enough to ensure that impacts to communities are genuinely being minimized and discussed between companies and communities, as there is still a large case being made that communities affected by biofuel development (viz. especially from 1G biofuels) have not been able to adequately reap benefits or otherwise see maturation of social capital.

The deprivation of community rights can manifest itself in various ways, which may include compulsory acquisition of land for cropping or refinery operations, lack of consultation with communities before establishing operations (and discussing impacts), deprivation of adequate benefits commensurate to any disturbance caused by the operation, or inability/unwillingness to negotiate in cases of conflict (e.g., over operational impacts). In all cases, there has been a lack of mutual understanding and agreement between corporations and communities, and the resulting disagreement is often perceived to be an act of disempowerment of the community as it is often presumed that the relative power of affirmative action lies almost completely with the corporation.

Many communities may be disempowered to take affirmative action to stand for their rights and negotiate fairly, although it is debatable as to which higher-level stakeholders—governments and/or corporations—are primarily responsible for not recognizing this disempowerment, or it may be an inherent problem compounded by lack of education or societal expectation to be able to assert rights. There are also cases where communities which have been characterized as disempowered have been able to legally assert their rights and, through gathering external momentum and influence, been able to stop biofuel projects, despite their overall disposition to adequately maintain those rights being fragile and undetermined [62]. Biofuel corporations may not pay enough attention to affected communities in cases where bureaucracy is extensive and the local government is not supportive of measures to ensure a healthy and fair relationship between the company and communities [63,64]. In other cases, however, there have been some successes in small communities with biofuel producers through cooperation and active empowerment by the biofuel company [65].

Effective consultation is key in ensuring that the expectations of all affected stakeholders are well aware of the social (and other sustainability) aspects which affect them. Consultation is also useful in creating an agreement on particular courses of action, including the promotion of socioenvironmental services and reassurance of social justice [66]. Consulting with different stakeholder groups not only is important for agreeing on actions but also characterizing critical concerns [67] and gauging acceptance across the supply chain [68].

10.6.3 Food versus Fuel

The "food versus fuel" debate is an uncertain but prevalent ongoing argument which is between the allocations of either:

- Food crops between supplying that crop for the food market, or sending it for biofuel production; and
- Land which may be used to grow either food market crops, or crops (food or otherwise) for biofuel production.

A large part of the debate stems from 1G biofuels which use crops which would normally be destined for food markets, for example, sugar (from sugarcane). The rise in demand for biofuels has created tension in agricultural circles between supplying for food or fuel markets. Although fuel markets may have a tendency to prove more economically lucrative, the pressure and subsequent diversion as such has had the potential to cause food prices to rise. The effect of rising food prices can compromise adequate food supply in developing or underdeveloped societies, particularly for those who are not directly connected to biofuel operations. In addition, the choice of two distinct markets to sell the same crop may create conflict due to the inability of some farmers to be able to sell their crop for biofuel production, even if it was desired. Finally, areas which have survived on subsistence farming for a long time before considering cropping for biofuels may give rise to social conflicts if biofuel companies do not adequately balance the needs of the community based on their previous subsistence lifestyle against that of producing biofuels with crops (even if the cropping area is expanded).

Ethanol is widely recognized as a food versus fuel issue [69], and in large part, the alleviation of pressure on food markets will need to be through avoiding conflicts of resources and demands with traditional food agriculture. This includes avoiding the use of land that may be prime for food agriculture (e.g., by considering the use of land with poor soils), and moving past 1G ethanol production and more towards either cellulosic 2G ethanol or further generations (e.g., algal bioethanol). An article in *Biofuels Digest* [70] suggests six technological considerations (see Table 10.4) which can also help alleviate the food versus fuel dilemma by improving the biofuel output from crops and the resources required to grow them, thus reducing resource intensity and allowing those relieved resources to be dedicated back to food market crops.

TABLE 10.4 Suggestions for Avoiding the Food versus Fuel Debate [70]

Suggestion	Description
Feedstock diversification	For fuel feedstocks, consideration of cellulosic feedstocks rather than direct food resources. For food manufacturers, diversifying to ingredient substitutes which may be unfavorable for biofuel production
Increasing yield per unit feedstock	Focus on continuous improvement to improve the yield of biofuel per unit mass of feedstock, and relevant incentives to move to best practice technology
Reducing water intensity	Focus on reduction of water use in refining or cropping, as well as creating synergies between refineries and other local water users, especially in water-starved regions
Improving feedstock traits and yields	Consideration of crops which are more resistant to environmental pressures, for example, heat, pesticides, etc. Such improvements may be realized through bioengineering
Utilizing wastelands	Use of land which is unsuitable for food agriculture. Consideration of algal biofuels
Improving results from photosynthesis	Possible use of bioengineering to improve solar energy uptake into plants to improve overall energy yields

10.6.4 Miscellaneous Social Issues

Some of the other social issues which are important to consider in biorefinery operations include:

- *Health impacts.* Biofuel operations—both cropping and refining—may impart health impacts on the local community, including dust deposition and odor. These effects may be further exacerbated if cane field burning is practiced. Companies need to be aware of both potential acute and chronic health effects that may arise from their operations, in addition to monitoring the local population for nonvisible signs of health issues that may have been introduced through the biofuel operation.
- *Social amenity.* This includes visual amenity of the biofuel operation, as well as noise, vibration and traffic generated due to the introduction of the operation. If personnel at a biorefinery are encouraged to stay in the local community, the issue of social disorder should also be monitored.

As well as considering the issues discussed in this section, it is important to note that even if an issue is merely perceived by a community, the fact that the perception exists should still be of serious concern to the company. Lack of education may result in these false perceptions, but the presence of a negative perception is already a source of conflict and may be enough to generate further, serious social issues.

10.7 MULTICRITERIA AND MULTIACTOR ASSESSMENT

The nature of sustainability, including a number of indicators which are not directly comparable with each other, demands that in order to ensure an integrated, holistic analysis, it is necessary to consider evaluation methods which can investigate multiple candidates based on multiple indicators according to a number of criteria. For this reason, multicriteria decision analysis (MCDA) or multicriteria analysis (MCA) systems are the most appropriate and sensible approach to achieving this goal. Indeed, the literature on the use of MCDA in sustainability assessment is growing steadily, owing to the ability to account for the three capital domains in a balanced and integrated fashion, overcoming difficulties encountered with more traditional, single-domain tools such as LCA [71].

There are numerous MCDA methods which may be applied to sustainability assessments; some of the most common ones are listed in Table 10.5 according to their approach methodology in determining the optimal scenario or option. By and large, MCDA methods are mathematically based, and consist of a number of salient components:

- *Criteria* and/or *indicators*: measurements which constitute the final evaluation of each scenario being considered. These are usually quantitative measures, but they can also be ordered qualitative measures (which are then usually cast to a numerical scale). Criteria may be a single, flat-level collection, or may be hierarchical (collections of subcriteria), and this organization has important implications on how the final assessment is completed. The measurement of each indicator is not usually against a common unit, so it is important

TABLE 10.5 List of Selected MCDA Methods Classified by Methodological Approach [71–74]

Method Abbreviation	Name
Outranking methods	
ELECTRE	Elimination and choice expressing reality
PROMETHEE	Preference ranking organization method for enrichment of evaluations
Utility-based models	
MAUT	Multiattribute utility theory
SMART	Simple multiattribute rated technique
AHP	Analytic hierarchy process
TOPSIS	Technique for the order of prioritization by similarity to ideal solution
Other methods	
NAIADE	Novel approach to imprecise assessment and decision environment
MACBETH	Measuring attractiveness by a categorical based evaluation technique
Programming approaches	
MOP	Multiple objective programming

to define the comparative behavior of each indicator as part of the overall assessment process. It is also often important to ensure that indicators do not measure the same aspect more than once in order to minimize unintended bias or accounting for the same contributing factor more than once.

- *Weighting factors*: for every criterion, a weighting factor is assigned to represent the relative importance of one criterion over others. A balanced perspective which treats all criteria equally simply means all weighting factors in the set are the same; however, customizing weighting factors allows for the inclusion of real-world biases as well as variations on scenarios (different stakeholder points of view, hypothetical or investigative scenarios, sensitivity testing). Weighting factors essentially dictate the "best fit" profile of the desired scenario. Weighting factors may either be additive to unity (within a given set of criteria, all weighting factors must add up to 1, such as where the sustainability capitals are weighted for final assessment [75]), or they may be based on a multiplicative comparison scale (the weighting factors do not need to sum to a given total, such as the weighting scale in Fig. 10.5 adapted from Ref. [76] as appearing in Ref. [77]).
- *Normalization, comparison, and evaluation method*: the comparison and assessment of each scenario is the key component of an MCDA method. The final result from the evaluation of each scenario may either order the scenarios in terms of satisfying the "best fit" (relative orientation, or ranking), or it may rate each scenario in terms of performance against a given scale (absolute orientation).

The ability to holistically determine the optimal or even properly rank scenarios within each capital domain of sustainability is not without difficulties, let alone when considering

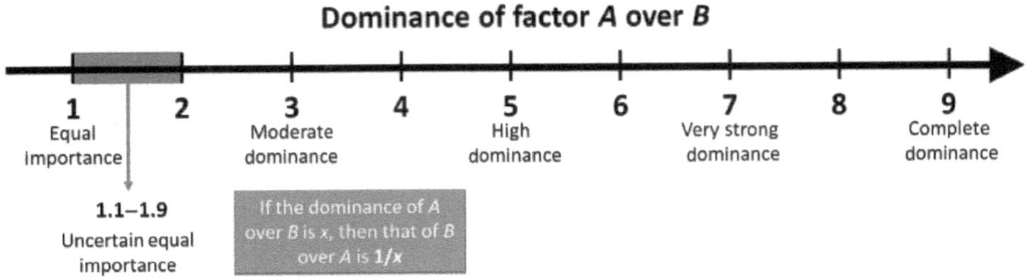

FIG. 10.5 Example of multiplicative weighting scale for an AHP MCDA. *Data from Saaty TL. The analytic hierarchy process. New York, NY: McGraw-Hill; 1988 as appearing in Macharis C, Verbeke A, De Brucker K. The strategic evaluation of new technologies through multicriteria analysis: the ADVISORS case. Res Transp Econ 2004;8:443–62, Economic Impacts of Intelligent Transportation Systems: Innovations and Case Studies.*

an even broader sustainability point of view. For example, with LCA and environmental assessment, potentials indicators are an attempt to give a more standard and aggregated viewpoint of impacts rather than simply trying to understand sets of LCIs; however, this has already been shown to have some drawbacks. Similarly, there have been attempts to standardize environmental indicator measurement by using embedded energy (emergy) [78] or the balance of potentially useful energy (exergy) [79]. Another attempt at standardizing the measurement of environmental impacts is suggested by the US EPA, known as the waste reduction algorithm [80,81]. One of the approaches used to attempt to standardize social indicator measurement is to define relative social capital measures and relate these to societal carrying capacities; however, it should be noted that these measures are usually only binary indicators of sustainability (sustainable or not sustainable) rather than absolute [82].

There are many examples of MCDA for determining the most sustainable scenario in the field of biorefineries. At a more simplistic level, one way to overcome the problem of inherently incompatible measurements across different criteria is to use class-based ordinal measures, or logic-based scales, in order to classify measurements of different indicators according to equal perceptions of broad magnitude, such as in the sustainability impact assessment method described in Ref. [83]. A similar example is in the work of Perimenis et al. [84], where each indicator is described on a five-point scale; depending on the indicator, the assignment of grades is either based on a qualitative classifier, or on a quantitative range of values. In this way, the final score for a scenario is based on weighted averages of the grades constituting each capital domain (including the technical domain). The tendency to use class-based rating scales to equalize incomparable factors is common for broad measurement applications, such as meeting standards for certification, visionary planning, or measurements of maturity. Such systems are also considerably less mathematically complex and more accessible and apparent to a wider range of stakeholders, which is why they are also seen more commonly in management or policy/political situations. However, such systems are inappropriate for more complex assessment, especially that for a more technical or project planning application, requiring a finer degree of comparison and appreciation of the differences between scenarios.

Among the several different methods of MCDA available, the predominant ones in use in the literature on biorefineries typically consist of three types: mathematical programming models (similar to MAUT or MOP), hierarchical normalization approaches (like the analytic hierarchy process, AHP), and outranking tools (such as elimination and choice expressing reality (ELECTRE) and preference ranking organization method for enrichment of evaluations (PROMETHEE)). The salient characteristics that contrast between these three kinds of approaches are summarized in Table 10.6, accompanied with selected studies where such MCDA approaches have been used.

10.7.1 Mathematical Programming Models

Mathematical programming models which are developed completely from sets of component equations, constraints, and optimization functions are different from the established methodologies where the decision processes are less transparent; moreover, mathematical programming approaches can be more easily conducted without requiring access to particular software tools. This allows for some unique approaches to evaluating sustainability, such as temporally dynamic models [86]. Since such approaches are nearly always specific to a study, they are not readily transferrable without significantly restructuring the mathematical analysis, which limits their portability and extensibility. In addition, with such approaches, due to the risk of overwhelming complexity, there is a greater propensity to either reduce traditionally nonquantifiable indicators to numerical scales, or ignore them completely. There is also a risk that these models may compensate for the performance (or lack thereof) across different factors, even where such trade-offs may be undesirable. Indeed, while a weak sustainability approach which allows trade-offs between capitals may be desired, nearly all programming approaches give no allowance to enforce critical reserves of sustainability capital, which may result in, for example, a conclusion where the best scenario may have excellent

TABLE 10.6 Summary of Selected MCDA Methods With Potential Applications in Sustainability Assessment of Biorefineries

Criteria	Mathematical Programming (MAUT, MOP)	Hierarchical Normalization (AHP)	Outranking Methods (PROMETHEE)
Weighting typology	Uniform compensability	Weighting factors	Weighting factors
Compensability/ sustainability strength	Possible (very weak)	Possible (weak)	Generally impossible (strong)
Complexity of implementation	Very high	Moderate	High
Ability to reuse or extend	Low	High	High
Examples	Andiappan et al. [85]; Chatterjee et al. [86]; Santibañez-Aguilar et al. [87]; You et al. [88]	Cobuloglu and Büyüktahtakın [89]; Macharis et al. [90]; Schaidle et al. [75]; Turcksin et al. [91]	Ziolkowska [92,93]

Data from Cinelli M, Coles SR, Kirwan K. Analysis of the potentials of multi criteria decision analysis methods to conduct sustainability assessment. Ecol Indic 2014;46:138–48 and Polatidis H, Haralambopoulos DA, Munda G, Vreeker R. Selecting an appropriate multicriteria decision analysis technique for renewable energy planning. Energy Sources B 2006;1:181–93.

economic performance but at the great expense of environmental capital. In such cases, additional constraints built into the model to avoid more extreme unacceptable scenarios have an overall limited effectiveness.

10.7.2 Hierarchical Normalization Methods

To avoid complete trade-offs across the entire indicator set, separating indicators into different subcategories and aggregating these in turns may yield a better result. In this way, the comparison and aggregation of indicators can be better controlled, which allows a stronger sustainability approach overall. In this respect, methodologies, such as AHP, which utilize a hierarchical indicator structure and normalization may be suitable. The normalization/aggregation of indicators has a further level of control in the form of weighting factors, which can be implemented at any level of aggregation along the hierarchy. Criteria hierarchies for methods such as AHP can be fairly flat (e.g., in Schaidle et al. [75]) or contain multiple tiers (e.g., in Macharis et al. [90]). Wang et al. [74] mentions that AHP is a widely popular, comprehensive method for multicriteria sustainable energy decision-making considerations. This is most likely due to the ease of understanding the weighting system (see Fig. 10.5) which is suitable even for a nontechnical audience, and indeed the modified AHP process suited for multiple stakeholders and used for an assessment of biofuels in Belgium [91] has a key component where the weighting factors for sets of indicators are requested from various stakeholders using a common graphical interface. AHP has also been adapted to attempt to account for uncertainties in precision as well as temporal variation [89]. Although there is a risk of rank reversal [71], AHP and associated methods are a fair compromise between implementation complexity, sustainability strength, and accessibility/transparency of the assessment process.

10.7.3 Outranking Methods

Insisting on a strong sustainability approach requires a more complex MCDA process as, by this time, criteria are more difficult to compare between each other due to the inability of one factor to compensate for another easily. To obtain the best scenario in a selection of candidates that best fit according to the acceptability of each criterion, outranking methods such as ELECTRE or PROMETHEE may be suggested. Outranking methods are suggested in cases where the comparability between criteria is complex and thus makes the ranking of scenarios without such tools nearly impossible [74]. Since criteria cannot be traded off against one another, the need to produce accurate, consistent rating scales for each criteria becomes less important. PROMETHEE, additionally, offers the ability to specify exactly the normalization profile for every indicator. The latter allowance for different profiles consists of:

- High and low thresholds of measurement in order to confine the otherwise unbound continuous range of an indicator (i.e., define limits of acceptability); and
- Nonlinear normalization of an indicator: for example, between the lowest and highest values in a range for an indicator, the values in between do not have to correspond to a linear trend.

The combination of thresholds and the normalization profile is known as the preference function for a given criterion or indicator. PROMETHEE can be extended to take into account uncertainties [92].

PROMETHEE generates results that show the alignment of the scenarios against the indicators, showing which indicators support certain scenarios, which indicators have synergies (or otherwise) with others, and the relative magnitudes of such discriminatory conclusions.

The use of PROMETHEE in the biofuel literature is limited; an example of its application to investigate sustainability in biofuel production according to different feedstocks is the work by Ziolkowska [92], although the application is quite limited in the number of indicators used in the assessment and is especially limited in social indicators. Given the amount of customization possible at the indicator level, the complexity of implementing PROMETHEE is much more than other MCDA methods. At its basis, PROMETHEE is a pairwise comparison method much like AHP. However, the integration method based on the preference functions and weighting factors is much more complex; such details are often not included in results reporting, but due to the difficulties in understanding the outranking process given that indicators do not generally compensate for each other, at times the results of PROMETHEE may be seen as not as transparent as, for example, the results from AHP. Like AHP, PROMETHEE has the risk of rank reversal when additional factors are introduced to an existing assessment. Finally, it should be stressed that PROMETHEE does not necessarily directly promote which is the best scenario from a given set; rather, the methodology shows the relationships between scenarios and each of the constituent criteria which, combined with the knowledge of synergizing or antagonistic criteria and the magnitudes of such relationships, will allow decision makers to make more informed decisions, as well as possibly allowing the ability to identify coupled risks and opportunities. For integrated sustainability assessment, the PROMETHEE method would appear to be the most promising despite its complexity, mainly due to its suitability for strong sustainability, which is better suited for allowing an equal amount of decision-making power through all constituent criteria, including those which are not normally quantified. While AHP may be simpler overall to implement and has some ability to prevent undesired compensations among indicators and deliver definitive rank scores of scenarios at the same time, organizing the indicator hierarchy to counter such undesired compensations may render it overly complex, or superficially insure against it. In addition, complete ranking might be desirable from a decision-making point of view, but the ability to understand scenario and criteria relationships in PROMETHEE may actually be more beneficial to decision makers in lieu of their being delivered an absolute result.

10.7.4 Assessment Involving Multiple Actors

The assessment of scenarios against the expectations of multiple stakeholders (multiple actors) is by and large an extension of the MCDA tools, with each criterion extending along an additional dimension to include the preferences of each type of relevant actor. Analysis including multiple actors, such as that for sustainable transport decisions discussed by Macharis et al. [90] and shown in Fig. 10.6, clearly emphasizes the stakeholder analysis process and the resulting capture and inclusion of multiple sets of preferences, though the method in which such preferences are integrated into the decision-making process is varied.

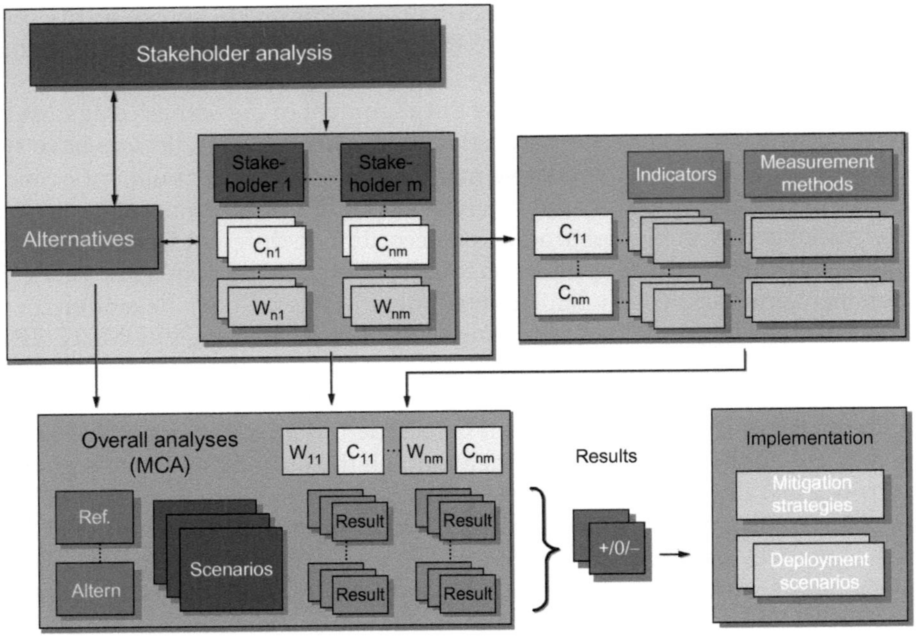

FIG. 10.6 Multiactor multicriteria analysis (MAMCA) methodology [90].

The aggregation of multiple sets of preferences is usually given less rigorous treatment compared to aggregating indicators, and could be as simple as an average of all preferences to give the final weightings [94]. A similar approach may be adopted in accounting for multiple stakeholders in PROMETHEE. In other cases, the indicators are divided among the stakeholder groups so that they rate different aspects that are specific to each of those groups, as shown in the example hierarchy of indicators in Fig. 10.7. This last kind of approach removes the need for specific extensions that take into account the preferences of multiple stakeholders, although care needs to be taken that the factors being considered are important to each set of stakeholders without unintentionally disempowering a group in the overall assessment.

In these ways, the extensions to the methodology are not complex and the different preferences of actors are not intricately integrated into the assessment process. It is possible to assign different biases to sets of preferences so that a certain group of stakeholder's preferences has a larger influence on the assessment compared to others, but such a bias would require substantially greater justification compared to the same biasing among indicators. In addition, a very important aspect to consider for multiactor assessment is accurately reflecting the preferences of different stakeholders in the assessment process; this requires both fair surveying processes to obtain the preferences of each group of stakeholders, and ensuring that stakeholders understand how to specify their preferences and what effects these will have on the assessment process.

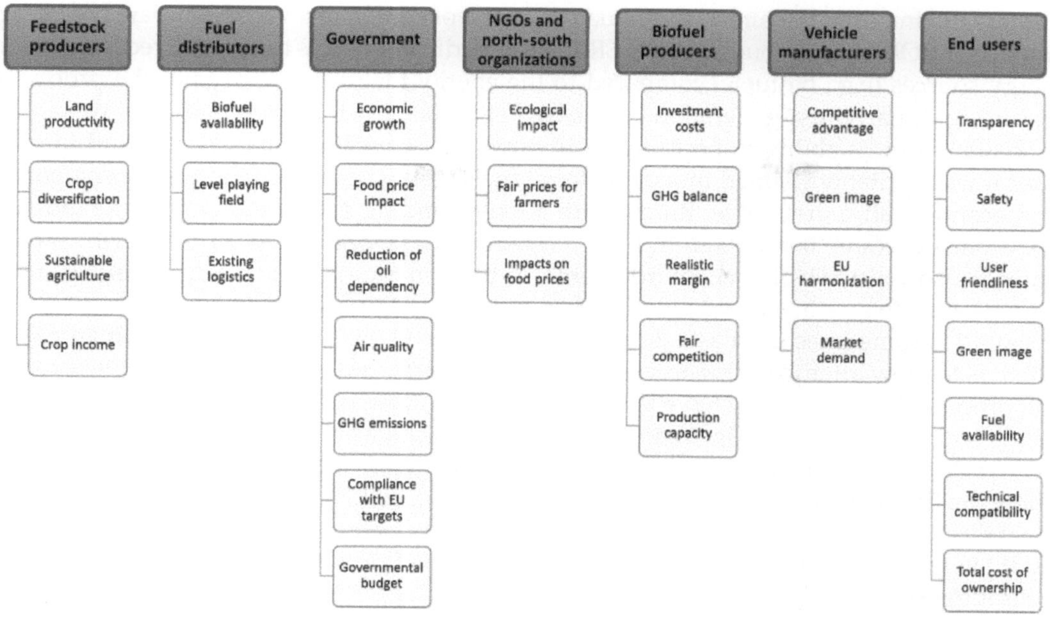

FIG. 10.7 Indicator framework for multicriteria multiactor assessment of biofuel use in Belgium, showing the different indicators in specific stakeholder groups [91].

10.8 ASSESSMENT PERSPECTIVES AND DEVELOPMENT

So far in this chapter, there has been a broad discussion about some of the pertinent sustainability issues related to biofuels, including their production operation in a given environment and community. It has also been acknowledged that sustainability is multifaceted and difficult to measure using traditional approaches that tend to adopt uniform scaling or overly reductionist methods; in this manner, MCDA methods can help to overcome some of these challenges.

This section will discuss some of the perspectives about sustainability assessment of biorefineries, before the end of the section which will discuss a possible guiding framework for the integrated sustainability assessment of biorefineries. The different perceptions of sustainability assessment include a combination of methods which have been attempted or relied upon in place of holistic assessment methods in order to qualify the sustainability of a biofuel.

10.8.1 Energy Recovery and Investment (EROEI)

An important environmental issue of biofuels is mainly borne from LCA-based studies, which is to ensure the ability for biofuels to make an earnest contribution towards alleviating the risks of climate change (rather than using fossil fuels). A prominent argument which was made against the early claims of the "biofuel as an energy panacea" was a precursor to LCA studies, and in particular aimed to evaluate the ratio of energy recovered compared to energy invested (EROEI) for various biofuels. EROEI is broadly defined as the energy recovered from energy sources (e.g., biofuels) compared to the energy that must be used in the process of producing it. Early first-generation corn-based ethanol used for automobile fuel was found to be limited in its ability to perform environmentally better than fossil fuels, and in some cases it was veritably claimed that using fossil fuels would have been more environmentally (and economically) preferable. Brazil, with a long-standing and dominant biofuel industry based on sugarcane, has proven to have a much more favorable EROEI than that for corn, which is also assisted by lower environmental impacts in transportation categories (e.g., moving feedstock to a refinery, or biofuel from a refinery to the end use).

EROEI is sometimes used as a definitive test for the viability of a biofuel; with an EROEI which is too low after factoring in various energy-consuming aspects of the biofuel production process, the biofuel may have to be discarded or reviewed. The EROEI for a selection of energy sources is shown in Fig. 10.8. The deterministic value of EROEI is useful for qualifying the viability of a proposed biofuel operation, as well as allowing further understanding as to where the flows of energy are located about a process. EROEI by itself is not useful for a sustainability assessment measure, although it can be a useful metric. The measurement of EROEI in terms of energy investment needs to be defined clearly before any such assessment to limit the scope of contributory energy sources.

10.8.2 Governance

An interesting aspect of the treatment of social sustainability within the field of bioenergy in general is the inclusion of governance within the social sphere. Governance, in a broad sense,

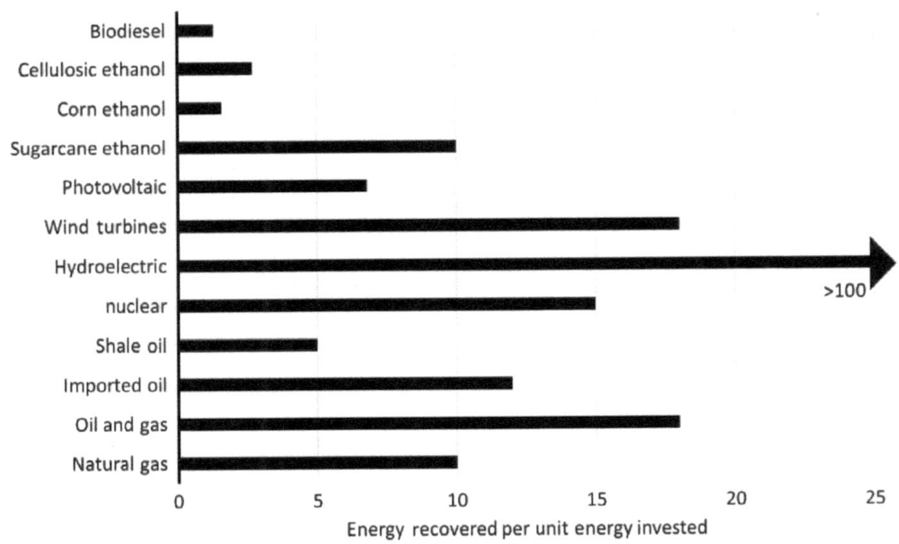

FIG. 10.8 Various energy sources in the United States compared by the value of energy recovered over energy invested (EROEI) [95].

may be described as "the manner in which power is exercised in the management of a country's economic and social resources for development" [96]. However, the term has been extended to broadly include the processes and their maturity to ensure fair, just and accountable decision making in all disciplinary spheres for various actors, not just including governments, for example, corporations, industries, and even civil societies. Due to this broadened understanding of the term, governance usually commands an important influence on the other two domains of sustainability in terms of ensuring the presence, appropriateness, maturity and development of decision making and accountability processes (e.g., measurements, reporting frameworks, decision support pathways, etc.). Governance has more appropriately been suggested as a separated complement to sustainability rather than forming part of the social domain. An example of this separation is in the sustainability guidelines of the Mining and Minerals for Sustainable Development [97], which has been endorsed by several of the major mining companies; yet another subtle example of this is the Global Reporting Initiative's G4 Reporting Guidelines, where standard management approach disclosures enforce the reporting of an organization's management of reporting and accountability processes across all indicators but separate from the indicators themselves and from the social disclosures [98].

In the assessment of biofuels and their operations, governance is usually accepted as part of social sustainability, and is generally not well recognized as being connected to nonsocial aspects of sustainability, despite several aspects of poor governance being extremely important to all sustainability aspects, including:

- Accountability processes, bureaucracy, and institutional support [64]
- Management of tension between agricultural and energy sectors in order to avoid negative socioeconomic impacts [99]
- Transparency and corruption
- Ability to meet commitments to global and national sustainability

Good governance can be fostered through assessment methodologies through the requirement of adequate, justifying documentation in assessment components with a generally open-minded approach to the assessment process. For corporations, using the correct tools and adopting good operating principles, which may be better than those of the affected societies, are essential for fostering effective governance and thus are precursors for holistic sustainability assessment, as well as better understanding and integration of operations with communities and other affected stakeholders.

10.8.3 Certification Schemes

In order to enforce sustainability on the biofuel process, particularly to ensure the processes for social sustainability, certification schemes for biofuels or bioenergy have been created.

One prime example of a certification scheme for sugarcane (which can be extended to sugarcane-produced biofuels) is the Bonsucro Certification System [100]. Bonsucro certification consists of a set of sustainability standards, a guidance for meeting the standards and an auditing system to verify compliance against the standards. There are two certification standards for Bonsucro: one mainly concerns the production of sugarcane and its auxiliary products, whereas another one concerns the entire product supply chain. The production standard encompasses six broad principles with component certification obligations for a sugarcane operation. Fig. 10.9 depicts the principles and key component obligations for Bonsucro certified production.

The intention of standards such as Bonsucro is to encourage both compliance as well as accountability, since auditing against a standard will ensure that adequate monitoring of impacts and systems will be necessary to prove compliance. On the other hand, certification schemes are largely nondriven and rely on companies to enact the relevant processes to ensure operation within compliance of standards. Fortin and Richardson [101] contend that certification schemes such as Bonsucro or even the RSB are simply enablers to scrutinize an operation's sustainability credentials, but are not as effective in enforcing the standards. In other cases, standards may not enforce the necessary requirements to improve practices from historical cases of inadequate land ownership and lack of access to natural resources. In cases like these, standards like Bonsucro may offer a veil of false legitimacy for companies to perpetuate outdated and unethical practices, despite its purported independence from sociopolitical influence [102].

10.8.4 Characterization of Sustainability Assessments

Although sustainability assessment of biofuel production from a multicapital point of view is not a completely novel concept, it is important to understand some of the key characteristics and dimensions of assessment in order to achieve a broad, holistic assessment method which will be suitable for fully understanding the impacts of biofuels on various stakeholders. In particular, "sustainability" assessments must actually address sustainability concerns from a multifaceted, multicapital standpoint, and not just be presumed to be constraints or guidelines for formulating technical and economic options [103,104].

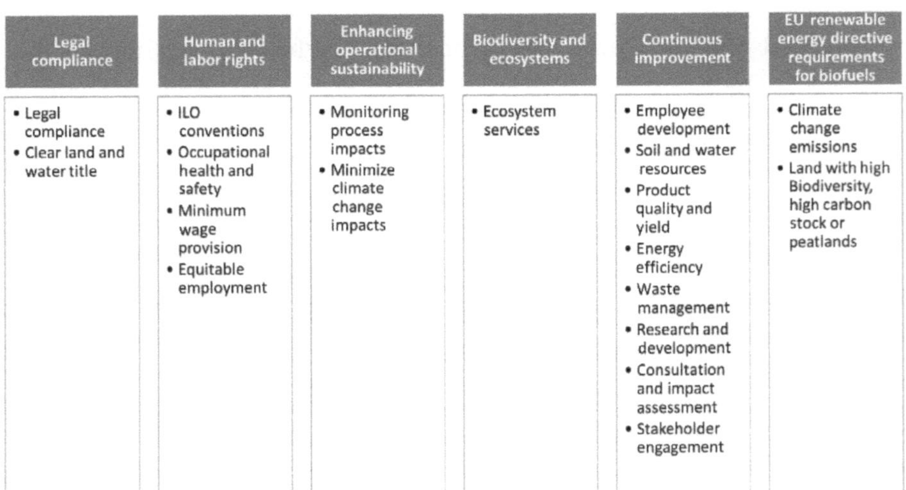

FIG. 10.9 The Bonsucro Production Standard [100].

Sustainability assessment processes can be designed for specific applications or generalized (like frameworks) to encompass a set of related processes. In the case of the sustainability impact assessment of biorefineries, the assessments can be generalized depending on the life-cycle stage of the refinery, as well as the operational context (e.g., corporate level or engineering level). The management of the different sustainability issues which may affect the production life cycle at different stages and logical business levels means that, while an overarching framework is useful for ensuring continuity and consistency, the correct kind of approach for assessment must be adopted for the correct level of organizational detail. For example, food security is recognized as a very important issue in managing biofuel projects; however, addressing such issues is usually managed at a state or national level, rather than at the operation. A summary of the importance of matching the correct managerial response to the correct kinds of sustainability issues can be exemplified in the management topography by Kunz et al. [105], as shown in Fig. 10.10. The framework shows that attempting to assign incorrect levels of management to make decisions at the wrong logical levels can lead to paralysis in decision making through overwhelming complications, or paralysis in tactical progress due to lack of specificity. Although this framework was developed in order to guide approaches to water management, it is very much adaptable to other multileveled, complex engineering systems. The larger challenge is, as shown in the framework, to ensure alignment

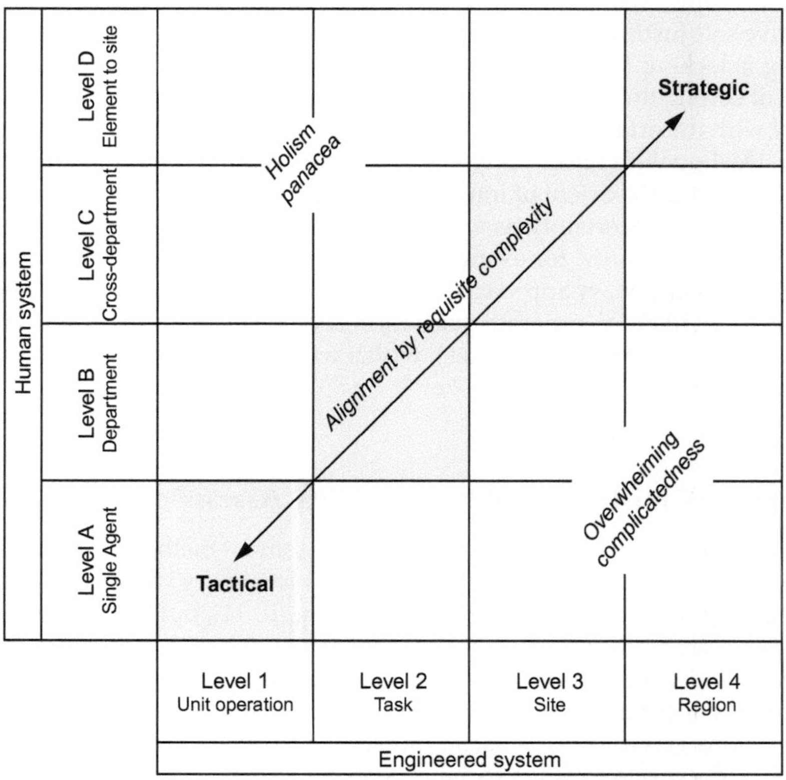

FIG. 10.10 Multilevel framework matching problem complexity with appropriate management action [105].

and consistency across the decision-making spectrum, viz. strategic to tactical. It is not effective to express high-level commitments or statements about sustainability and not be able to have that considered at the operation or community; similarly, policy-level decisions may address matters of concern which cannot be greatly influenced at the community or operational level, but a lack of consultation across levels compounded by a lack of consistent conceptual approach will inevitably result in conflicts.

Current predominant approaches to measuring biorefinery or biofuel production sustainability generally manifests itself in sets of measurable, numerical indicators and possibly aggregation methods. However, many social and some environmental indicators become very difficult to measure quantitatively, which will necessitate embracing uncertainty and nonnumerical outcomes in holistic assessment. In some cases, indicators can be created which attempt to relate the extent of otherwise intangible social indicators to risks or quantifiable indirect effects [106]. The indicator framework for biofuels proposed by Sacramento-Rivero [106] attempts to offer a fully quantifiable, ratio-based assessment method to characterize sustainability. Table 10.7 shows the proposed indicators, which have been organized according to locational source of sustainability impacts rather than by sustainability capital domain.

Reducing and operationalizing sustainability assessment into figures mainly allow for more extensive scopes of assessment that would be otherwise more considerably complex, for example the stochastic supply chain model created by Osmani and Zhang [107] for cellulosic bioethanol supply chains. Such models can only be, at best, understood as rough guides and not effective sole justifications for implementation, particularly as a great degree of social understanding is lacking from such results.

A key part of setting the correct context for sustainability assessment is an understanding of how to deal with uncertainty, as well as deciding on the relative merits of strong and weak sustainability. Dealing with uncertainty in assessment does not mean an inability to measure or correctly account for the extent of impacts or benefits, but implies that assessors should be prepared to document uncertainty as well as generally adopt a considered approach to potentially resolving uncertainty, for example, sensitivity analysis, adaptive management, scenario analysis or precautionary approach [108]. The strength of sustainability—between weak and strong—is more likely to generate discussion about critical resources (i.e., inventorying, characterization of these critical resources), which will be useful for later stages of a sustainability assessment.

10.8.5 Framework for Integrated Sustainability Assessment

The main shortcoming of several sustainability assessment methods is the inability to characterize social sustainability aspects. Social aspects of biorefinery sustainability may manifest themselves at any stage of the project life cycle (see Fig. 10.11), but the certainty at which impacts can be assessed and the nature of those impacts vary with stage, and generally with greater certainty and wider ranges of issues as the project matures. By the time the project is within the stages nearing real-world implementation of the biorefinery, there should be serious considerations given towards social impacts. An understanding of the importance of life-cycle stage and expectations for social aspects will assist an assessment team in contextualizing the various sustainability indicators.

TABLE 10.7 List of Indicators Comprising the Sustainability Assessment Framework From Sacramento-Rivero [106]

Indicator	Ideal Value	Critical Value
Feedstock		
Normalized biotechnological valorization potential	Feedstock sources exhibit the most desirable features according to predefined criteria	Feedstock sources exhibit the minimum expected features according to predefined criteria
Raw materials consumption	Rate of regeneration of renewable raw materials is infinitely greater than consumption	Rate of regeneration of renewable raw materials is equal to consumption
Fraction of revenue for feedstock	Cost of process feedstock is infinitely lower than the value of products	Value of products equals the cost of process feedstock
Process		
Freshwater use reduction	Reduction of freshwater consumption is achieved (relative to a reference system) at a magnitude infinitely better than a critical reduction target	Freshwater consumption represents a reduction on freshwater use (relative to a reference system) equal to an acknowledged target
Modified gross margin	The cost of production is infinitely lower than net sales	The cost of production is equal to the net sales
Products		
Nonrenewable energy share	Share of nonrenewable energy consumption in the system is infinitely better than an acknowledged target	Share of nonrenewable energy consumption in the system is equal to an acknowledged target
Reduction of fossil energy use		Biorefinery requires the same amount of fossil energy as a reference system
Raw materials cost ratio	Raw materials cost is infinitely lower than those in a reference system	Raw materials cost is equal to the cost of process inputs
Environment		
Reduction of baseline emissions	Baseline emissions are reduced (relative to a reference system) infinitely many times more than an acknowledged target	Baseline emissions are reduced (relative to a reference system) by an acknowledged target
Sustainable land use	No net land use change is required	Land use change caused by biorefinery is at the limit of causing net biodiversity loss
Corporate		
EHS Management System compliance	Biorefinery is EHS Management System compliant with maximum accreditation	Biorefinery is recognized as EHS Management System compliant with lowest level of accreditation
Health and safety compliance	Biorefinery has a recognized OHS system in place and is maximally compliant	Biorefinery has a recognized OHS system in place and is minimally compliant
Recordable incident rate	No recordable incidents in the annual operation of the biorefinery	Maximum allowed incidents in the annual operation of the biorefinery
Rate of fatal work injuries	No fatal work injuries in the annual operation of the biorefinery	Maximum allowed fatal work injuries in the annual operation of the biorefinery
Social investment	Philanthropic and community development projects, and number of employees per volume of net sales are both at predefined maximum values	Philanthropic and community development projects, and number of employees per volume of net sales are both at predefined minimum values

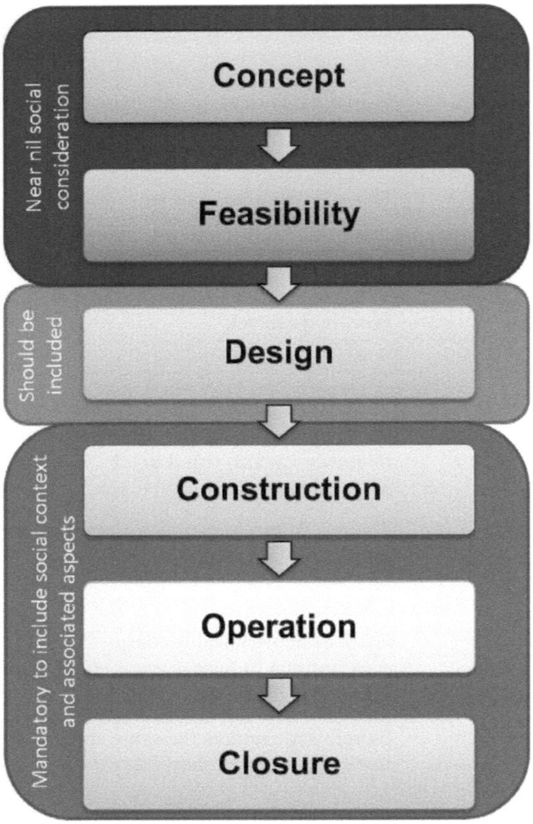

FIG. 10.11 Project life-cycle stages, including general indications of earnest consideration of social sustainability aspects.

Since social aspects are usually strongly tied to a particular area or region, it is necessary to introduce the local context into sustainability assessment. As well, sustainability assessment does not need to render traditional technical-economic assessment or LCA obsolete, but rather the results of the latter methods can complement and feed into the integrated sustainability assessment. Thus, a high-level framework for the sustainability assessment of biorefineries is proposed and presented graphically in Fig. 10.12, with major steps explained in Table 10.8. The proposed framework is adaptable to all stages of the project life cycle, and the steps should not only be guiding towards progressive sustainability characterization of the various scenarios, but also generate worthwhile discussion between members of an assessment team about various aspects that may be key to the overall decision-making process.

As an important part of sustainability assessment is the indicator set, judicious consideration should be given to selecting the critical indicator set that is useful to the objectives of the assessment, are discriminatory and material. Azapagic and Perdan [109] propose eight guiding keywords in selecting indicators for characterizing sustainability, as shown in Fig. 10.13.

Reference systems are important in LCAs as they form the default alternative scenario (or base case), often used to compare against the proposed concept plant in the study. For biofuels, it is common to set the reference system as a fossil-fuel based equivalent scenario,

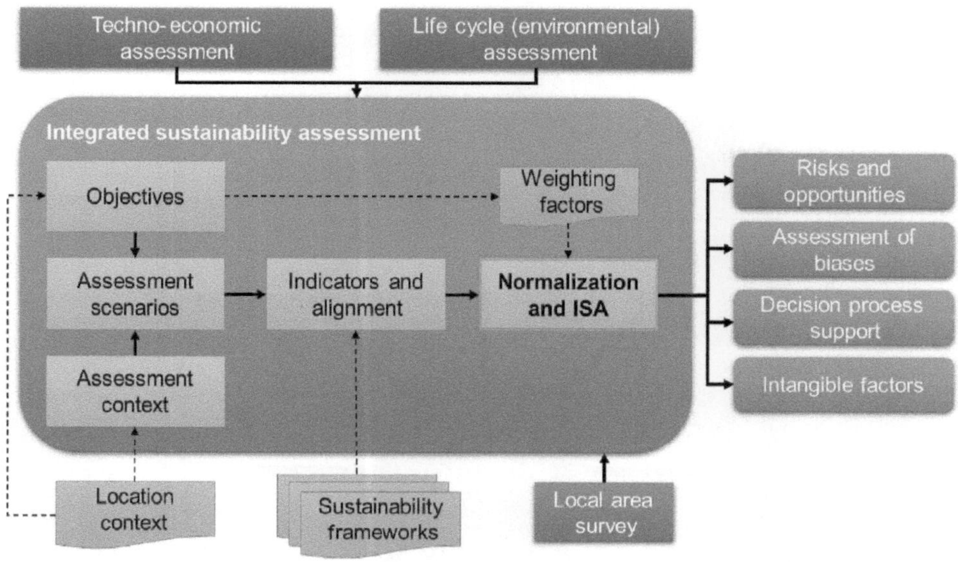

FIG. 10.12 Proposed framework for integrated sustainability assessment of biorefineries.

TABLE 10.8 Description of Components of Proposed Integrated Sustainability Assessment Framework (See Fig. 10.12)

Framework Step	Description
Objectives	The goals of the assessment need to be established. Are alternative configurations being compared? Are there critical or sensitive variables or indicators that must be kept in mind especially for this assessment (e.g., highly sensitive environmental or cultural assets, water-starved region)?
Assessment context	The assessment context needs to be established with respect to the geographical area or community that is being affected by the biorefinery operation. This includes an understanding of local climate, environmental state, community demographics, local governance, accessibility and other pertinent background or baseline factors. A lot of the assessment context may be established through a local area survey
Assessment scenarios	From the combined understanding of the objectives of the assessment and the context within which the potential biorefinery operation will take place, the alternative options or scenarios for the assessment will be characterized. Scenarios can include both alternative operational configurations as well as reference systems
Indicators and alignment	The sustainability indicators for assessment will be elicited through an understanding of selected sustainability frameworks, usually comprising an established global sustainability framework and the corporate sustainability framework. Indicator elicitation may also be guided by general selection principles and materiality (see Fig. 10.13), in order to have the best chance of discriminating between scenarios, but also to best understand how indicator results will be related back to sustainability goals or requirements (e.g., reporting)
Normalization and integrated sustainability assessment (ISA)	The assessment of the scenarios against the proposed list of indicators takes place with respect to both (a) normalization clusters, and (b) weighting factors. Normalization will attempt to equalize the characterization of various indicators so to allow adequate and equitable decision power for each indicator (subject to weighting). However, this process will also necessitate decisions where attempts to normalize some indicators will be difficult or impossible, and how they will be accounted for in the final results

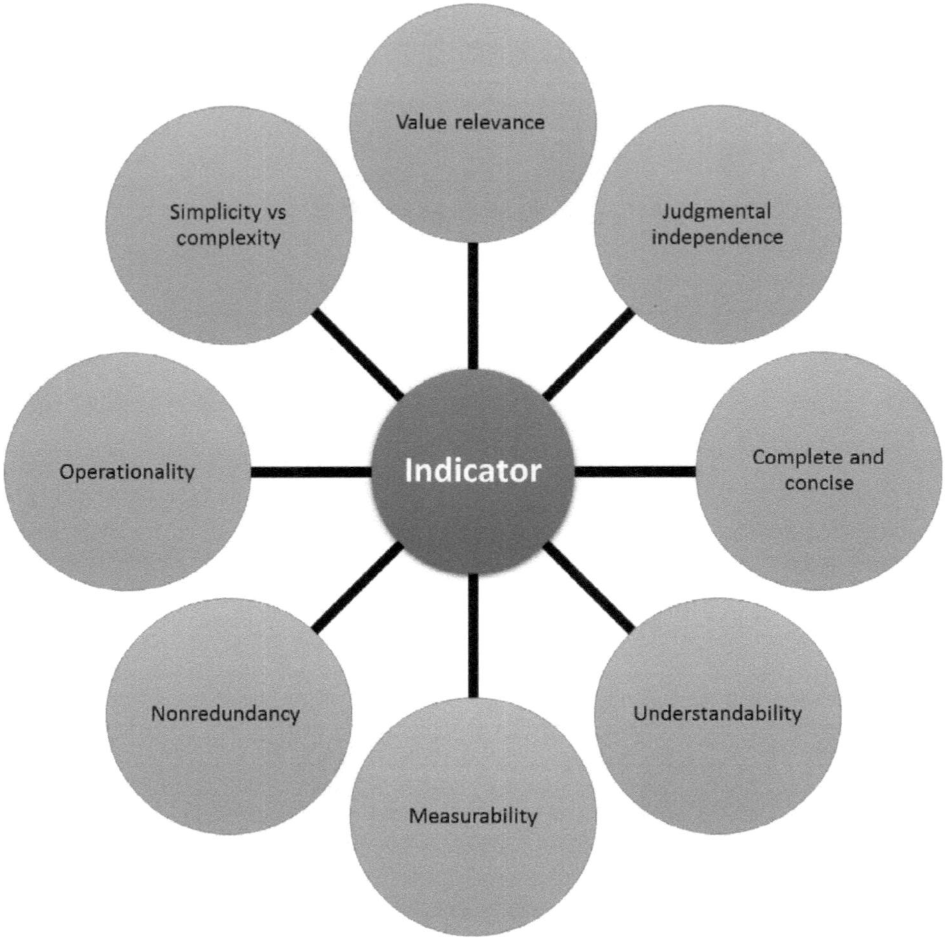

FIG. 10.13 Guideline keywords for selecting indicators for characterizing sustainability issues [109].

which provides the same services as afforded by the proposed system under study. In reference to project management, such reference systems may also be referred to as "business as usual" scenarios. Reference systems are also necessary when using impact potentials as a measure, since it is more relevant to compare potentials rather than understand them as standalone numbers.

For the integrated sustainability assessment, the reference system may be defined similarly, although the context will include several additional aspects compared to previous ones. In addition, depending on the aims of the assessment and project, the reference system may be defined in subtly different manners:

- The reference system may involve comparing the system under study with that of a fossil fuel–based system that supplies the same services. The fossil fuel–based system will be based in the same location (or proximate or equivalent) as the system under study, which means the objective is to compare a traditional fossil fuel method along with a proposed improved biofuel method of supplying a service (e.g., automobile fuel).

- The reference system may involve a similar set-up to the previously mentioned fossil fuel–based system, but the location at which the fossil fuels are produced to fulfill the equivalent service may not be the same as that of the proposed biofuel system.
- The reference system may be a "do nothing" scenario, by which the systems being compared are the system under study and the same location but no new activity is proposed (cf. the previous proposal where the reference system still involves the creation of a fossil fuel–based operation to fulfill the provision of services). The provision of services in the reference system equivalent to the system under study may be supplied from another context, or it may simply "fail" to provide such services, which can have impacts associated with that. The use of this kind of reference system may, in addition to highlighting increased pressures on energy security or reliance on fossil fuels, also reveal the potential losses of opportunities or social investments, for example, jobs, education, etc. Such a reference system is very similar to the "business as usual" or "do nothing" scenarios, which are used as similar defaults in business cases.

The normalization step of the framework may be achieved by reducing all indicators into a common measurable basis, but this kind of reductionist approach has been shown previously to be broadly unproductive. If such an approach is still desired, the limitations and assumptions associated with adopting that approach should be well documented and accompany results of the assessment. In particular, the indicators or aspects where a reductionist approach may prove to be quite awkward may be documented in more detail in the output on intangible factors or bias assessment. Alternatively, normalization may be achieved through other means which do not involve common-base reduction; for example, outranking methods (such as PROMETHEE) promote measurement independence of indicators (criteria), while at the same time maintaining a strong sustainability approach which avoids trade-offs between various factors (which may prove inequitable or ethically unfavorable). Methods like PROMETHEE produce results which are not absolute and can be referenced in several ways with respect to the constituent indicators (or indicator clusters). The uniqueness of these various approaches give rise to different reporting forms for the major outputs of the integrated sustainability assessment process:

- *Risks and opportunities.* The assessment process may have revealed particular risks to certain scenarios which were analyzed; alternatively, instead of risks the assessment process may have shown that there are valuable opportunities which deserve more detailed investigation.
- *Assessment of biases.* Through the process of deciding weighting factors and even indicator selection, apart from obvious connections to the assessment objectives, the aspects of noneven weighting or the particular selection of indicators should be discussed and documented to support decisions which are proposed from the assessment, promoting full accountability and transparency in the assessment process.
- *Decision process support.* Ultimately, a decision on the various candidate scenarios will be made from the assessment process. The decision support in this case complements existing decision-making methods or even the results from established techno-economic and life-cycle assessments; alternatively, the decision support may reveal and explain differences between the sustainability and alternative assessment processes.
- *Intangible factors.* This includes a discussion of indicators or sustainability aspects which are difficult or impossible to adequately include in the normalized sustainability assessment. It is important not to ignore intangible factors in sustainability assessment, as their implicated impacts may manifest themselves in more tangible ways if not monitored.

10.9 CONCLUSIONS AND PERSPECTIVES

This chapter has discussed the sustainability assessment of biorefineries beyond LCA, which has been a well-established tool for measuring the environmental impacts of biorefinery operations. While LCA has been shown to be well grounded and effective in the characterization of environmental burdens, it lacks the ability to be adapted to a local context, as well as the extensibility to effectively measure nonenvironmental concerns for biofuel operations. LCA thus forms a necessary but insufficient part of an effective, overall assessment of biorefinery operations.

The discussion of various sustainability issues comprising the triple bottom line capital domains—economic, environmental, and social—shows that not only is LCA relatively ineffective at capturing some of these issues, but such issues can have major implications on the sustainability of biorefineries. A large part of recognizing the various sustainability issues is their connection to the local context, that is, affected ecosystems, natural resources, communities (including their human and legal rights) and social structures.

Since sustainability is multifaceted and difficult to normalize effectively using a simplistic approach, it lends itself very easily to various forms of MCDA. In particular, outranking MCDA methods hold the best promise for ensuring a fairly strong sustainability paradigm, but at the same time allowing relative indicator independence and thus preservation of equitable decision power. The trade-off for these characteristic strengths is the relative complexity of using outranking methods, such as PROMETHEE.

The final part of this chapter discussed various perspectives around sustainability assessment or attempts to effectively include and account for sustainability of biofuel production processes. Strong governance forms the basis of ensuring effective and equitable sustainability assessment and general oversight of the sustainability of a biorefinery. While certification schemes have gone some way to attempting to enforce sustainability standards on biofuels or other byproducts, there are doubts as to whether the relative passivity of meeting standards does little more than give some scrutiny to corporate practice rather than actually drive sustainability objectives. Sustainability assessment should be suited to the type of decision that needs to be made (level of specificity or corporate level within a company), but it should also be based on consistent frameworks and guiding principles. It must also be careful to be inclusive of multiple stakeholder groups, including avoiding disempowering affected communities. To this extent, a framework for integrated sustainability assessment, based on a structured approach of considering local conditions and encouraging discussion and documentation of underlying assumptions, biases and intangible/unquantifiable factors for full accountability, has been proposed and is adaptable to biorefinery assessment at any level of the project life cycle.

References

[1] RSB. RSB principles and criteria for sustainable biofuels production (no. RSB-STD-01-001 v2). Geneva; 2010.
[2] Tuazon D, Corder GD, McLellan BC. Sustainable development: a review of theoretical contributions. Int J Sustain Future Hum Secur 2013;1:40–8.

[3] Brundtland GH, WCED. Our common future. Oxford: Oxford University Press; 1987.
[4] Elkington J. Cannibals with forks: the triple bottom line of 21st century business. Oxford: Capstone; 1997.
[5] Forum for the Future. The five capitals model; 2005. [WWW document]. URL, http://www.forumforthefuture.org/our-approach/tools-and-methodologies/5capitals [accessed 11.03.14].
[6] Carew AL, Mitchell CA. Teaching sustainability as a contested concept: capitalizing on variation in engineering educators' conceptions of environmental, social and economic sustainability. J Clean Prod 2008;16:105–15.
[7] Mitchell CA, Carew AL, Clift R. The role of the professional engineer and scientist in sustainable development. In: Azapagic A, Perdan S, Clift R, editors. Sustainable development in practice: case studies for engineers and scientists. Chichester: John Wiley & Sons; 2004. p. 29–55.
[8] Moran CJ, Kunz NC. Sustainability as it pertains to minerals and energy supply and demand: a new interpretative perspective for assessing progress. J Clean Prod 2014;84:16–26.
[9] Liew WH, Hassim MH, Ng DKS. Review of evolution, technology and sustainability assessments of biofuel production. J Clean Prod 2014;71:11–29.
[10] Azapagic A. Sustainability considerations for integrated biorefineries. Trends Biotechnol 2014;32:1–4.
[11] Dias MOS, Junqueira TL, Cavalett O, Cunha MP, Jesus CDF, Mantelatto PE, et al. Cogeneration in integrated first and second generation ethanol from sugarcane. Chem Eng Res Des 2013;91:1411–7.
[12] Plevin RJ, Delucchi MA, Creutzig F. Using attributional life cycle assessment to estimate climate-change mitigation benefits misleads policy makers. J Ind Ecol 2014;18:73–83.
[13] de Haes HAU, Heijungs R, Suh S, Huppes G. Three strategies to overcome the limitations of life-cycle assessment. J Ind Ecol 2004;8:19–32.
[14] Turconi R, Boldrin A, Astrup T. Life cycle assessment (LCA) of electricity generation technologies: overview, comparability and limitations. Renew Sustain Energy Rev 2013;28:555–65.
[15] Sala S, Farioli F, Zamagni A. Progress in sustainability science: lessons learnt from current methodologies for sustainability assessment. Int J Life Cycle Assess 2012;18:1653–72.
[16] Wu R, Yang D, Chen J. Social life cycle assessment revisited. Sustainability 2014;6:4200–26.
[17] Brent A, Labuschagne C. Social indicators for sustainable project and technology life cycle management in the process industry. Int J Life Cycle Assess 2006;11:3–15.
[18] Ekvall T, Finnveden G. Allocation in ISO 14041—a critical review. J Clean Prod 2001;9:197–208.
[19] Stamp A, Althaus HJ, Wäger PA. Limitations of applying life cycle assessment to complex co-product systems: the case of an integrated precious metals smelter-refinery. Resour Conserv Recycl 2013;80:85–96.
[20] Ardente F, Cellura M. Economic allocation in life cycle assessment. J Ind Ecol 2012;16:387–98.
[21] Cherubini F, Strømman AH, Ulgiati S. Influence of allocation methods on the environmental performance of biorefinery products—a case study. Resour Conserv Recycl 2011;55:1070–7.
[22] Luo L, van der Voet E, Huppes G, Haes HAU. Allocation issues in LCA methodology: a case study of corn stover-based fuel ethanol. Int J Life Cycle Assess 2009;14:529–39.
[23] Reap J, Roman F, Duncan S, Bras B. A survey of unresolved problems in life cycle assessment. Int J Life Cycle Assess 2008;13:374–88.
[24] Finnveden G. On the limitations of life cycle assessment and environmental systems analysis tools in general. Int J Life Cycle Assess 2000;5:229–38.
[25] Florin MJ, van de Ven GWJ, van Ittersum MK. What drives sustainable biofuels? A review of indicator assessments of biofuel production systems involving smallholder farmers. Environ Sci Policy 2014;37:142–57.
[26] O'Shea T, Golden JS, Olander L. Sustainability and earth resources: life cycle assessment modeling. Bus Strat Environ 2013;22:429–41.
[27] Levasseur A, Lesage P, Margni M, Deschênes L, Samson R. Considering time in LCA: dynamic LCA and its application to global warming impact assessments. Environ Sci Technol 2010;44:3169–74.
[28] McBride AC, Dale VH, Baskaran LM, Downing ME, Eaton LM, Efroymson RA, et al. Indicators to support environmental sustainability of bioenergy systems. Ecol Indic 2011;11:1277–89.
[29] Kami Delivand M, Gnansounou E. Life cycle environmental impacts of a prospective palm-based biorefinery in Pará State-Brazil. Bioresour Technol 2013;150:438–46.
[30] Efroymson RA, Dale VH, Kline KL, McBride AC, Bielicki JM, Smith RL, et al. Environmental indicators of biofuel sustainability: what about context? Environ Manage 2012;51:291–306.

[31] Searchinger T, Heimlich R, Houghton RA, Dong F, Elobeid A, Fabiosa J, et al. Use of U.S. croplands for biofuels increases greenhouse gases through emissions from land-use change. Science 2008;319:1238–40.
[32] Fargione J, Hill J, Tilman D, Polasky S, Hawthorne P. Land clearing and the biofuel carbon debt. Science 2008;319:1235–8.
[33] Sanchez ST, Woods J, Akhurst M, Brander M, O'Hare M, Dawson TP, et al. Accounting for indirect land-use change in the life cycle assessment of biofuel supply chains. J R Soc Interface 2012;9:1105–19.
[34] Ahlgren S, Lucia LD. Indirect land use changes of biofuel production—a review of modelling efforts and policy developments in the European Union. Biotechnol Biofuels 2014;7:35.
[35] Bird DN, Zanchi G, Pena N. A method for estimating the indirect land use change from bioenergy activities based on the supply and demand of agricultural-based energy. Biomass Bioenergy 2013;59:3–15.
[36] Kim S, Dale BE, Heijungs R, Azapagic A, Darlington T, Kahlbaum D. Indirect land use change and biofuels: mathematical analysis reveals a fundamental flaw in the regulatory approach. Biomass Bioenergy 2014;71:408–12.
[37] Humalisto NH. Climate policy integration and governing indirect land-use changes—actors in the EU's biofuel policy-formulation. Land Use Policy 2015;45:150–8.
[38] Liska AJ, Perrin RK. Indirect land use emissions in the life cycle of biofuels: regulations vs science. Biofuel Bioprod Biorefin 2009;3:318–28.
[39] Fritsche UR, Sims REH, Monti A. Direct and indirect land-use competition issues for energy crops and their sustainable production—an overview. Biofuel Bioprod Biorefin 2010;4:692–704.
[40] Mathews JA, Tan H. Biofuels and indirect land use change effects: the debate continues. Biofuel Bioprod Biorefin 2009;3:305–17.
[41] Renó MLG, del Olmo OA, Palacio JCE, Lora EES, Venturini OJ. Sugarcane biorefineries: case studies applied to the Brazilian sugar–alcohol industry. Energy Convers Manage 2014;86:981–91.
[42] Richards PD, Walker RT, Arima EY. Spatially complex land change: the indirect effect of Brazil's agricultural sector on land use in Amazonia. Global Environ Change 2014;29:1–9.
[43] Obidzinski K, Andriani R, Komarudin H, Andrianto A. Environmental and social impacts of oil palm plantations and their implications for biofuel production in Indonesia. Ecol Soc 2012;17:25.
[44] Ramdani F, Hino M. Land use changes and GHG emissions from tropical forest conversion by oil palm plantations in Riau province, Indonesia. PLoS ONE 2013;8:e70323.
[45] Smit HH, Meijaard E, van der Laan C, Mantel S, Budiman A, Verweij P. Breaking the link between environmental degradation and oil palm expansion: a method for enabling sustainable oil palm expansion. PLoS ONE 2013;8: e68610.
[46] Gelfand I, Sahajpal R, Zhang X, Izaurralde RC, Gross KL, Robertson GP. Sustainable bioenergy production from marginal lands in the US Midwest. Nature 2013;493:514–7.
[47] Thomsen TP, Ahrenfeldt J, Thomsen ST. Assessment of a novel alder biorefinery concept to meet demands of economic feasibility, energy production and long term environmental sustainability. Biomass Bioenergy 2013;53:81–94.
[48] Eranki PL, Manowitz DH, Bals BD, Izaurralde RC, Kim S, Dale BE. The watershed-scale optimized and rearranged landscape design (WORLD) model and local biomass processing depots for sustainable biofuel production: integrated life cycle assessments. Biofuel Bioprod Biorefin 2013;7:537–50.
[49] Thiffault E, Endres J, McCubbins JS, Junginger M, Lorente M, Fritsche U, et al. Sustainability of forest bioenergy feedstock supply chains: local, national and international policy perspectives. Biofuel Bioprod Biorefin 2015;9:283–92.
[50] Manning P, Taylor G, Hanley ME. Bioenergy, food production and biodiversity—an unlikely alliance? GCB Bioenergy 2015;7:570–6.
[51] Van Meerbeek K, Appels L, Dewil R, Calmeyn A, Lemmens P, Muys B, et al. Biomass of invasive plant species as a potential feedstock for bioenergy production. Biofuel Bioprod Biorefin 2015;9:273–82.
[52] Committee on Water Implications of Biofuels Production in the United States. Water implications of biofuels production in the United States. Washington, DC: National Research Council; 2008.
[53] Tsao CC, Campbell JE, Mena-Carrasco M, Spak SN, Carmichael GR, Chen Y. Increased estimates of air-pollution emissions from Brazilian sugar-cane ethanol. Nat Clim Change 2012;2:53–7.
[54] Golberg A, Vitkin E, Linshiz G, Khan SA, Hillson NJ, Yakhini Z, et al. Proposed design of distributed macroalgal biorefineries: thermodynamics, bioconversion technology, and sustainability implications for developing economies. Biofuel Bioprod Biorefin 2014;8:67–82.

[55] Pieprzyk B, Rojas Hilje P, Kortlüke N. The substitution of marginal oil with biofuels. Biofuel Bioprod Biorefin 2013;7:525–36.
[56] Gnansounou E, Dauriat A. Techno-economic analysis of lignocellulosic ethanol: a review. Bioresour Technol 2010;101:4980–91.
[57] Salvo A, Huse C. Build it, but will they come? Evidence from consumer choice between gasoline and sugarcane ethanol. J Environ Econ Manage 2013;66:251–79.
[58] Eggert H, Greaker M. Promoting second generation biofuels: does the first generation pave the road? Energies 2014;7:4430–45.
[59] Acheampong E, Campion BB. The effects of biofuel feedstock production on farmers' livelihoods in Ghana: the case of *Jatropha curcas*. Sustainability 2014;6:4587–607.
[60] Cansino J, Cardenete M, González-Limón J, Román R. Economic impacts of biofuels deployment in Andalusia. Renew Sustain Energy Rev 2013;27:274–82.
[61] Favretto N, Stringer LC, Dougill AJ. Towards improved policy and institutional coherence in the promotion of sustainable biofuels in Mali: towards improved policy and institutional coherence. Environ Policy Gov 2015;25:36–54.
[62] Gingembre M. Resistance or participation? Fighting against corporate land access amid political uncertainty in Madagascar. J Peasant Stud 2015;42:561–84.
[63] Ciptaningtyas AL. Biodiesel governance not ready for now government says [WWW document]. Middle East North Africa Financial Network; 2015. URL, http://www.menafn.com/qn_news_story_s.aspx?storyid=1094189596&title=Biodiesel-Governance-Not-Ready-For-Now-Government-Says [accessed 06.12.15].
[64] Jupesta J, Harayama Y, Parayil G. Sustainable business model for biofuel industries in Indonesia. Sustain Acc Manage Policy J 2011;2:231–47.
[65] Stattman SL, Mol APJ. Social sustainability of Brazilian biodiesel: the role of agricultural cooperatives. Geoforum 2014;54:282–94.
[66] Venghaus S, Selbmann K. Biofuel as social fuel: introducing socio-environmental services as a means to reduce global inequity? Ecol Econ 2014;97:84–92.
[67] Cruse RM, Herndl CG, Polush EY, Shelley MC. An assessment of cellulosic ethanol industry sustainability based on industry configurations. J Soil Water Conserv 2012;67:67–74.
[68] Chin HC, Choong WW, Wan Alwi SR, Mohammed AH. Issues of social acceptance on biofuel development. J Clean Prod 2014;71:30–9.
[69] Wharton School of Pennsylvania. The transportation nexus: ethanol is a "food vs. fuel" issue. [WWW document]. Knowledge@Wharton; 2013 URL, http://knowledge.wharton.upenn.edu/article/the-transportation-nexus-ethanol-is-a-food-vs-fuel-issue/ [accessed 06.02.15].
[70] Lane J. 6 methods to avoid the food v fuel dilemma [WWW document]. Biofuels Digest; 2012. URL, http://oilprice.com/Energy/Energy-General/6-Methods-to-Avoid-the-Food-v-Fuel-Dilemma.html [accessed 06.02.15].
[71] Cinelli M, Coles SR, Kirwan K. Analysis of the potentials of multi criteria decision analysis methods to conduct sustainability assessment. Ecol Indic 2014;46:138–48.
[72] De Montis A, De Toro P, Droste-Franke B, Omann I, Stagl S. Criteria for quality assessment of MCDA methods. In: Presented at the 3rd biennial conference of the European society for ecological economics. Vienna: ESEE; 2000.
[73] Polatidis H, Haralambopoulos DA, Munda G, Vreeker R. Selecting an appropriate multi-criteria decision analysis technique for renewable energy planning. Energy Source B 2006;1:181–93.
[74] Wang JJ, Jing YY, Zhang CF, Zhao JH. Review on multi-criteria decision analysis aid in sustainable energy decision-making. Renew Sustain Energy Rev 2009;13:2263–78.
[75] Schaidle JA, Moline CJ, Savage PE. Biorefinery sustainability assessment. Environ Prog Sustain Energy 2011;30:743–53.
[76] Saaty TL. The analytic hierarchy process. New York, NY: McGraw-Hill; 1988.
[77] Macharis C, Verbeke A, De Brucker K. The strategic evaluation of new technologies through multicriteria analysis: the ADVISORS case. Res Transp Econ 2004;8:443–62. Economic Impacts of Intelligent Transportation Systems: Innovations and Case Studies.
[78] Ren J, Manzardo A, Mazzi A, Fedele A, Scipioni A. Emergy analysis and sustainability efficiency analysis of different crop-based biodiesel in life cycle perspective. Sci World J 2013;2013. 918514.

[79] Maes D, Van Passel S. Advantages and limitations of exergy indicators to assess sustainability of bioenergy and biobased materials. Environ Impact Assess Rev 2014;45:19–29.
[80] Moncada J, Tamayo JA, Cardona CA. Integrating first, second, and third generation biorefineries: incorporating microalgae into the sugarcane biorefinery. Chem Eng Sci 2014;118:126–40.
[81] Moncada J, Matallana LG, Cardona CA. Selection of process pathways for biorefinery design using optimization tools: a Colombian case for conversion of sugarcane bagasse to ethanol, poly-3-hydroxybutyrate (PHB), and energy. Ind Eng Chem Res 2013;52:4132–45.
[82] McElroy MW, Jorna RJ, van Engelen J. Sustainability quotients and the social footprint. Corp Soc Responsib Environ Manage 2008;15:223–34.
[83] Gnansounou E. Assessing the sustainability of biofuels: a logic-based model. Energy 2011;36:2089–96.
[84] Perimenis A, Walimwipi H, Zinoviev S, Müller-Langer F, Miertus S. Development of a decision support tool for the assessment of biofuels. Energy Policy 2011;39:1782–93.
[85] Andiappan V, Ko ASY, Lau VWS, Ng LY, Ng RTL, Chemmangattuvalappil NG, et al. Synthesis of sustainable integrated biorefinery via reaction pathway synthesis: economic, incremental environmental burden and energy assessment with multiobjective optimization. AIChE J 2015;61:132–46.
[86] Chatterjee R, Sharma V, Mukherjee S, Kumar S. Quantitative assessment of biodiesel production: a model-based scenario of sustainable development. J Renew Sustain Energy 2014;6:013129.
[87] Santibañez-Aguilar JE, González-Campos JB, Ponce-Ortega JM, Serna-González M, El-Halwagi MM. Optimal planning and site selection for distributed multiproduct biorefineries involving economic, environmental and social objectives. J Clean Prod 2014;65:270–94.
[88] You F, Tao L, Graziano DJ, Snyder SW. Optimal design of sustainable cellulosic biofuel supply chains: multiobjective optimization coupled with life cycle assessment and input–output analysis. AICHE J 2012;58:1157–80.
[89] Cobuloglu HI, Büyüktahtakın İE. A stochastic multi-criteria decision analysis for sustainable biomass crop selection. Exp Syst Appl 2015;42:6065–74.
[90] Macharis C, Turcksin L, Lebeau K. Multi actor multi criteria analysis (MAMCA) as a tool to support sustainable decisions: state of use. Decis Support Syst 2012;54:610–20.
[91] Turcksin L, Macharis C, Lebeau K, Boureima F, Van Mierlo J, Bram S, et al. A multi-actor multi-criteria framework to assess the stakeholder support for different biofuel options: the case of Belgium. Energy Policy 2011;39:200–14.
[92] Ziolkowska JR. Optimizing biofuels production in an uncertain decision environment: conventional vs. advanced technologies. Appl Energy 2014;114:366–76.
[93] Ziolkowska JR. Evaluating sustainability of biofuels feedstocks: a multi-objective framework for supporting decision making. Biomass Bioenergy 2013;59:425–40.
[94] Ren J, Fedele A, Mason M, Manzardo A, Scipioni A. Fuzzy multi-actor multi-criteria decision making for sustainability assessment of biomass-based technologies for hydrogen production. Int J Hydrogen Energy 2013;38:9111–20.
[95] Tao L, Tan EC, McCormick R, Zhang M, Aden A, He X, et al. Techno-economic analysis and life-cycle assessment of cellulosic isobutanol and comparison with cellulosic ethanol and n-butanol. Biofuel Bioprod Biorefin 2014;8:30–48.
[96] The World Bank. Managing development: the governance dimension (discussion paper). Washington, DC: The World Bank; 1991.
[97] WBCSD, IIED. Breaking new ground: mining, minerals, and sustainable development: the report of the MMSD project. London: Earthscan; 2002.
[98] Global Reporting Initiative. G4 sustainability reporting guidelines—reporting principles and standard disclosures. Amsterdam: Global Reporting Initiative; 2013.
[99] Stattman SL, Hospes O, Mol APJ. Governing biofuels in Brazil: a comparison of ethanol and biodiesel policies. Energy Policy 2013;61:22–30.
[100] Bonsucro. Guidance for the production standard. 4th ed. London: Bonsucro; 2014.
[101] Fortin E, Richardson B. Certification schemes and the governance of land: enforcing standards or enabling scrutiny? Globalizations 2013;10:141–59.
[102] Selfa T, Bain C, Moreno R. Depoliticizing land and water "grabs" in Colombia: the limits of Bonsucro certification for enhancing sustainable biofuel practices. Agric Hum Values 2014;31:455–68.

[103] Vlysidis A, Binns M, Webb C, Theodoropoulos C. A techno-economic analysis of biodiesel biorefineries: assessment of integrated designs for the co-production of fuels and chemicals. Energy 2011;36:4671–83.
[104] Zhu LD, Hiltunen E, Antila E, Zhong JJ, Yuan ZH, Wang ZM. Microalgal biofuels: flexible bioenergies for sustainable development. Renew Sustain Energy Rev 2014;30:1035–46.
[105] Kunz NC, Moran CJ, Kastelle T. Implementing an integrated approach to water management by matching problem complexity with management responses: a case study of a mine site water committee. J Clean Prod 2013;52:362–73.
[106] Sacramento-Rivero JC. A methodology for evaluating the sustainability of biorefineries: framework and indicators. Biofuel Bioprod Biorefin 2012;6:32–44.
[107] Osmani A, Zhang J. Economic and environmental optimization of a large scale sustainable dual feedstock lignocellulosic-based bioethanol supply chain in a stochastic environment. Appl Energy 2014;114:572–87.
[108] Hacking T, Guthrie P. A framework for clarifying the meaning of triple bottom-line, integrated and sustainability assessment. Environ Impact Assess Rev 2008;28:73–89.
[109] Azapagic A, Perdan S. An integrated sustainability decision-support framework: part I: problem structuring. Int J Sustain Dev World 2005;12:98–111.

Index

Note: Page numbers followed by *f* indicate figures and *t* indicate tables.

F

M